$44.25

PROGRESS IN THEORETICAL ORGANIC CHEMISTRY

VOLUME 1

THEORY AND PRACTICE OF MO CALCULATIONS ON ORGANIC MOLECULES

PROGRESS IN THEORETICAL ORGANIC CHEMISTRY

VOLUME 1

THEORY AND PRACTICE OF MO CALCULATIONS ON ORGANIC MOLECULES

I.G. CSIZMADIA

Department of Chemistry
University of Toronto
Toronto, Ontario
Canada M5S 1A1

ELSEVIER SCIENTIFIC PUBLISHING COMPANY
Amsterdam – Oxford – New York 1976

ELSEVIER SCIENTIFIC PUBLISHING COMPANY
335 Jan van Galenstraat
P.O. Box 211, Amsterdam, The Netherlands

AMERICAN ELSEVIER PUBLISHING COMPANY, INC.
52 Vanderbilt Avenue
New York, New York 10017

ISBN: 0-444-41468-1

Printed in The Netherlands

PREFACE

There is only one way to learn something .... by doing it. This means that the road to knowledge in theoretical organic chemistry is built from quantum chemical computations on organic molecules. It is hoped that after an experimental organic chemist has finished this book he will be ready to begin. Even if he feels he is not ready, he should start and after some computational experience has been gained the content of this introductory book will mean considerably more.

TO MY PARENTS

ACKNOWLEDGEMENTS

This book has emerged from a series of seminars, lectures and graduate courses the author has delivered during the period 1965-1975 at three Canadian Universities:

University of Alberta, Edmonton
Queen's University, Kingston
University of Toronto, Toronto

The author is indebted to Professors O. P. Strausz and S. Wolfe as well as to Drs. E. Lown and P. G. Mezey for reading parts of the manuscript during the early period of its preparation.

The assistance of John D. Goddard during the "final" period of manuscript preparation is gratefully acknowledged. Without his conscientious efforts it would have been impossible to finish this manuscript on time.

Thanks are also due to Frank Safian for the art work, to John Glover for the photography and to Patti Epstein for typing the manuscript.

Most of the computed results, used to illustrate the various applications of MO theory, emerged from research projects financially supported by the National Research Council of Canada.

TABLE OF CONTENTS

SECTION A

INTRODUCTION

CHAPTER I

INTRODUCTORY REMARKS

1. The Role of Theories and Models

1. The Role of Theories and Molecules

The following scheme encompasses in a very general way all scientific activity

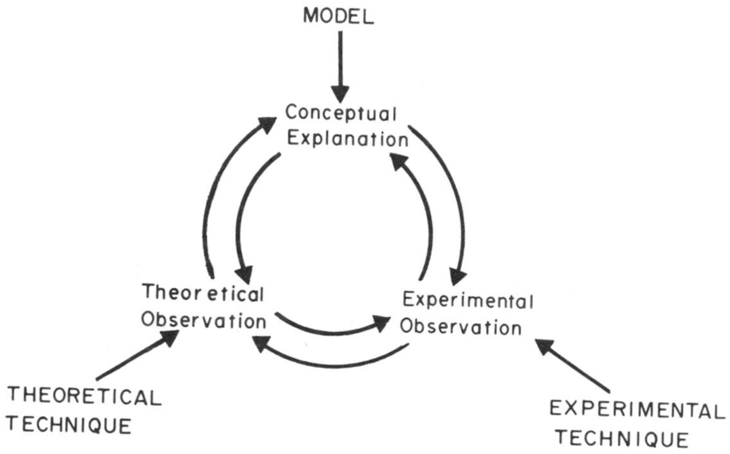

Figure I-1. A schematic illustration of the interrelationship of experiment, theory and modelling.

It should be noted that THEORY and EXPERIMENT are equally fundamental in any branch of science that has reached a rigorous stage. MODELS on the other hand are built either on theoretical and/or experimental observations and are expected to provide a conceptual explanation for the phenomenon investigated.

When its chemical implications are sought we can identify the "THEORETICAL TECHNIQUE" with "Quantum Theory" including all its various branches. The "MODEL" may be identified with all the "Rules" we have in chemistry such as the "Selection Rules", the "Woodward-Hoffmann Rules" and the like. The application of any THEORETICAL TECHNIQUE to a particular chemical problem is carried out through computation which in turn produces the Theoretical (i.e. numerical) Observations in very much the same way as the application of an EXPERIMENTAL TECHNIQUE leads to qualitative (i.e. non-numeric) or quantitative (i.e. numerical) results. The application of a MODEL to the same chemical problem produces qualitative (i.e. non-numerical) results which may be regarded as concepts or conceptual explanations and possessing them we usually declare that we have some understanding of the problem.

There are at least three things to be noticed. One is that the results obtained from theory are always quantitative from models always qualitative, while experimental results may be either qualitative or quantitative. The second thing to notice is that in both theory and models we are dealing with intellectual constructs while in any chemical experiment one is working with actual chemical substances. The third point is that there exists a whole spectrum of intellectual constructs between theory and models and it need not be trivially obvious into which category a construct fits.

At this stage it is necessary to point out that the above viewpoint, in which all of our scientific activities are unified on an equal basis, by no means enjoys uniform acceptance in the chemical community. In fact, we may classify most of the confessed opinions into the following four categories.

1. Only MODELS and therefore Conceptual Explanations are of any importance and therefore of any real relevance to chemistry because THEORY is largely incomprehensible and consequently irrelevant.
2. THEORY is important in so far as it supports a useful MODEL.
3. THEORY is of primary importance and a MODEL is acceptable only in so far as it is in close agreement with the Theoretical Observations.
4. THEORY is the only scientifically acceptable and therefore relevant method that complements EXPERIMENT. Consequently all MODELS are more or less useless and thus irrelevant.

Perhaps most experimentalists' point of view could be fitted in the first three categories while most theoreticians' views could be accommodated by the last three. To the average chemist, however, the relatively moderate views of (2) and (3) are most appealing and the first and last statements appear extreme; the first one being at the ultra conservative extreme while the last is at the radical extreme. At this stage it may be appropriate to make an operational distinction between theory and a model. Such a distinction may be made in terms of their applicability. If we apply a Theory to investigate a question that asks WHAT we may compute an answer within any quantitative theory. For example: WHAT is the most stable geometry of NH_3? We may obtain r(N-H) = 1.01 Å and <HNH = 107°. Alternatively we may ask: WHAT is the barrier to pyramidal inversion in NH_3? We may compute 6 Kcal/mole. However no theory can answer a question: WHY is ammonia pyramidal or WHY is its barrier to pyramidal inversion 6 Kcal/mole? As soon as we try to obtain an answer to this type of question, wittingly or unwittingly we are using a model (cf. Figure I-2.).

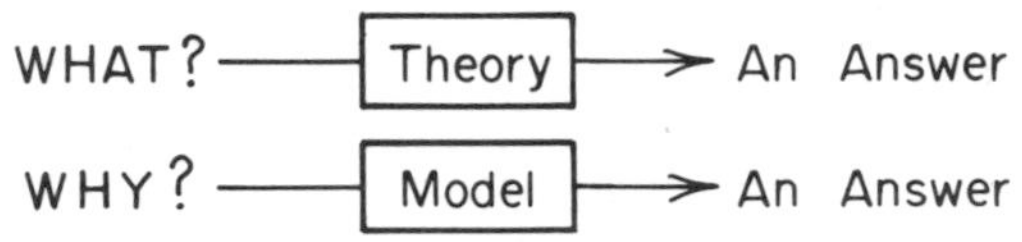

Figure I-2. The differing utility of a theory and a model.

The fundamental problem is that with the question WHY we would like to know what is causing what. For example we may say that NH_3 is pyramidal because such a geometry is guaranteed by its electron distribution. Immediately then we may ask WHY is the electron density so distributed. Then one might suggest that the position of the nuclei is causing such an electron distribution. Even at this stage we may ask WHY is the balance between nuclear and electronic forces such that the pyramidal geometry is favoured. In other words a question WHY leads to an infinite number of questions until one arrives at the ultimate or "primary cause". In contrast to this situation once an answer is given to a question WHAT no further question of WHAT may follow. Consequently people in category No. 4 are of the opinion that WHAT constitutes a scientific question while WHY constitutes a philosophical or theological question.

After summarising the rationale behind the extreme viewpoint (4), it is enough to say that quite likely you, the reader, may identify your point of view with one of the four (1-4) statements above. Irrespective of your point of view you probably wish to find out more about the THEORETICAL TECHNIQUES used for the generation of Theoretical Observations which is the subject of the subsequent chapters.

CHAPTER II

MATHEMATICAL INTRODUCTION

It is advantageous to treat orbitals as vectors and to deal with the mathematical problem of generating molecular orbitals (MO) from atomic orbitals (AO) as a transformation from one set of vectors to another. In this sense molecular orbital theory may be viewed as an exercise in Linear Algebra and this presentation of molecular orbital theory therefore begins with a review of vectors, vector spaces and their associated manipulations.

1. Vectors and Vector Spaces

A vector space is a set of mathematical objects ϕ, χ, η... called vectors. Although we may think of a vector ϕ as a geometrical object characterized by a length and a direction, however, it is best to abandon this fixed idea because we shall consider other than geometrical objects, such as mathematical functions to be vectors.

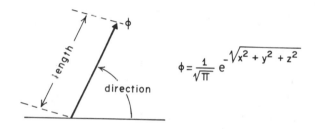

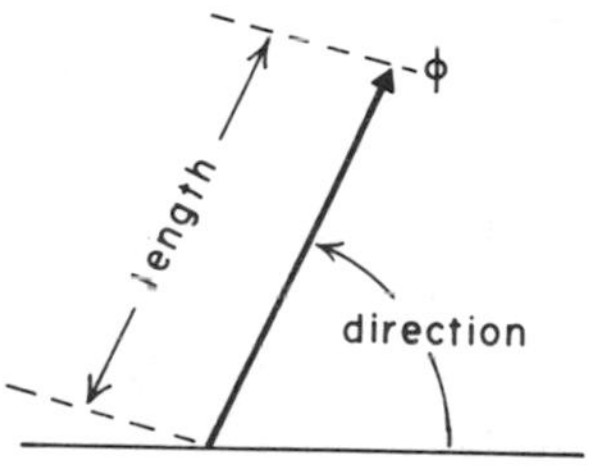

$$\phi = \frac{1}{\sqrt{\pi}} e^{-\sqrt{x^2+y^2+z^2}}$$

Figure II.1 The vector as geometrical object and as mathematical function.

Consequently, if we wish to consider abstract vector spaces, which is the most convenient way to discuss quantum chemistry then it is best to consider vectors as abstract mathematical objects. However, from time to time geometric illustrations will be given to facilitate understanding.

For any pair of vectors that belong to the vector space there is a unique sum. The addition of vectors is both commutative and associative

$$\phi + \chi = \chi + \phi \qquad \text{(commutative)} \qquad \{II-1\}$$

$$\phi + (\chi + \eta) = (\phi + \chi) + \eta \qquad \text{(associative)} \qquad \{II-2\}$$

There is one unique vector in every vector space called origin and zero vector: O and for every vector ϕ there exists an inverse vector $-\phi$. These special vectors are subject to the following addition laws

$$\phi + O = \phi \qquad \{II-3\}$$

$$\phi + (-\phi) = O \qquad \{II-4\}$$

For each scalar a and each vector ϕ there exists a multiple vector which is the product of a and ϕ. The multiplication of scalars is both distributive and associative

$$a(\phi + \chi) = a\phi + a\chi \qquad \{II-5\}$$

$$(a + b)\phi = a\phi + b\phi \qquad \text{(distributive)} \qquad \{II-6\}$$

$$(ab)\phi = a(b\phi) \qquad \text{(associative)} \qquad \{II-7\}$$

The minimum number of vectors required to define the vector space is termed the dimension of the vector space. It is easy to provide geometrical illustrations of one, two and three dimensional vector spaces.

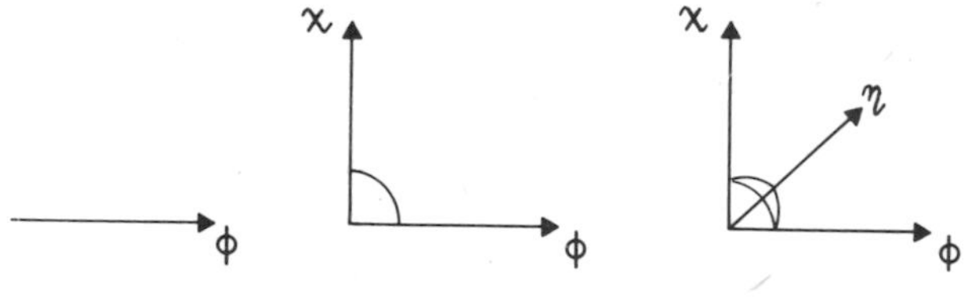

Figure II.2 Geometrical illustration of 1, 2 and 3 dimensional vector spaces.

To define a vector space it is necessary to have a set of linearly independent vectors. If the equation

$$k_1 \eta_1 + k_2 \eta_2 + \ldots\ldots + k_n \eta_n = 0 \qquad \{II-8\}$$

is valid only for <u>all</u> $k_i = 0$ the set of vectors $\{\eta_i\}$ <u>is linearly independent</u> (if any $k_i \neq 0$ the vectors are linearly dependent).

A linearly independent set of vectors which spans a vector space represents the <u>basis of the vector space</u>. Any other vector in this vector space is linearly dependent on the basis set (i.e. not all $k_i = 0$) and may therefore be expressed as a <u>linear combination of the basis vectors</u>

$$\phi = C_1\eta_1 + C_2\eta_2 + \ldots. + C_n\eta_n = \sum_{i=1}^{n} C_i\eta_i \qquad \{II-9\}$$

where the set of numbers $c_1, c_2 \ldots. c_n$ are the <u>components</u> of the vector over the <u>basis set</u> $\{\eta\}$.

A geometrical illustration of this principle for a two dimensional vector space is given in the following figure.

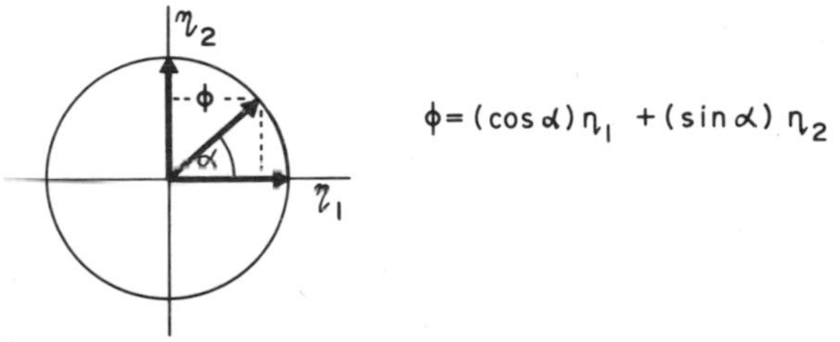

Figure II.3 The concept of linear dependence in a two dimensional vector space.

Since it is impossible for both components (sin α and cos α) to be zero at any value of α, ϕ is linearly dependent on the set of {η} which is <u>the basis set of the vector space</u>.

The basis vectors may be arranged in a row (row vector) and abbreviated as $\langle\eta|$

$$\langle\eta| = (\eta_1, \eta_2 \ldots. , \eta_n) \qquad \{II-10\}$$

or in a column (column vector) and denoted as $|\eta>$

$$|\eta> = \begin{pmatrix} \eta_1 \\ \eta_2 \\ \cdot \\ \cdot \\ \cdot \\ \cdot \\ \eta_n \end{pmatrix} \qquad \{II\text{-}11\}$$

The transpose (denoted by a prime) of a column vector is a row vector:

$$|\eta>' \equiv \begin{pmatrix} \eta_1 \\ \eta_2 \\ \cdot \\ \cdot \\ \cdot \\ \cdot \\ \eta_n \end{pmatrix}' = (\eta_1 \eta_2 \; \cdots\cdots \; \eta_n) \equiv <\eta| \qquad \{II\text{-}12\}$$

and vice versa.

If the basis vectors are not real but complex (such as complex functions) the corresponding relationship holds for the adjoint (denoted by a dagger) which is the transpose complex conjugate (the complex conjugate is denoted by an asterisk)

$$|\Psi>^{\dagger} = \begin{pmatrix} \Psi_1 \\ \Psi_2 \\ \cdot \\ \cdot \\ \cdot \\ \cdot \\ \cdot \\ \Psi_n \end{pmatrix}^{\dagger} = (\Psi_1{}^* \; \Psi_2{}^* \; \cdots\cdots \; \Psi_n{}^*) \equiv <\Psi| \qquad \{III\text{-}13\}$$

With this notation at hand it is possible to develop a vector notation for the linear combination of the basis vectors as defined in equation {II-9}

$$\phi = C_1\eta_1 + C_2\eta_2 + \dots + C_n\eta_n = (C_1C_2 \dots C_n) \begin{pmatrix} \eta_1 \\ \eta_2 \\ \cdot \\ \cdot \\ \cdot \\ \cdot \\ \eta_n \end{pmatrix} \qquad \{II\text{-}14\}$$

or in its equivalent form

$$\phi = \eta_1C_1 + \eta_2C_2 + \dots + \eta_nC_n = (\eta_1\eta_2 \dots \eta_n) \begin{pmatrix} C_1 \\ C_2 \\ \vdots \\ C_n \end{pmatrix} \qquad \{II\text{-}15\}$$

It probably should be pointed out at this stage that this is exactly the method (as indicated in equations {II-9}, {II-11} and {II-15} used to generate molecular orbitals from atomic orbitals where the basis set $\{\eta\}$ stands for the set of AO used and ϕ represents one particular molecular orbital. The expansion coefficients (i.e. the coefficients of the linear combination) are labelled as $C_1C_2 \dots C_n$ and these are in fact the components of the particular MO: ϕ over the chosen AO basis $\{\eta\}$.

2. Inner Product and Orthogonality

Molecular orbital theory involves special types of vector spaces where the basis set (the atomic orbitals) consists of scalar-valued continuous functions of space (i.e. x, y, z) defined in an interval {a,b} for each one of the independent variables (x, y, z). This example may be referred to as a vector space of continuous functions on {a,b}. It may be noted that using a Cartesian coordinate system the interval is $\{-\infty, +\infty\}$ for all three independent variables.

For such a vector space the inner product of any two basis vectors is defined as the following definite integral

$$S_{ij} = \int_{-\infty}^{+\infty}\int_{-\infty}^{+\infty}\int_{-\infty}^{+\infty} \eta_i(x,y,z)\eta_j(x,y,z)\ dx\ dy\ dz \qquad \{II\text{-}16\}$$

If the set of orbitals $\{\eta\}$ used are complex valued functions the above definition should be generalized as follows

$$S_{ij} = \int_{-\infty}^{+\infty}\int_{-\infty}^{+\infty}\int_{-\infty}^{+\infty} \eta_i^*(x,y,z)\eta_j(x,y,z)\ dx\ dy\ dz \qquad \{II\text{-}17\}$$

This integral (which is frequently called the overlap integral) is abbreviated in the following fashion in Dirac's notation

$$S_{ij} \equiv \langle\eta_i|\eta_j\rangle \qquad \{II\text{-}18\}$$

where the $\langle\eta_i|$ includes the complex conjugate as specified in equation {II-17}.

When S_{ij} is zero then the two vectors η_i and η_j are said to be orthogonal. Although orthogonality is a more general concept than perpendicularity, however it may be helpful to review a geometrical example.

As indicated by the geometrical relationship shown in Figure II-4 the inner product:

$$S \equiv \langle\phi|\chi\rangle = |\phi|.|\chi|.\cos\theta \qquad \{II\text{-}19\}$$

will vanish when the two vectors are perpendicular to each other (i.e. when the angle θ is 90° or 270°).

In quantum chemistry orthogonalization of the basis i.e. the basis vectors (e.g. the AO) facilitates the calculation and is therefore of some importance. The most frequently used methods are the symmetric (Löwdin) orthogonalization and the Schmidt orthogonalization. The geometrical illustration of these two methods of orthogonalization is shown in Figure II-5.

In actual calculations one usually works with more than two basis vectors (i.e. AO) and the whole set needs to be orthogonalized so that any pair of vectors in the set will be

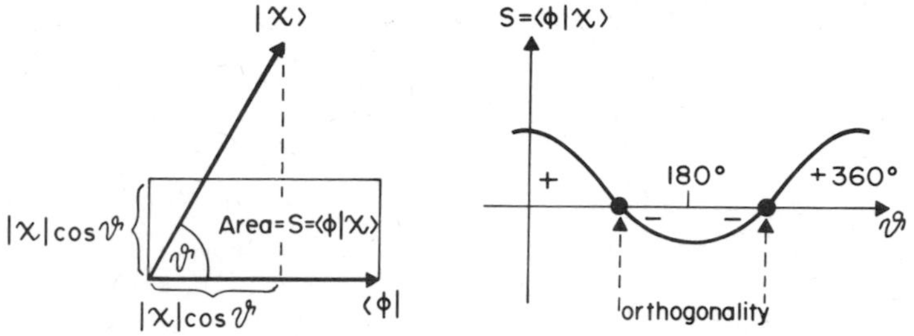

Figure II.4 Geometrical analogue of orthogonality.

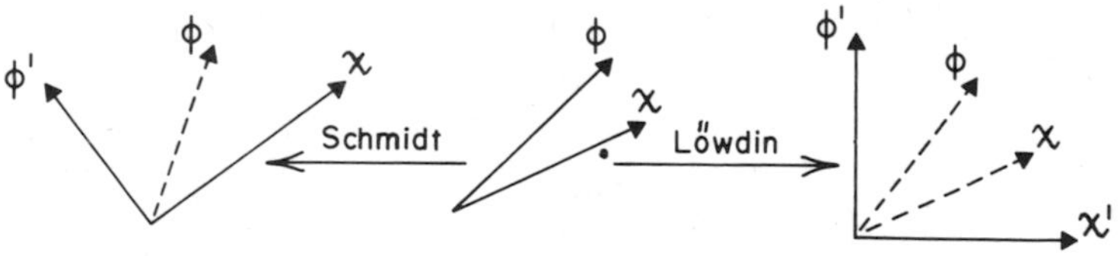

Figure II.5 Geometrical illustration of Schmidt and Löwdin type orthogonalization.

orthogonal. In such a case one vector (such as an atomic core: 1s) needs to be chosen as fixed and all the other vectors (orbitals) are orthogonalized to that particular vector in the method of Schmidt orthogonalization. In the Löwdin method of orthogonalization each and every vector is changed until the whole set is orthogonal and therefore this procedure is frequently called the symmetric orthogonalization.

If the components of χ are the same as those of ϕ (i.e. $\phi = \chi$) then the inner product is called the norm (S = N)

$$N = \langle\phi|\phi\rangle = |\phi|^2 \qquad \{II-20\}$$

the square root of the norm may be associated with the length of the vector. A vector is normalized if its norm is unity. A vector may be normalized if its norm is not zero.

$$\phi_{normalized} = \frac{1}{\sqrt{\langle\phi|\phi\rangle}}\,\phi_{unnormalized} \qquad \{II-21\}$$

Calculations in quantum chemistry are always simpler if the orbitals are normalized so that it is not surprising that normalization is usually performed.

An orthonormal set of vectors is a set of vectors each of which is normalized and is orthogonal to every other vector in the set.

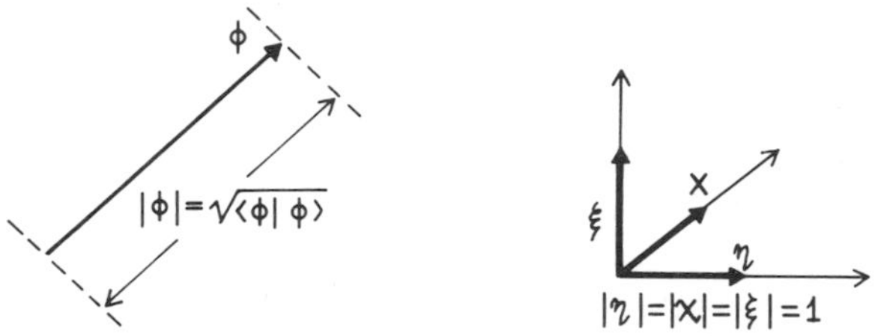

Figure II.6 Geometrical illustration for the norm of a vector and for an orthonormal set of vectors.

The overlap between members of an orthonormal set of vector is denoted by the Kronecker delta

$$\langle\phi_i|\phi_j\rangle = \delta_{ij} \qquad \{II-22\}$$

The Kronecker delta (δ_{ij}) is a widely used discontinuous function with the following properties

$$\delta_{ij} = 1 \quad (\text{if } i = j)$$
$$\delta_{ij} = 0 \quad (\text{if } i \neq j) \qquad \{II-23\}$$

Consider an n-dimensional vector space defined by an orthonormal basis set $\chi_1,\chi_2 \ldots\ldots,\chi_n$. This means that the inner product (i.e. overlap) between the members of the basis set is equal to the Kronecker delta:

$$S_{ij} = \langle\chi_i|\chi_j\rangle = \delta_{ij} \qquad \{II-24\}$$

Now let us take two vectors ϕ and Ψ defined in this n-dimensional vector space in terms of the orthonormal basis set $\{\chi\}$.

$$|\psi\rangle = \sum_{j=1}^{n} \chi_j b_j \quad (\chi \text{ is written as a row vector}) \qquad \{II-25\}$$

$$\langle\phi| = \sum_{i=1}^{n} a_i\chi_i \quad (\chi \text{ is written as a column vector}) \qquad \{II-26\}$$

The inner product of $\langle\phi|\psi\rangle$ may therefore be expressed in terms of the orthonormal basis set:

$$\langle\phi|\Psi\rangle = \langle\sum_{i=1}^{n} a_i\chi_i| \sum_{j=1}^{n} \chi_j b_j\rangle = \sum_{i=1}^{n}\sum_{j=1}^{n} \langle a_i\chi_i|\chi_j b_j\rangle =$$

$$= \sum_{i=1}^{n} \sum_{j=1}^{n} a_i \langle \chi_i | \chi_j \rangle b_j = \sum_{i=1}^{n} \sum_{j=1}^{n} a_i S_{ij} b_j \qquad \{II-27\}$$

However, in the present case $S_{ij} = \delta_{ij}$ as specified by equation {II-24} therefore we may have the following simplified expression for $\langle\phi|\psi\rangle$ since $\delta_{ij} \neq 0$ if and only if i=j (i.e. $\delta_{ii}=1$).

$$\langle\phi|\psi\rangle = \sum_{i=1}^{n} \sum_{j=1}^{n} a_i \delta_{ij} b_j = \sum_{i=1}^{n} a_i b_i = a_1 b_1 + a_2 b_2 + \ldots\ldots a_n b_n \qquad \{II-28\}$$

The last part of this equation is the usual expression one defines the inner product in terms of vector components, i.e. in terms of the coefficients of the linear combination.

If the vectors are not real but complex then the appropriate complex-conjugate should be used in equations {II-26-28} as indicated below.

$$\langle\phi| = \sum_{i=1}^{n} a_i^* \chi_i^* \qquad \{II-26a\}$$

$$\langle\phi|\psi\rangle = \sum_{i=1}^{n} \sum_{j=1}^{n} a_i^* S_{ij} b_j \qquad \{II-27a\}$$

$$\langle\phi|\psi\rangle = \sum_{i=1}^{n} a_i^* b_i = a_1^* b_1 + a_2^* b_2 + \ldots\ldots + a_n^* b_n \qquad \{II-28a\}$$

3. Special Vector Spaces

There are two special vector spaces which are of some importance in quantum chemistry, viz. the Euclidean vector space and the Hermitian vector space. The basis vectors of the Euclidean vector space are real and the basis vectors of the Hermitian vector space are complex so that the Euclidean vector space is a special case of the Hermitian vector space. The characteristics of the two spaces are best illustrated in terms of the properties of the inner product of two vectors:

Euclidean Vector Space	Hermitian Vector Space
a) The inner product is real: $\langle\phi\|\chi\rangle = a_1b_1 + a_2b_2 + \ldots.$	a) The inner product is complex $\langle\phi\|\chi\rangle = (a_1{}^*b_1 + a_2{}^*b_2 + \ldots.$ {II-29a,b}
b) The inner product is symmetric (i.e. it is equal to its own transpose) $\langle\phi\|\chi\rangle' \equiv \langle\chi\|\phi\rangle = \langle\phi\|\chi\rangle$	b) The inner product is Hermitian (i.e. it is equal to its own adjoint) $\langle\phi\|\chi\rangle^\dagger \equiv \langle\chi\|\phi\rangle = \langle\phi\|\chi\rangle$ {II-30a,b}
c) The inner product is bilinear $\langle\phi + \chi\|\psi\rangle = \langle\phi\|\psi\rangle + \langle\chi\|\psi\rangle$	c) The inner product is bilinear $\langle\phi + \chi\|\psi\rangle = \langle\phi\|\psi\rangle + \langle\chi\|\psi\rangle$ {II-31a,b}
d) The norm is positive $N = \langle\phi\|\phi\rangle > 0$	d) The norm is positive $N = \langle\phi\|\phi\rangle > 0$ {II-32a,b}

The following figure illustrates by 3 dimensional geometrical analogues the Euclidean (real) vector space.

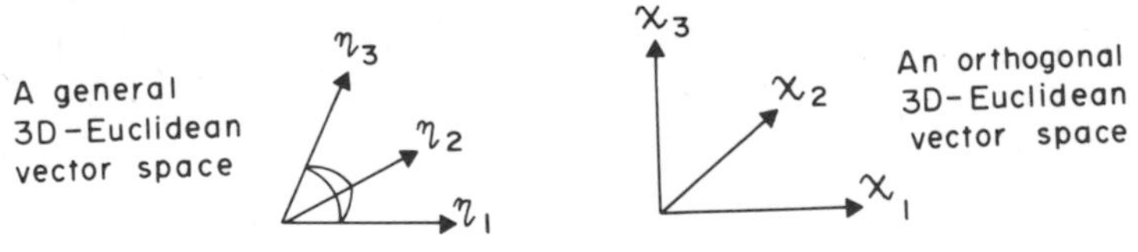

Figure II.7 Geometrical illustrations for Euclidean vector spaces.

4. Matrices

When real or complex scalars are arranged in a rectangular array the resulting object is called a matrix. A matrix consisting of m-rows and n-columns is an m x n matrix. The i, jth element of the matrix is found at the intersection of the ith row and the jth column.

$$\underline{\underline{A}} = \begin{pmatrix} a_{11} & a_{12} & a_{13} & a_{14} \\ a_{21} & a_{22} & a_{23} & a_{24} \\ a_{31} & a_{32} & a_{33} & a_{34} \end{pmatrix} \quad \text{a 3 x 4 matrix} \qquad \{II-33\}$$

A square matrix has the same number of rows and columns.

$$\underline{\underline{B}} = \begin{pmatrix} b_{11} & b_{12} & b_{13} \\ b_{21} & b_{22} & b_{23} \\ b_{31} & b_{32} & b_{33} \end{pmatrix} \quad \text{a 3 x 3 matrix} \qquad \{II-34\}$$

In this case there is a unique distinction between diagonal and off diagonal elements. The diagonal elements are those that fall on the principal diagonal of the square (i.e. b_{ii}) while all other elements (i.e. b_{ij}, $i \neq j$) are called off-diagonal elements.

The sum of the diagonal elements of a square matrix is usually referred to as the trace of the matrix (abbreviated as tr) and it is of some importance in quantum chemistry

$$\text{tr } \underline{\underline{B}} = \sum_{i=1}^{n} b_{ii} \qquad \{II-35\}$$

The multiplication of matrices is of great importance in computational quantum chemistry. Suppose we wish to multiply matrix $\underline{\underline{B}}$ with matrix $\underline{\underline{A}}$ defined in equations {II-34} and {II-33} respectively to obtain a product matrix $\underline{\underline{C}}$

$$\underline{\underline{C}} = \underline{\underline{B}}\,\underline{\underline{A}} \qquad \{II-36\}$$

The elements of $\underline{\underline{C}}$ are defined as:

$$c_{ij} = \sum_{k=1}^{n} b_{ik} a_{kj} \qquad \{II\text{-}37\}$$

This means that the elements of the ith row in $\underline{\underline{B}}$ are multiplied by the corresponding elements of the jth column in $\underline{\underline{A}}$ and the sum of these corresponds to the i,j^{th} element of the product matrix $\underline{\underline{C}}$. This is illustrated for $c_{2,3}$ in the next two equations

$$\begin{pmatrix} c_{11} & c_{12} & c_{13} & c_{14} \\ c_{21} & c_{22} & \boxed{c_{23}} & c_{24} \\ c_{31} & c_{32} & c_{33} & c_{34} \end{pmatrix} = \begin{pmatrix} b_{11} & b_{12} & b_{13} \\ \boxed{b_{21} \quad b_{22} \quad b_{23}} \\ b_{31} & b_{32} & b_{33} \end{pmatrix} \begin{pmatrix} a_{11} & a_{12} & \boxed{a_{13}} & a_{14} \\ a_{21} & a_{22} & \boxed{a_{23}} & a_{24} \\ a_{31} & a_{32} & \boxed{a_{33}} & a_{34} \end{pmatrix} \qquad \{II\text{-}38\}$$

$$c_{23} = b_{21}a_{13} + b_{22}a_{23} + b_{23}a_{33} \qquad \{II\text{-}39\}$$

From the foregoing it should be evident that not any two matrices may be multiplied together and even two matrices which may be multiplied together in one particular order might not be multiplied in the reverse order. For example $\underline{\underline{A}} \cdot \underline{\underline{B}}$ cannot be multiplied but $\underline{\underline{B}} \cdot \underline{\underline{A}}$ may be multiplied. This is because $\underline{\underline{B}}$ has so many columns as many rows are in $\underline{\underline{A}}$.

$$\begin{matrix} \underline{\underline{C}} & = & \underline{\underline{B}} & \cdot & \underline{\underline{A}} \\ (3x4) & & (3x3) & & (3x4) \\ (nxm) & & (nxk) & & (kxm) \end{matrix} \qquad \{II\text{-}40\}$$

Even if one deals with square matrices the order which they are multiplied together is quite important since $\underline{\underline{A}}.\underline{\underline{B}}$ is not necessarily the same as $\underline{\underline{B}}.\underline{\underline{A}}$:

$$\underline{\underline{A}}.\underline{\underline{B}} \neq \underline{\underline{B}}\ \underline{\underline{A}} \qquad \{II\text{-}41\}$$

There are some special matrices that are frequently used in computational quantum chemistry. These are the following.

<u>A Diagonal Matrix</u> is a square matrix with non-zero diagonal elements and with all other elements being zero.

$$\underline{\underline{D}} \quad \begin{pmatrix} d_{11} & & & O \\ & d_{22} & & \\ & & \ddots & \\ O & & & d_{nn} \end{pmatrix} \qquad \{II-42\}$$

<u>A Unit Matrix</u> is a special case of $\underline{\underline{D}}$ with unit diagonal elements. Naturally all off-diagonal elements are zero

$$\underline{\underline{1}} = \begin{pmatrix} 1 & 0 & 0 \\ 0 & 1 & 0 \\ 0 & 0 & 1 \end{pmatrix} \qquad \{II-43\}$$

Multiplication of a matrix with unit matrix leaves the matrix unchanged. Also in this case the order of multiplication is immaterial.

$$\underline{\underline{B}} \cdot \underline{\underline{1}} = \underline{\underline{1}} \cdot \underline{\underline{B}} = \underline{\underline{B}} \qquad \{II-44\}$$

<u>A Null Matrix</u> is a matrix (may or may not be a square matrix) which has only zero elements

$$\underline{\underline{O}} = \begin{pmatrix} 0 & 0 & 0 \\ 0 & 0 & 0 \\ 0 & 0 & 0 \end{pmatrix} \qquad \{II-45\}$$

The addition and multiplication that involves a null matrix must obey the following relationships

$$\underline{\underline{A}} + \underline{\underline{O}} = \underline{\underline{A}} \qquad \text{(for any dimensions)} \qquad \{II-46\}$$

$$\underline{\underline{A}} \cdot \underline{\underline{O}} = \underline{\underline{O}}\cdot\underline{\underline{A}} = \underline{\underline{O}} \quad \text{(for square matrices)} \qquad \{II-47\}$$

There are certain matrix operations that are used frequently in computational quantum chemistry

The Complex Conjugate. If the elements of Matrix $\underline{\underline{A}}$ are complex then the complex conjugate of the matrix ($\underline{\underline{A}}^*$) may be generated by taking the complex conjugate of the elements.

$$(\underline{\underline{A}}^*)_{ij} = a_{ij}^* \qquad \{II-48\}$$

If the elements of $\underline{\underline{A}}$ are real then

$$\underline{\underline{A}}^* = \underline{\underline{A}} \qquad \{II-49\}$$

Consequently in this particular case $\underline{\underline{A}}$ is a real matrix. The transpose of a matrix is obtained by interchanging the rows and columns

$$\underline{\underline{A}} = \begin{pmatrix} a_{11} & a_{12} & a_{13} \\ a_{21} & a_{22} & a_{23} \end{pmatrix} \qquad \underline{\underline{A}}' = \begin{pmatrix} a_{11} & a_{21} \\ a_{21} & a_{22} \\ a_{13} & a_{23} \end{pmatrix} \qquad \{II-50a,b\}$$

Here a 2x3 matrix is changed to a 3x2 matrix. For a square matrix (nxn) the dimension of the matrix remains the same.

$$\underline{\underline{A}}' = \begin{pmatrix} a_{11} & a_{12} & a_{13} \\ a_{21} & a_{22} & a_{23} \\ a_{31} & a_{32} & a_{33} \end{pmatrix}' = \begin{pmatrix} a_{11} & a_{21} & a_{31} \\ a_{12} & a_{22} & a_{32} \\ a_{13} & a_{23} & a_{33} \end{pmatrix} \qquad \{II-51\}$$

Note that the diagonal elements remained unchanged and the i,j indices interchanged in such a way that the 2,3 element of $\underline{\underline{A}}'$ is a_{32}.

In the case of a symmetric arrangement when

$$a_{ij} = a_{ji} \tag{II-52}$$

the transpose of a matrix is the same as the original matrix

$$\underline{\underline{A}}' = \underline{\underline{A}} \tag{II-53}$$

This type of matrix is usually called a <u>symmetric</u> matrix. If the elements of the matrix are real then the matrix is normally referred to as a <u>real symmetric</u> matrix which has the following property

$$(\underline{\underline{A}}^*)' = \underline{\underline{A}}$$

or {II-54a,b}

$$a_{ji}^* = a_{ij}$$

This matrix is very frequently used in quantum chemistry. The transpose complex conjugate is usually called <u>the adjoint</u> and denoted by a dagger:

$$\underline{\underline{A}}^\dagger = (\underline{\underline{A}}^*)'$$

<u>The Inverse of a Matrix</u> is defined by the following identity:

$$\underline{\underline{A}}^{-1} \cdot \underline{\underline{A}} = \underline{\underline{1}} \tag{II-55}$$

<u>For example</u>:

$$\underline{\underline{A}} = \begin{pmatrix} 1 & 2 \\ 3 & 4 \end{pmatrix}; \quad \underline{\underline{A}}^{-1} = \begin{pmatrix} -2 & 1 \\ 3/2 & -1/2 \end{pmatrix} \tag{II-56a,b}$$

$$\underline{\underline{A}}\underline{\underline{A}}^{-1} = \begin{pmatrix} 1 & 2 \\ 3 & 4 \end{pmatrix}\begin{pmatrix} -2 & 1 \\ 3/2 & -1/2 \end{pmatrix} = \begin{pmatrix} 1 & 0 \\ 0 & 1 \end{pmatrix} = \underline{\underline{1}} \tag{II-57}$$

The method of matrix inversion will not be discussed here since those matrices whose inverse are normally required fall into a very special category. These are the orthogonal ($\underline{\underline{\theta}}$) and the unitary ($\underline{\underline{U}}$) matrices whose inverse are the transpose and the adjoint respectively

$$\underline{\underline{\theta}}^{-1} = \underline{\underline{\theta}}' \qquad \{II-58\}$$

$$\underline{\underline{U}}^{-1} = \underline{\underline{U}}^{\dagger} \qquad \{II-59\}$$

The following relationships should be observed in connection with the transpose and the inverse of the product of two matrices:

$$(\underline{\underline{B}}\ \underline{\underline{A}})' = \underline{\underline{A}}' . \underline{\underline{B}}' \qquad \{II-60\}$$

$$(\underline{\underline{B}}\ \underline{\underline{A}})^{-1} = \underline{\underline{A}}^{-1} . \underline{\underline{B}}^{-1} \qquad \{II-61\}$$

These properties can be used to discover under what conditions can one interchange the order of matrix multiplication

Usually $$\underline{\underline{A}} \cdot \underline{\underline{B}} \neq \underline{\underline{B}} \cdot \underline{\underline{A}} \qquad \{II-62\}$$

However, if both $\underline{\underline{A}}$ and $\underline{\underline{B}}$ are real diagonal matrices then:

$$\underline{\underline{A}}\ \underline{\underline{B}} = \underline{\underline{B}}\ \underline{\underline{A}} \qquad \{II-63\}$$

Proof:

The i,j elements of $\underline{\underline{A}}\ \underline{\underline{B}}$ and $\underline{\underline{B}}\ \underline{\underline{A}}$ are given by

$$(\underline{\underline{A}}\ \underline{\underline{B}})_{ij} = \sum_k A_{ik}B_{kj} \qquad (\underline{\underline{B}}\ \underline{\underline{A}})_{ij} = \sum_k B_{ik}A_{kj} \qquad \{II-64\}$$

Since $\underline{\underline{A}}$ and $\underline{\underline{B}}$ are diagonal, the terms in the summations in {II-64} are 0 unless i = j = k. Thus

$$(\underline{\underline{A}}\ \underline{\underline{B}})_{ij} = (\underline{\underline{B}}\ \underline{\underline{A}})_{ij} = 0 \text{ unless } i = j \qquad \{II-65\}$$

$$(\underline{\underline{A}}\ \underline{\underline{B}})_{ii} = \sum_k A_{ik}B_{ki} = A_{ii}B_{ii} \qquad \{II-66a\}$$

$$(\underline{\underline{B}}\ \underline{\underline{A}})_{ii} = \sum_k B_{ik}A_{ki} = B_{ii}A_{ii} = A_{ii}B_{ii} = (\underline{\underline{A}}\ \underline{\underline{B}})_{ii} \qquad \{II-66b\}$$

5. Transformations

A transformation converts a region of an n-dimensional vector space into an m dimensional vector space where $m \leq n$. This is frequently achieved by matrix multiplication

$$\phi = \eta \underline{\underline{C}} \tag{II-67}$$

or, in detail, (using a square transforming matrix i.e. m=n):

$$\underset{1\mathrm{x}n}{(\phi_1 \; \phi_2 \; \cdots \; \phi_n)} = \underset{1\mathrm{x}n}{(\eta_1 \; \eta_2 \; \cdots \; \eta_n)} \underset{n\mathrm{x}n}{\begin{pmatrix} c_{11} & c_{12} & \cdots\cdots & c_{1n} \\ c_{21} & c_{22} & \cdots\cdots & c_{2n} \\ \vdots & \vdots & & \vdots \\ c_{n1} & c_{n2} & \cdots\cdots & c_{nn} \end{pmatrix}} \tag{II-68}$$

In quantum chemistry the molecular orbitals (ϕ) are always obtained by a transformation from some set of basis functions such as atomic orbitals (η) and the transforming matrix ($\underline{\underline{C}}$) is then referred to as the coefficient matrix. The number of occupied MO must always be less than (or sometimes equal to) the number of basis orbitals ($m \leq n$).

$$(\underbrace{\phi_1\phi_2\cdots\phi_m}_{\substack{m\\ \text{occupied}\\ \text{MO}}}\underbrace{\cdots\cdots\phi_n}_{\substack{(n-m)\\ \text{virtual}\\ \text{MO}}}) = (\underbrace{\eta_1\eta_2\cdots\cdots\eta_n}_{\substack{n\\ \text{AO}}}) \begin{pmatrix} c_{11} & c_{12}\cdots\cdots c_{1m} & \cdots\cdots c_{1n} \\ c_{21} & c_{22}\cdots\cdots c_{2m} & \cdots\cdots c_{2n} \\ \vdots & \vdots \qquad\;\; \vdots & \vdots \\ \underbrace{c_{n1} & c_{n2}\cdots\cdots c_{nm}}_{\substack{m\\ \text{columns}}} & \underbrace{\cdots\cdots c_{nn}}_{\substack{(n-m)\\ \text{columns}}} \end{pmatrix} \tag{II-69}$$

It is reasonable that if we wish to know the properties of transformations we must examine the properties of transforming matrices. Two transformations are most important in quantum chemistry, viz., the orthogonal and the unitary transformations.

An orthogonal transformation preserves the length (normalization) and orthogonality of vectors in Euclidean vector space. A unitary transformation preserves normalization and orthogonality of vectors in Hermitian vector space. The matrices that perform these transformations are called orthogonal and unitary. The inverse of an orthogonal matrix is equal to the transpose of the original orthogonal matrix. The inverse of a unitary matrix is equal to its adjoint (transpose complex conjugate).

Table II-1 Summary of Special Transforming Matrices in Special Vector Spaces

Vector Space	Transforming Matrix	Inverse of Transforming Matrix
Euclidean	Orthogonal: $\underline{\underline{\Theta}}$	$\underline{\underline{\Theta}}^{-1} = \underline{\underline{\Theta}}'$
Hermitian	Unitary: $\underline{\underline{U}}$	$\underline{\underline{U}}^{-1} = \underline{\underline{U}}^{+}$

This means that if $\phi = \chi\underline{\underline{\Theta}}$ instead of $\chi = \phi\underline{\underline{\Theta}}^{-1}$ one may write $\chi = \phi\underline{\underline{\Theta}}'$.

It may be convenient to think of the orthogonal and unitary transformations as the rotation of a real (Euclidean) or complex (Hermitian) vector space respectively. For this reason the following example is relevant to an orthogonal transformation (rotation of a real - Euclidean - vector space). Since most MO theory is based on real (rather than complex) AO this exercise is quite important.

We consider the rotation of a vector in a two dimensional vector space and the rotation of the two dimensional space itself. If we carry out these two operations in sequence we recover the initial situation because the second operation will undo the effect of the first. Thus the two operations are inverse to one another.

Before elaborating let us note the following trigonometric relationships:

$$\cos(\alpha + \beta) = \cos\beta \cos\alpha - \sin\beta \sin\alpha$$

$$\sin(\alpha + \beta) = \cos\beta \sin\alpha + \sin\beta \cos\alpha \qquad \{II\text{-}70a\text{-}d\}$$

$$\cos(\beta - \alpha) = \cos\beta \cos\alpha + \sin\beta \sin\alpha$$

$$\sin(\beta - \alpha) = \cos\beta \sin\alpha - \sin\beta \cos\alpha$$

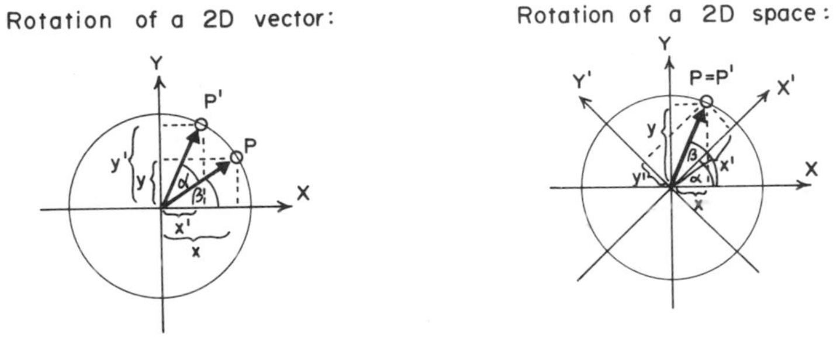

Figure II.8 Rotation in two dimensions (note that the term "rotation of 2D space" means rotation of the basis - i.e. the coordinates -).

$$x' = r\cos(\alpha+\beta) = \underbrace{r.\cos\beta}_{x}.\cos\alpha - \underbrace{r.\sin\beta}_{y}.\sin\alpha \qquad \{II\text{-}71a\}$$

$$x' = r.\cos(\beta-\alpha) = \underbrace{r.\cos\beta}_{x}\cos\alpha + \underbrace{r\sin\beta}_{y}\sin\alpha \qquad \{II\text{-}71b\}$$

$$y' = r \sin(\alpha+\beta) = \underbrace{r.\cos\beta}_{x} \sin\alpha + \underbrace{r.\sin\beta}_{y}.\cos\alpha \qquad \{II\text{-}72a\}$$

$$y' = r.\sin(\beta-\alpha) = \underbrace{r.\sin\beta}_{x} \cos\alpha - \underbrace{r.\cos\beta}_{y} \sin\alpha \qquad \{II\text{-}72b\}$$

$$x' = x \cos\alpha - y \sin\alpha \qquad x' = x \cos\alpha + y \sin\alpha \qquad \{II\text{-}73a,b\}$$

$$y' = x \sin\alpha + y \cos\alpha \qquad y' = y \cos\alpha - x \sin\alpha \qquad \{II\text{-}74a,b\}$$

$$\underbrace{\begin{pmatrix} x' \\ y' \end{pmatrix}}_{\underline{r}'} = \underbrace{\begin{pmatrix} \cos\alpha & -\sin\alpha \\ \sin\alpha & \cos\alpha \end{pmatrix}}_{\underline{\underline{\theta}}} \underbrace{\begin{pmatrix} x \\ y \end{pmatrix}}_{\underline{r}} \qquad \{II\text{-}75a\}$$

$$\underbrace{\begin{pmatrix} x' \\ y' \end{pmatrix}}_{\underline{r}'} = \underbrace{\begin{pmatrix} \cos\alpha & \sin\alpha \\ -\sin\alpha & \cos\alpha \end{pmatrix}}_{\underline{\underline{\theta}}'} \underbrace{\begin{pmatrix} x \\ y \end{pmatrix}}_{\underline{r}} \qquad \{II\text{-}75b\}$$

$$\underline{r}' = \underline{\underline{\theta}}\,\underline{r} \qquad \underline{r}' = \underline{\underline{\theta}}'\underline{r} \qquad \{II\text{-}76a,b\}$$

where $\underline{\underline{\theta}}'$ is the transpose of $\underline{\underline{\theta}}$.

It is evident that the two transforming matrices are transpose to each other. However if they are to be orthogonal they must also be the inverse of each other. In this case their product must yield a unit matrix:

$$\underbrace{\begin{pmatrix} \cos\alpha & -\sin\alpha \\ \sin\alpha & \cos\alpha \end{pmatrix}}_{\underline{\underline{\theta}}} \underbrace{\begin{pmatrix} \cos\alpha & \sin\alpha \\ -\sin\alpha & \cos\alpha \end{pmatrix}}_{\underline{\underline{\theta}}'} =$$

$$\begin{pmatrix} \{\cos^2\alpha + \sin^2\alpha\} & \{\cos\alpha\sin\alpha - \cos\alpha\sin\alpha\} \\ \{\sin\alpha\cos\alpha - \sin\alpha\cos\alpha\} & \{\sin^2\alpha + \cos^2\alpha\} \end{pmatrix} = \underbrace{\begin{pmatrix} 1 & 0 \\ 0 & 1 \end{pmatrix}}_{\underline{\underline{1}}} \qquad \{II-77\}$$

consequently: $\underline{\underline{\theta}}' = \underline{\underline{\theta}}^{-1}$ {II-78}

because $\underline{\underline{\theta}}\underline{\underline{\theta}}' = \underline{\underline{1}} = \underline{\underline{\theta}}\underline{\underline{\theta}}^{-1}$ {II-79}

Since $\underline{\underline{\theta}}^{-1}$ is the inverse of $\underline{\underline{\theta}}$ it will undo the operation of $\underline{\underline{\theta}}$ (or vice versa) i.e. $\underline{\underline{\theta}}\underline{\underline{\theta}}^{-1} = \underline{\underline{\theta}}^{-1}\underline{\underline{\theta}} = \underline{\underline{1}}$. The resulting unit matrix indicates that the final state is the same as the initial state if we rotate the vector first, by an angle α, and then rotate the vector space by the same angle α.

6. Operators and their Matrix Representative

An operator is an instruction or series of instructions which change a vector space. This may be represented symbolically in the following fashion

$$\Psi = \hat{A}\phi \qquad \{II-80\}$$

The operator $\hat{A}$ is a linear operator if it obeys the following relations:

$$\hat{A}(\phi + \chi) = \hat{A}\phi + \hat{A}\chi \qquad \{II-81\}$$

$$\hat{A}(c\phi) = c\hat{A}\phi \qquad \{II-82\}$$

where c is a scalar, real or complex.

Let $\Psi = \hat{A}\phi$ where both Ψ and ϕ belong to the orthonormal vector space $\{\eta_i\}$. We generate the inner product between ϕ and Ψ in terms of $\{\eta_i\}$ by expanding ϕ in terms of $\{\eta\}$(i.e. expressing ϕ as a linear combination of the set $\{\eta\}$):

$$\phi = \sum_{i=1}^{n} \eta_i c_i \qquad \{II\text{-}83\}$$

$$\langle\phi|\Psi\rangle = \langle\phi|\hat{A}\phi\rangle \equiv \langle\phi|\hat{A}|\phi\rangle$$

$$= \langle\sum_{i=1}^{n} c_i\eta_i|\hat{A}|\sum_{j=1}^{n} \eta_j c_j\rangle \qquad \{II\text{-}84\}$$

$$= \sum_{i=1}^{n}\sum_{j=1}^{n} c_i^* \underbrace{\langle\eta_i|\hat{A}|\eta_j\rangle}_{a_{ij}} c_j$$

Observe that $\langle\phi|\hat{A}|\phi\rangle$ is a special notation for the inner product between ϕ and $\hat{A}\phi$ and is frequently called the "expectation value of the operator $\hat{A}$

$$\langle\phi|\Psi\rangle = \sum_{i=1}^{n}\sum_{j=1}^{n} c_i^* a_{ij} c_j \qquad \{II\text{-}85\}$$

$$= (c_1^* c_2^* \; .... \; c_n^*) \begin{pmatrix} a_{11} & a_{12} & .... & a_{1n} \\ a_{21} & a_{22} & .... & a_{2n} \\ \vdots & \vdots & & \vdots \\ a_{n1} & a_{n2} & .... & a_{nn} \end{pmatrix} \begin{pmatrix} c_1 \\ c_2 \\ \vdots \\ c_n \end{pmatrix}$$

Thus one may write:

$$\langle\phi|\Psi\rangle = \underline{C}^{\dagger}.\underline{\underline{A}}.C$$

$$(1x1) \quad (1xn).(nxn).(nx1) \qquad \{II\text{-}86\}$$

where the square matrix $\underline{\underline{A}}$ has the following definition

$$\underline{\underline{A}} = \begin{pmatrix} a_{11} & a_{12} & \cdots & a_{1n} \\ a_{21} & a_{22} & \cdots & a_{2n} \\ \vdots & \vdots & & \vdots \\ a_{n1} & a_{n2} & \cdots & a_{nn} \end{pmatrix}$$

$$= \begin{pmatrix} \langle\eta_1|\hat{A}|\eta_1\rangle & \langle\eta_1|\hat{A}|\eta_2\rangle & \cdots & \langle\eta_1|\hat{A}|\eta_n\rangle \\ \langle\eta_2|\hat{A}|\eta_1\rangle & \langle\eta_2|\hat{A}|\eta_2\rangle & \cdots & \langle\eta_2|\hat{A}|\eta_n\rangle \\ \vdots & \vdots & & \vdots \\ \langle\eta_n|\hat{A}|\eta_1\rangle & \langle\eta_n|\hat{A}|\eta_2\rangle & \cdots & \langle\eta_n|\hat{A}|\eta_n\rangle \end{pmatrix} \{II\text{-}87\}$$

Here the matrix $\underline{\underline{A}}$ represents the operator $\hat{A}$ in the basis $\{\eta_i\}$ or we may say that $\underline{\underline{A}}$ is the matrix representative of operator $\hat{A}$ in the basis $\{\eta_i\}$. Since operator $\hat{A}$ may have a matrix representative in any basis it is advisable to indicate the basis as a superscript: $\underline{\underline{A}}^{\eta}$ with <u>matrix elements</u> $\langle\eta_i|\hat{A}|\eta_j\rangle$. If $\{\eta_i\}$ represents AO and $\hat{A}$ is the kinetic energy operator then $\underline{\underline{A}}$ collects the kinetic energy values associated with the set of AO used.

The adjoint of an operator is defined in terms of its matrix elements such that:

$$\langle\eta_i|\hat{A}^{\dagger}|\eta_j\rangle \equiv \langle\eta_j|\hat{A}|\eta_i\rangle^* \qquad \{II\text{-}88\}$$

<u>A Hermitian Operator</u> is a linear operator that is self adjoint

$$\hat{H}^{\dagger} = \hat{H} \qquad \{II\text{-}89\}$$

and in an n-dimensional vector space it has n distinct eigen vectors* (orthonormal set) and n real eigen values*. A <u>Unitary Operator</u> is a linear operator, the adjoint of which is its inverse

$$U^{\dagger} = U^{-1} \qquad \{II\text{-}90\}$$

and in an n-dimensional vector space it has n distinct eigen vectors and n eigen values of ± 1. Again the situation is simpler in a real vector space where a Hermitian Operator becomes a Symmetric Operator and the Unitary Operator will be an Orthogonal Operator.

Table II-2 A Summary of Special Operators in Special Vector Spaces

	Vector Space	
	Hermitian	Euclidean
OPERATOR	Hermitian	Symmetric
	Unitary	Orthogonal

7. <u>Similarity Transformations</u>

Since one may have a matrix representative over any basis, e.g. the kinetic energy or potential energy operator over the AO basis, we may need to obtain a <u>similar</u> matrix representative over another basis, such as the MO basis. Let

---

*To be defined in Chapter II, Section 8.

$$\psi = \phi \underline{\underline{U}} \text{ i.e. } (\psi_1 \psi_2 \; .... \; \psi_n) = (\phi_1 \phi_2 \; .... \; \phi_n) \begin{pmatrix} U_{11} & U_{12} & .... & U_{1n} \\ U_{21} & U_{22} & .... & U_{2n} \\ \vdots & \vdots & & \vdots \\ U_{n1} & U_{n2} & .... & U_{nn} \end{pmatrix}$$

{II-91}

where both $\{\phi_i\}$ and $\{\psi_i\}$ are the basis vectors of two n-dimensional orthonormal vector spaces.
Correspondingly

{II-92}

$$\psi^{+} = \underline{\underline{U}}^{+} \phi^{\dagger} \text{ i.e. } \begin{pmatrix} \psi_1^* \\ \psi_2^* \\ \vdots \\ \psi_n^* \end{pmatrix} = \begin{pmatrix} U_{11}^* & U_{21}^* & .... & U_{n1}^* \\ U_{12}^* & U_{22}^* & .... & U_{n2}^* \\ \vdots & \vdots & & \vdots \\ U_{1n}^* & U_{nn}^* & .... & U_{nn}^* \end{pmatrix} \begin{pmatrix} \phi_1^* \\ \phi_2^* \\ \vdots \\ \phi_n^* \end{pmatrix}$$

Now we wish to express $\underline{\underline{A}}^{\psi}$ in terms of $\underline{\underline{A}}^{\phi}$ because $\underline{\underline{A}}^{\phi}$ is given and the unitary matrix $\underline{\underline{U}}$ that connects $\underline{\psi}$ and $\underline{\phi}$ is available

$$\underline{\underline{A}}^{\psi} = \begin{pmatrix} \langle\psi_1|\hat{A}|\psi_1\rangle & \langle\psi_1|\hat{A}|\psi_2\rangle & .... & \langle\psi_1|\hat{A}|\psi_n\rangle \\ \langle\psi_2|\hat{A}|\psi_1\rangle & \langle\psi_2|\hat{A}|\psi_2\rangle & .... & \langle\psi_2|\hat{A}|\psi_n\rangle \\ \vdots & \vdots & & \vdots \\ \langle\psi_n|\hat{A}|\psi_1\rangle & \langle\psi_n|\hat{A}|\psi_2\rangle & & \langle\psi_n|\hat{A}|\psi_n\rangle \end{pmatrix} \quad \{\text{II-93}\}$$

We have shown in the previous section (No. 6) that a given matrix element may be expressed in the following way:

$$\langle\psi_k|\hat{A}|\psi_\ell\rangle = \sum_{i=1}^{n} \sum_{j=1}^{n} U_{ki}{}^* a^{\phi}{}_{ij} U_{j\ell} \qquad \{II-94\}$$

$$= (U_{k1}{}^* U_{k2}{}^* \ \ldots \ U_{kn}{}^*) \begin{pmatrix} a_{11} & a_{12} & \cdots & a_{1n} \\ a_{21} & a_{22} & \cdots & a_{2n} \\ \vdots & \vdots & & \vdots \\ a_{n1} & a_{n2} & \cdots & a_{nn} \end{pmatrix} \begin{pmatrix} U_{11} \\ U_{21} \\ \vdots \\ U_{n1} \end{pmatrix}$$

or

$$a^{\psi}{}_{k1} = U_k \underline{\underline{A}}^{\phi} U_1 \qquad \{II-95\}$$

therefore the whole matrix may be written as

$$\begin{pmatrix} a_{11}{}^{\psi} & a_{12}{}^{\psi} & \cdots & a_{1n}{}^{\psi} \\ a_{21}{}^{\psi} & a_{22}{}^{\psi} & \cdots & a_{2n}{}^{\psi} \\ \vdots & \vdots & & \vdots \\ a_{n1}{}^{\psi} & a_{n2}{}^{\psi} & \cdots & a_{nn}{}^{\psi} \end{pmatrix} = \qquad \{II-96\}$$

$$\begin{pmatrix} U_{11}{}^* & U_{21}{}^* & \cdots & U_{n1}{}^* \\ U_{12}{}^* & U_{22}{}^* & \cdots & U_{n2}{}^* \\ \vdots & \vdots & & \vdots \\ U_{1n}{}^* & U_{2n}{}^* & \cdots & U_{nn}{}^* \end{pmatrix} \begin{pmatrix} a_{11}{}^{\phi} & a_{12}{}^{\phi} & \cdots & a_{1n}{}^{\phi} \\ a_{21}{}^{\phi} & a_{22}{}^{\phi} & \cdots & a_{2n}{}^{\phi} \\ \vdots & \vdots & & \vdots \\ a^{\phi}{}_{n1} & a^{\phi}{}_{n2} & \cdots & a^{\phi}{}_{nn} \end{pmatrix} \begin{pmatrix} U_{11} & U_{12} & \cdots & U_{1n} \\ U_{21} & U_{22} & \cdots & U_{2n} \\ \vdots & \vdots & & \vdots \\ U_{n1} & U_{n2} & \cdots & U_{nn} \end{pmatrix}$$

or

$$\underline{\underline{A}}^{\psi} = \underline{\underline{U}}^{\dagger} \ \underline{\underline{A}}^{\phi} \ \underline{\underline{U}} \qquad \{II-97\}$$

This equivalence is usually referred to a Similarity Transformation. If we work in a real (i.e. Euclidean) vector space then

$$\underline{\underline{A}}^{\psi} = \underline{\underline{\mathbf{O}}}'\ \underline{\underline{A}}^{\phi}\ \underline{\underline{\mathbf{O}}} \qquad \{II\text{-}98\}$$

i.e. $\underline{\underline{U}}$(unitary) $\longrightarrow$ $\underline{\underline{\mathbf{O}}}$(orthogonal) matrix as we saw before.

8. Eigen-Problem Equation

The eigen problem equation is of great importance in molecular orbital theory and is usually written in the following way:

$$\hat{Q}\phi_i = \phi_i \varepsilon_i \qquad \{II\text{-}99\}$$

where ϕ_i is the ith eigen vector and ε_i is the ith eigen value of operator $\hat{Q}$. For an n-dimensional space the eigen-problem equation may be written as

$$\hat{Q}\phi = \phi\ \underline{\underline{\varepsilon}} \qquad \{II\text{-}100\}$$

$$(1xn) \qquad (1xn) \quad (nxn)$$

which corresponds to

$$\hat{Q}(\phi_1\phi_2 \cdots \phi_n) = (\phi_1\phi_2 \cdots \phi_n) \begin{pmatrix} \varepsilon_1 & & & O \\ & \varepsilon_2 & & \\ & & \ddots & \\ O & & & \varepsilon_n \end{pmatrix} \qquad \{II\text{-}101\}$$

Taking the inner product with ϕ_i on both sides of equation {II-99} one obtains

$$\langle\phi_i|\hat{Q}\phi_i\rangle = \langle\phi_i|\phi_i\rangle\varepsilon_i \qquad \{II\text{-}102\}$$

or from equation {II-100} in matrix notation it is written as follows

$$\underline{\underline{Q}}^{\phi} = \underline{\underline{S}}^{\phi}\ \underline{\underline{\varepsilon}} \qquad \{II\text{-}103\}$$

$$\text{then} \qquad \underline{\underline{Q}}^{\phi} = \underline{\underline{\varepsilon}}$$

$$\text{if and only if} \qquad \underline{\underline{S}}^{\phi} = \underline{\underline{1}} \qquad \{II\text{-}104a,b\}$$

This implies that the matrix representative of an operator over the eigen vector space is a diagonal matrix.

If $\{\phi_i\}$ represents the MO, which are unknown, we have to obtain $\underline{\underline{Q}}^{\phi}$ by means of a similarity transformation from the known $\underline{\underline{Q}}^{\eta}$ where $\{\eta_i\}$ is the chosen set of AO.

$$\phi = \eta\, \underline{\underline{C}} \quad \text{(row vector notation)} \qquad \{II\text{-}105\}$$

$$\phi\dagger = C\dagger\eta\dagger \quad \text{(column vector notation)} \qquad \{II\text{-}106\}$$

By carrying out the appropriate substitution in the eigen problem equation one obtains the following relationship

$$\underline{\underline{Q}}^{\phi} = \underline{\underline{C}}\dagger\, \underline{\underline{Q}}^{\eta}\, \underline{\underline{C}} = \underline{\underline{C}}\dagger\, \underline{\underline{S}}^{\eta}\, \underline{\underline{C}}\, \underline{\underline{\varepsilon}} \qquad \{II\text{-}107\}$$

In this equation the form of a similarity transformation is apparent (transforming from AO to MO). If we expand $\underline{\phi}$ in terms of an orthonormal AO basis set $\{\chi\}$ noting that

$$\phi = \chi\, \underline{\underline{U}} \quad \text{and} \quad \phi\dagger = \underline{\underline{U}}\dagger\, \chi\dagger \qquad \{II\text{-}108a,b\}$$

then

$$\{II\text{-}109\}$$

$$\underline{\underline{Q}}^{\phi} = \underline{\underline{U}}\dagger\underline{\underline{Q}}^{\chi}\underline{\underline{U}} = \underline{\underline{U}}\dagger\, \underline{\underline{S}}^{\chi}\, \underline{\underline{U}}\, \underline{\underline{\varepsilon}} = \underline{\underline{U}}\dagger\, \underline{\underline{1}}\, \underline{\underline{U}}\, \underline{\underline{\varepsilon}} = \underline{\underline{U}}\dagger\, \underline{\underline{U}}\, \underline{\underline{\varepsilon}} = \underline{\underline{1}}\, \underline{\underline{\varepsilon}} = \underline{\underline{\varepsilon}}$$

Thus

$$\boxed{\underline{\underline{U}}\dagger\underline{\underline{Q}}^{\chi}\underline{\underline{U}} = \underline{\underline{\varepsilon}}}$$

Since $\{\chi\}$ is an orthonormal basis, $\underline{\underline{S}}^{\chi} = \underline{\underline{1}}$.

Since $\underline{\underline{U}}$ is a unitary matrix $\underline{\underline{U}}\dagger = \underline{\underline{U}}^{-1}$ and $\underline{\underline{U}}\dagger\underline{\underline{U}} = \underline{\underline{1}}$.

Since, in practice, one works with real (i.e. not complex) orbitals $\underline{\underline{U}}$ is an orthogonal matrix (and $\underline{\underline{U}}\dagger$ is its transpose) while $\underline{\underline{Q}}^{\chi}$ becomes a real symmetric matrix. Consequently the diagonalization of a real symmetric matrix is of utmost importance because the matrix ($\underline{\underline{U}}$) that diagonalizes $\underline{\underline{Q}}^{\chi}$ *via* a similarity transformation is the matrix that transforms the orthogonal AO basis set $\{\chi\}$ to the MO basis set $\{\phi\}$.

9. Jacobi's Method of Matrix Diagonalization

Although there are a number of methods to diagonalize a real symmetric matrix, Jacobi's method is probably the most widely used and computer programs, written to carry out MO calculations, frequently perform diagonalization by this method. It is therefore reviewed here.

The method is based on the successive plane (i.e. 2x2) rotations discussed previously. In fact an infinite number of rotations are required to achieve a "perfect" diagonalization but, for working accuracy, a finite number of steps is satisfactory. Jacobi's procedure may be illustrated in the following sequence where the original real symmetric matrix is $\underline{\underline{Q}}_o$, the matrix obtained after rotation in the first plane is $\underline{\underline{Q}}_1$, etc., and the final, diagonal, matrix is $\underline{\underline{\varepsilon}}$.

$$\underline{\underline{Q}}_o \rightarrow \underline{\underline{Q}}_1 \rightarrow \ldots . \rightarrow \underline{\underline{Q}}_{k-1} \rightarrow \underline{\underline{Q}}_k \rightarrow \ldots . \rightarrow \underline{\underline{\varepsilon}} \qquad \{II\text{-}110\}$$

The matrix obtained in the kth step has the following relationship to the previous matrix

$$\underline{\underline{Q}}_k = \underline{\underline{U}}_k^{\dagger} \, \underline{\underline{Q}}_{k-1} \, \underline{\underline{U}}_k \qquad \{II\text{-}111\}$$

where $\underline{\underline{U}}_k$ is an orthogonal matrix (and $\underline{\underline{U}}_k^{\dagger}$ is its transpose) associated with the rotation in the kth plane.

If the kth step involves rotation in the r,s plane then the orthogonal matrix ($\underline{\underline{U}}_k$) has the form similar to that discussed in Section 5 for the 2x2 rotation

$$\underline{\underline{U}}_k = \begin{pmatrix} 1 & 0 & \ldots & 0 & \ldots\ldots & 0 & \ldots & 0 & 0 \\ 0 & 1 & \ldots & 0 & \ldots\ldots & 0 & \ldots & 0 & 0 \\ \vdots & \vdots & & \vdots & & \vdots & & \vdots & \vdots \\ 0 & 0 & \ldots & \cos\alpha & \ldots & -\sin\alpha & \ldots & 0 & 0 \\ \vdots & \vdots & & \vdots & & \vdots & & \vdots & \vdots \\ 0 & 0 & \ldots & \sin\alpha & \ldots & \cos\alpha & \ldots & 0 & 0 \\ \vdots & \vdots & & \vdots & & \vdots & & \vdots & \vdots \\ 0 & 0 & \ldots & 0 & \ldots\ldots & 0 & \ldots & 0 & 1 \end{pmatrix} \begin{matrix} \\ \\ \\ \ldots r \\ \\ \ldots s \\ \\ \\ \end{matrix} \qquad \{II\text{-}112\}$$

$$\begin{matrix} & & & r & & s & & & \end{matrix}$$

In general the elements of this orthogonal matrix are:

$$\left.\begin{aligned} U_{rr} &= U_{ss} = \cos\alpha \\ U_{rs} &= -U_{sr} = -\sin\alpha \\ U_{ii} &= U_{jj} = 1 \quad (i \text{ or } j \neq r \text{ or } s) \\ U_{ij} &= U_{ji} = 0 \quad (i \text{ or } j \neq r \text{ or } s) \end{aligned}\right\} \qquad \{II\text{-}113a\text{-}d\}$$

Because the purpose of the plane rotation is to eliminate the off-diagonal matrix elements the angle of rotation (α) must be chosen in such a way that the r,sth element of the matrix $\underline{\underline{Q}}_k$, i.e. $q_{rs}^{\,k}$, will be equal to zero.

By substituting $\underline{\underline{U}}_k$ and $\underline{\underline{U}}_k^{\dagger}$ into the equation

$$\underline{\underline{Q}}_k = \underline{\underline{U}}_k^{\dagger}\, \underline{\underline{Q}}_{k-1}\, \underline{\underline{U}}_k \qquad \{II\text{-}114\}$$

one obtains the following diagonal and off-diagonal elements of $\underline{\underline{Q}}_k$ which are associated with the r,s plane

$$q_{rr}^{(k)} = q_{rr}^{(k-1)} \cos^2\alpha + 2\, q_{rs}^{(k-1)} \cos\alpha \sin\alpha + q_{ss}^{(k-1)} \sin^2\alpha$$

{II-115}

$$q_{ss}^{(k)} = q_{rr}^{(k-1)} \sin^2\alpha - 2\, q_{rs}^{(k-1)} \cos\alpha \sin\alpha + q_{ss}^{(k-1)} \cos^2\alpha$$

{II-116}

$$q_{rs}^{(k)} = q_{sr}^{(k-1)} = [q_{ss}^{(k-1)} - q_{rr}^{(k-1)}] \cos\alpha \sin\alpha + q_{rs}^{(k-1)} [\cos^2\alpha - \sin^2\alpha]$$

{II-117}

As noted above the purpose of diagonalization is to eliminate the off-diagonal elements so that the last equation must be set equal to zero

$$[q_{ss}^{(k-1)} - q_{rr}^{(k-1)}] \cos\alpha \sin\alpha + q_{rs}^{(k-1)} [\cos^2\alpha - \sin^2\alpha] = 0$$

{II-118}

By noting the following trigonometric relationships

$$\cos\alpha \sin\alpha = (1/2)\sin 2\alpha$$

$$\cos^2\alpha - \sin^2\alpha = \cos 2\alpha$$

{II-119a,b}

we obtain the following simplified expression

$$-(1/2)[q_{rr}^{(k-1)} - q_{ss}^{(k-1)}] \sin 2\alpha - q_{rs}^{(k-1)} \cdot \cos 2\alpha = 0$$

{II-120}

or, combining the appropriate terms,

$$\tan 2\alpha \;=\; \frac{2q_{rs}^{(k-1)}}{q_{rr}^{(k-1)} - q_{ss}^{(k-1)}} \qquad \{II\text{-}121\}$$

It is customary to choose α within the range of $\pm\,45^{\circ}$ since in the limiting cases $2\alpha = \pm 90^{\circ}$. The 45° limit is therefore appropriate whenever the denominator becomes zero (i.e. $q_{rr}^{(k-1)} = q_{ss}^{(k-1)}$. The sign of the angle is always the same as the sign of the numerator. The numerator will not be zero unless the diagonalization is complete.

The final orthogonal matrix $\underline{U}$ must incorporate all the rotational matrices $\underline{\underline{U}}_k$ used in the process of diagonalization.

Since:

$$\underline{\underline{Q}}_1 = \underline{U}_1{}^{+}\underline{\underline{Q}}_0\underline{U}_1$$

$$\underline{\underline{Q}}_2 = \underline{U}_2{}^{+}\underline{\underline{Q}}_1\underline{U}_2 \;\;= \underline{U}_2{}^{+}\underline{U}_1{}^{+}\underline{\underline{Q}}_0\underline{U}_1\underline{U}_2 \qquad\qquad \{II\text{-}122\}$$

$$= \underline{\underline{Q}}_n = \underline{U}_n{}^{+}\underline{\underline{Q}}_{n-1}\underline{U}_n = \underline{U}_n{}^{+}\cdots\; \underline{U}_2{}^{+}\underline{U}_1{}^{+}\underline{\underline{Q}}_0\underline{U}_1\underline{U}_2 \;\cdots\; \underline{\underline{U}}_n$$

it is evident that the final $\underline{U}$ is the matrix product of all $\underline{\underline{U}}_k$.

$$\underline{U} \;=\; \underline{\underline{U}}_1\underline{\underline{U}}_2 \;\cdots\; \underline{\underline{U}}_n \qquad \{II\text{-}123\}$$

In practice each $\underline{\underline{U}}_k$ matrix is multiplied by the previous products of $\underline{\underline{U}}_k$ so that one obtains successively $\underline{\underline{U}}_1$, $(\underline{\underline{U}}_1\underline{\underline{U}}_2)$, $(\underline{\underline{U}}_1\underline{\underline{U}}_2\underline{\underline{U}}_3)$ and so on

EXAMPLE

$$\underline{\underline{Q}}_0 \;=\; \begin{pmatrix} 0.6532 & 0.2165 & 0.0031 \\ & 0.4105 & 0.0052 \\ & & 0.2132 \end{pmatrix} \begin{matrix} 2 \\ \vdots \\ \cdots\, 1 \end{matrix} \qquad \{II\text{-}124a\}$$

Rotation in the 1,2 plane (i.e. x,y plane)

$$\tan 2\alpha = \frac{2 \times 0.2165}{0.6532 - 0.4105} \qquad \{II\text{-}124b\}$$

$$\cos\alpha = 0.8628; \quad \sin\alpha = 0.5055$$

$$\underline{\underline{U}}_1 = \begin{pmatrix} 0.8628 & -0.5055 & 0 \\ 0.5055 & 0.8628 & 0 \\ 0 & 0 & 1 \end{pmatrix} \qquad \{II\text{-}124c\}$$

$$\underline{\underline{Q}}_1 = \underline{\underline{U}}_1^{\dagger}\underline{\underline{Q}}_o\underline{\underline{U}}_1 \qquad \{II\text{-}124d\}$$

$$\underline{\underline{Q}}_1 = \begin{pmatrix} 0.7800 & 0.0000 & 0.0053 \\ & 0.2836 & 0.0029 \\ & & 0.2132 \end{pmatrix} \begin{matrix} \overset{3}{\vdots} \\ \ldots.1 \end{matrix} \qquad \{II\text{-}125a\}$$

Rotation in 1,3 plane (i.e. x,z plane)

$$\tan 2\alpha \quad \frac{2 \times 0.0053}{0.7800 - 0.2123} \qquad \{II\text{-}125b\}$$

$$\cos\alpha = 1.0000; \quad \sin\alpha = 0.0094$$

(Note that q_{13} is small; therefore, α is small.)

$$\underline{\underline{U}}_2 = \begin{pmatrix} 1.0000 & 0 & -0.0094 \\ 0 & 1.0000 & 0 \\ 0.0094 & 0 & 1.000 \end{pmatrix} \qquad \{II\text{-}125c\}$$

$$\underline{\underline{Q}}_2 = \underline{\underline{U}}_2^{\dagger}\underline{\underline{Q}}_1\underline{\underline{U}}_2$$

$$\underline{\underline{Q}}_2 = \begin{array}{c} \\ \\ \\ \end{array}\begin{array}{ccc|c} & & 3 & \\ & & \vdots & \\ \end{array}$$

$$\underline{\underline{Q}}_2 = \begin{pmatrix} 0.7801 & 0.0000 & 0.0000 \\ & & \vdots \\ & 0.2836 & 0.0029 \\ & & 0.2132 \end{pmatrix} \; \ldots\ldots \; 2 \qquad \{\text{II-126a}\}$$

Rotation in 2,3 plane (i.e. y,z plane)

$$\tan 2\alpha \qquad \frac{2 \times 0.0029}{0.2836 - 0.2132} \qquad \{\text{II-126b}\}$$

$$\cos\alpha = 0.9991; \quad \sin\alpha = 0.0411$$

$$\underline{\underline{U}}_3 = \begin{array}{ccc} 1.0000 & 0 & 0 \\ & 0.9991 & -0.0411 \\ 0 & 0.0411 & 0.9991 \end{array} \qquad \{\text{II-126c}\}$$

$$\underline{\underline{Q}}_3 = \underline{\underline{U}}_3{}^{\dagger}\underline{\underline{Q}}_2\underline{\underline{U}}_3 \qquad \{\text{II-126d}\}$$

$$\underset{\text{eigen values}}{\underline{\underline{Q}}_3} = \begin{pmatrix} 0.7801 & 0.0000 & 0.0000 \\ & 0.2837 & 0.0000 \\ & & 0.2131 \end{pmatrix} = \underline{\underline{\varepsilon}} \qquad \{\text{II-127a}\}$$

The eigen vectors are the columns of the product of the three rotating matrices:

$$\underline{\underline{U}} = \underline{\underline{U}}_1\underline{\underline{U}}_2\underline{\underline{U}}_3 = \begin{pmatrix} 0.8628 & -0.5054 & 0.0126 \\ 0.5055 & 0.8618 & -0.0402 \\ 0.0094 & 0.0411 & 0.9991 \end{pmatrix} \qquad \{\text{II-128}\}$$

10. Determinants

To every square (nxn) matrix there may be associated a determinant symbolized by

$$\det|\underline{\underline{A}}| = \begin{vmatrix} a_{11} & a_{12} & \cdots & a_{1n} \\ a_{21} & a_{22} & \cdots & a_{2n} \\ a_{n1} & a_{n2} & \cdots & a_{nn} \end{vmatrix} \qquad \{II-129\}$$

The determinant is simply a function which may be associated with the elements of a matrix. If det $\underline{\underline{A}} = 0$ then $\underline{\underline{A}}$ is a <u>singular</u> matrix otherwise $\underline{\underline{A}}$ is said to be <u>nonsingular</u>.

Considering the well known cases of a 2x2 and a 3x3 matrix

$$\det|\underline{\underline{A}}| = \begin{vmatrix} a_{11} & a_{12} \\ a_{21} & a_{22} \end{vmatrix} \qquad \{II-130\}$$

$$= a_{11}a_{22} - a_{12}a_{21}$$

$$\det|\underline{\underline{A}}| = \begin{vmatrix} a_{11} & a_{12} & a_{13} \\ a_{21} & a_{22} & a_{23} \\ a_{31} & a_{32} & a_{33} \end{vmatrix}$$

$$= a_{11}a_{22}a_{33} - a_{11}a_{23}a_{32} - a_{12}a_{21}a_{33} \qquad \{II-131\}$$

$$+ a_{13}a_{21}a_{32} - a_{13}a_{22}a_{31} + a_{12}a_{23}a_{31}$$

In general, the determinant associated with an nxn matrix has n! terms in its expansion of which n!/2 are positive and n!/2 negative.

The general rules for expanding a determinant in terms of its elements are:

1. Form the n! products

$$a_{ij}a_{k\ell}a_{mn} \quad \cdots \quad a_{st} \qquad \{II\text{-}132\}$$

$$i \neq k \neq m \neq \quad \ldots \neq s$$
$$j \neq \ell \neq n \neq \quad \ldots \neq t$$

2. Arrange each product so the row indices are in natural order 1, 2, ..., n.
3. The column indices then appear as the n! different permutations of n integers, for example, the 3! permutations of the integers 1, 2, and 3 in the 3x3 case.
4. Let p be the least number of transpositions needed to put the column indices in their natural order then $(-1)^p$ is the parity of the transpositions required to effect this reordering. For the 3x3 case, the results are given in Table II-3.

Table II-3 Transposed Products and the Parity of a 3x3 Expansion

Product	p	$(-1)^p$
$a_{11}a_{22}a_{33}$	0	1
$a_{11}a_{23}a_{32}$	1	-1
$a_{12}a_{21}a_{33}$	1	-1
$a_{13}a_{21}a_{32}$	2	1
$a_{13}a_{22}a_{31}$	3	-1
$a_{12}a_{23}a_{31}$	2	1

The value of the determinant is simply the sum of the individual products multiplied by the parity as given by {II-131}. The procedure may be expressed symbolically as

$$\det|A| = \sum_{P} (-1)^p \hat{P} a_{1\ell_1} a_{2\ell_2} \cdots a_{n\ell_n} \qquad \{II\text{-}133\}$$

where $\hat{P}$ is an operator which forms n! different permutations of the column indices with the row indices in natural order.

$(-1)^p\hat{P}$ can be combined and symbolized as the antisymmetrizer, $\hat{A}$.

Two important results regarding determinants are (1) that the interchange of any two rows or columns of a determinant changes the sign of the determinant and thus (2) if two rows or columns of a determinant are identical the value of the determinant is zero. These results are important in the construction of correctly symmetrized many electron wavefunctions as will be discussed later.

CHAPTER III

QUANTUM MECHANICAL BACKGROUND

This short section makes no attempt at rigour or completeness but merely presents very briefly a few topics in quantum mechanics of particular relevance to molecular orbital-calculations.

1. The Heisenberg Uncertainty Principle

In its general form the Heisenberg uncertainty principle may be written

$$\Delta f \Delta g \;>\; 1/2\, |\langle[\hat{f},\hat{g}]\rangle| \qquad \{III-1\}$$

where Δf and Δg are the inexactitudes in the measurement of the dynamical variables f and g, while $\hat{f}$ and $\hat{g}$ are the associated operators. The notation $[\hat{f},\hat{g}]$ in {III-1} is termed the commutator of $\hat{f}$ and $\hat{g}$ and symbolizes the expression: $\hat{f}\hat{g} - \hat{g}\hat{f}$. Two familiar special cases of {III-1} are:

$$\Delta p_x \Delta x \gtrsim h \quad \text{and} \quad \Delta E \Delta t \gtrsim h \qquad \{III-2a,b\}$$

This inherent uncertainty in the physical world requires quantum mechanics for an adequate description.

2. The Postulates of Quantum Mechanics

Almost every text on elementary quantum mechanics presents a set of postulates forming the basis of quantum mechanics. There is no unique set of postulates. One such set which will suffice for our purposes is:

(1) The constituent particles of a given system may be in several arrangements. Each arrangement is called a state of the system. For every state of the system there is a wavefunction (Ψ) which describes the configuration of the particles of that particular state. The wavefunction has no physical meaning but its square (or more precisely $\Psi^*\Psi$) may be interpreted as measuring the probability of finding the particles in that state in the configurational space.

(2) For every dynamical quantity of classical physics such as distance, momentum, kinetic energy, potential energy there is a corresponding Quantum

Mechanical operator. The operator is characteristic for the system but it is independent of the state of the system.

$$\hat{\theta}\Psi = \theta\Psi$$

where Ψ is the wavefunction (or eigen-function or eigen-vector) of the operator $\hat{\theta}$ and θ is the eigen-value of operator $\hat{\theta}$ associated with Ψ. Usually there are several possible Ψ and therefore each Ψ_i has its own eigen-value θ_i. Solution Ψ_i is the eigen-function of the ith state of the given system and θ_i is the eigen-value of the ith state of the same system.

(4) Neither Ψ nor $\hat{\theta}$ has physical significance only θ represents a physical quantity. It is expected that there is a 1:1 correspondence between the values which are measured (θ_{exp}) and those obtained from the wave equation (θ_{calc}).

From our point of view, the most important operator will be the Hamiltonian operator (usually for a molecule) which has as its associated eigenvalues the quantum mechanically allowed energy levels of a system (molecule). Employing the bra-ket notation of Dirac:

$$\frac{\int \Psi_i^* \hat{H} \Psi_i d\tau}{\int \Psi_i^* \Psi_i d\tau} = \frac{\langle \Psi_i | \hat{H} | \Psi_i \rangle}{\langle \Psi_i | \Psi_i \rangle} = E_i \qquad \{III\text{-}3\}$$

The evaluation of such integrals and thus the solution of the above equation for organic molecules either rigorously or usually approximately will be the major goal of the rest of the course.

3. Atomic Units

Certain advantages are to be gained by working with quantities expressed in atomic units and these units will now be discussed.

Physical quantities, normally expressed in CGS or SI units, are computed in atomic units in quantum chemistry. This practice is very useful because the computed results are then independent of the values for the universal constants h (Planck's constant), e (the charge of the electron) and m_o (the rest mass of the electron). Even if the values of these constants are determined to a higher accuracy at some future date calculated quantum chemical values will remain unchanged if they are computed in atomic units.

The definition of atomic units involves the setting of m_o, $\hbar = h/2\pi$, and e equal to unity. This choice requires that charge separation ($\delta+$ and $\delta-$) in a molecule be expressed as a fraction or multiple of the charge of an electron. Furthermore, it immediately leads to the definition of the atomic unit of length: the Bohr radius or the Bohr atomic unit or simply the bohr.

$$a_o = \frac{(h/2\pi)^2}{m_o e^2} \qquad \{III-4\}$$

If all three quantities, involved in the above definition of a_o, are taken to be unity then clearly a_o itself is equal to 1 as well.

$$a_o \equiv 1 \text{ Bohr a.u.} \equiv 1 \text{ bohr} \qquad \{III-5\}$$

In Å or CGS or SI units a_o has approximately the following value

$$a_o = 0.529177 \text{ Å} = 0.529177 \times 10^{-8} \text{ cm} = 0.529177 \times 10^{-10} \text{ m} \qquad \{III-6\}$$

The atomic unit of energy can be derived from Coulombs law of electron-electron repulsion, i.e. charge times charge over distance:

$$E_o = \frac{e^2}{a_o} = \frac{m_o e^4}{(h/2\pi)^2} \qquad \{III-7\}$$

Again, since all three quantities involved in the definition of E_o are defined to be unity therefore E_o itself, the atomic unit of energy (Hartree a.u. or simply hartree), must also be equal to 1.

$$E_o \equiv 1 \text{ Hartree a.u.} \equiv 1 \text{ hartree} \qquad \{III-8\}$$

It should be mentioned that the energy computed in hartree is for a single particle, atom, or molecule in very much the same way as spectroscopic energy units are also used for a single particle:

$$1 \frac{\text{hartree}}{\text{particle}} = \frac{27.2117 \text{ e.V.}}{\text{particle}} = \frac{219475 \text{ cm}^{-1}}{\text{particle}} \qquad \{III-9\}$$

For an Avogadro number of particles the equivalent amount of energy in CGS or SI units is the following

$$1 \frac{\text{hartree}}{\text{particle}} = 627.52 \frac{\text{Kcal}}{\text{mole}} = 2625.51 \frac{\text{KJ}}{\text{mole}} \qquad \{III-10\}$$

For convenience, the numerical equivalence of all these energy units are summarized in Table III-1.

4. The Variational Theorem

Having briefly considered the fundamental principles of quantum mechanics and having introduced convenient atomic units it is important to consider the variational theorem. Historically, almost all calculations of molecular energy within the MO method have been variational in nature. The variational theorem states that any approximate wavefunction Φ will produce an energy ε which is higher than the exact energy E given by the exact wavefunction Ψ. As the approximate wavefunction approaches the exact ($\Phi \rightarrow \Psi$) then the approximate energy will also approach the exact ($\varepsilon \rightarrow E$). A formal proof of this theorem for the ground state of a system follows.

Table III-1

Conversion Table* for Frequently Used Energy Units

	hartree/particle	eV/particle	cm^{-1}/particle	kcal/mole	kJ/mole
hartree/particle	1	27.2117	219475	627.52	2625.51
eV/particle	3.675×10^{-2}	1	8066	23.06	96.48
cm^{-1}/particle	4.556×10^{-6}	1.239×10^{-4}	1	0.00286	0.0120
kcal/mole	1.5935×10^{-3}	4.3365×10^{-2}	349.65	1	4.1839
kJ/mole	3.808×10^{-4}	1.0365×10^{-1}	83.57	0.2390	1

*This table is one of equivalence and not of identity. For example, the cm^{-1} is not itself an energy unit but is related to energy through $E = hc\nu$.

Suppose the Hamiltonian of the chemical system is given. Then the Schrodinger equation has exact solutions Ψ_i (i = o, n; n → ∞) each with an associated energy E_i (for simplicity all states are assumed non-degenerate). The exact wavefunctions $\{\Psi_i\}$ form a complete orthonormal set and thus any approximate wavefunction, Φ, may be expanded to an arbitrary degree of precision in terms of these functions:

$$\Phi = \sum_{i=0}^{n} \Psi_i c_i \qquad \{III-11\}$$

with $$\sum_{i=0}^{n} c_i^* c_i = 1 \qquad \{III-12\}$$

This condition implies that the expansion will yield normalized

$$\langle\Phi|\Phi\rangle = \langle\sum_i^n c_i\psi_i | \sum_j^n \psi_i c_j\rangle$$

$$= \sum_i^n \sum_j^n c^*_i c_j \langle\psi_i\psi_j\rangle = \sum_i^n c^*_i c_i = 1 \qquad \{III-13\}$$

Evaluation of the expectation value of the Hamiltonian over the Φ yields

$$\varepsilon = \langle\Phi|\hat{H}|\Phi\rangle = \langle\sum_{k=0}^{n} c_k\Psi_k|\hat{H}|\sum_{\ell=0}^{n} \Psi_\ell c_\ell\rangle$$

$$= \sum_k^n\sum_\ell^n c_k^* c_\ell \langle\Psi_k|\hat{H}|\Psi_\ell\rangle$$

$$= \sum_k^n c_k^* c_k \langle\Psi_k|\hat{H}|\Psi_k\rangle \qquad \{III-14\}$$

$$= \sum_k^n c_k^* c_k E_k$$

Where use has been made of the orthogonality of wavefunctions of the same operator (in this case the Hamiltonian) with different eigenvalues.

Since

$$\sum_{k=0}^{n} c_k{}^*c_k = 1 \qquad \{III-12\}$$

E_o may be written

$$E_o = \sum_{k=0}^{n} c_k{}^*c_k\, E_o \qquad \{III-15\}$$

where $1 > c_k{}^*c_k > 0$. Subtracting {III-15} from both sides of {III-14} gives

$$\varepsilon - E_o = \sum_{k}^{n} c_k{}^*c_k\, E_k - \sum_{k=0}^{n} c_k{}^*c_k\, E_o$$

$$= \sum_{k}^{n} c_k{}^*c_k\, (E_k - E_o). \qquad \{III-16\}$$

Since $\sum_k c_k{}^*c_k > 0$ and $E_k - E_o > 0$, the right hand side of {III-16} is greater than zero and the same must be true for the left hand side as well:

$$\varepsilon - E_o > 0 \qquad \{III-17a\}$$

Thus

$$\varepsilon > E_o. \qquad \{III-17b\}$$

Any trial wavefunction yields an upper bound on the true energy. A wavefunction's quality may be judged on the basis of the energy it yields. The lower the energy the "better" the wavefunction in the sense of the variational theorem.

The above proof is, of course, a formal one since in practice the complete orthonormal set $\{\Psi_k\}$ of exact solutions is unknown. Therefore some systematic method of constructing the trial wavefunctions is required. Certain general conditions that any N-particle wavefunction must satisfy follow simply from the principle that in quantum mechanics identical particles are indistinguishable (i.e. it is not possible even in principle to attach labels to the particles).

5. General Remarks on the Construction of Wavefunctions

The wavefunction of a given state for an N-particle system describes the "distribution" or configuration of these N-particles in the given state, i.e.

$$\Phi(1,2, \ldots, N) \qquad \{III-18\}$$

A special case of an N particle wavefunction involves the case N = 1

$$\phi(1) = u(1) \qquad \{III-19\}$$

where the one electron function u has a special importance as defined by equation {III-20}

It should be recalled that elementary particles have spin characteristic of the particle. Depending on the type of particle we have a restriction on the wavefunction as given in Table III-2.

Table III-2 Spin Symmetry Restrictions

Particles		Spin	Description	
Examples	Type	Eigenvalue	Statistics	Wavefunction
α, $h\nu$	Bosons	$n\hbar$	Bose-Einstein	Symmetric
e,p,n	Fermions	$(n/2)\hbar$	Fermi-Dirac	Anti-symmetric

The one electron wavefunction u of equation {III-19} should include both a space function ϕ and a spin function (α or β).

$$u(1) = \begin{cases} \phi_1(1)\alpha(1) \\ \\ \phi_1(1)\beta(1) \end{cases} \qquad \{III-20\}$$

The one electron function u is a spin-orbital while the ϕ is a space orbital. The product of ϕ and α or β implies the simultaneous probability of space orbital ϕ and spin α or β.

Many-electron wavefunctions may be constructed as products of one electron functions or spin-orbitals

$$\Phi(1,2, \ldots., N) = u_1(1)u_2(2) \ldots. u_N(N) \qquad \{III-21\}$$

Since the electrons are indistinguishable any permutation of the labels of {III-21} should also be a satisfactory wavefunction thus yielding N! terms.

$$\Phi(1,2, \ldots., N) = \hat{A}\{u_1(1)u_2(2) \ldots. u_N(N)\} \qquad \{III-23\}$$

$$= \frac{1}{\sqrt{N!}} \sum_{\nu=1}^{N!} (-1)^{\nu}\hat{P}_{\nu} u_1(1)u_2(2) \ldots. u_N(N)$$

which was seen in Chapter II, Section 10 to be the formula for a determinant.

$$\Phi(1,2, \ldots., N) = \frac{1}{\sqrt{N!}} \begin{vmatrix} u_1(1) & \ldots. & u_N(1) \\ \vdots & & \vdots \\ u_1(N) & \ldots. & u_N(N) \end{vmatrix} \qquad \{III-24\}$$

The spin-orbitals may be replaced by the product of a space and spin function from which it follows that an orbital may appear only twice in a determinantal wavefunction once with α-spin and once with β-spin.

$$\Phi(1,2,\ \ldots\ 2M) = \qquad \{III-25\}$$

$$= \frac{1}{\sqrt{(2M)!}} \begin{vmatrix} \phi_1(1)\alpha(1) & \phi_1(1)\beta(1) & \ldots & \phi_{M-1}(1)\alpha(1) & \phi_M(1)\beta(1) \\ \vdots & \vdots & & \vdots & \vdots \\ \phi_1(2M)\alpha(2M) & \phi_1(2M)\beta(2M) & \ldots & \phi_{M-1}(2M)\alpha(2M) & \phi_M(2M)\beta(2M) \end{vmatrix}$$

This determinantal form of a wavefunction is referred to as a Slater determinant.

6. Introduction to Atoms

In general, atoms may be viewed as the simplest (i.e. monatomic) molecules. Moreover, many of the theoretical techniques with which we are concerned may be applied to atoms as well as to molecules. A brief consideration of atoms will serve to introduce some important concepts we will use later in our study of molecules.

(i) The Central Field Hamiltonians

The hydrogen atom is a simple mathematical problem in quantum mechanics but for any atom with more than one electron an exact solution is impossible. For two-electron atoms very elaborate approximate solutions of the Schrodinger equation have been set up (notably by E. Hylleraas about 1930) but these methods involving explicit use of interelectron co-ordinates are too difficult to have any general applicability to many electron atoms or molecules. A good starting point is provided by assuming that each electron moves in a central or spherically symmetrical force field produced by the nucleus and the other electrons. The instantaneous action of all the electrons of an atom on one of their number is replaced by the averaged charge distribution of each other electron. This charge distribution summed over the electrons of an atom is very nearly spherically symmetrical and in the central field approximation one takes a spherical average.

In the one electron central field problem (the hydrogen atom), the Schrödinger equation in CGS units is

$$\{-\frac{h^2}{8\pi e m_o}\nabla^2 - \frac{e^2}{r}\}\ \psi = E\psi \qquad \{III-26\}$$

where ∇^2 is the Laplacian operator

$$\nabla^2 = \frac{\partial^2}{\partial x^2} + \frac{\partial^2}{\partial y^2} + \frac{\partial^2}{\partial z^2} \qquad \{\text{III-27}\}$$

and the Hamiltonian operator in {III-26} may be identified as:

$$\hat{H}^{cgs} = -\frac{h^2}{8\pi^2 m_o} \nabla^2 - \frac{e^2}{r} \qquad \{\text{III-28}\}$$

As an exercise to show the simplifications obtained by using atomic units (Section 3) we will convert the Hamiltonian from CGS to atomic units.

Rewriting {III-28} using the definition of a_o {III-4} gives

$$\hat{H}^{cgs} = -\frac{1}{2} a_o e^2 \nabla^2 - \frac{e^2}{r} \qquad \{\text{III-29}\}$$

Conversion of $\hat{H}^{cgs}$ to $\hat{H}^{a.u.}$ then requires division by (e^2/a_o);

$$\hat{H}^{a.u.} = \frac{\hat{H}^{cgs}}{\frac{e^2}{a_o}} = -\frac{1}{2} \frac{a_o e^2}{\frac{e^2}{a_o}} \nabla^2 - \frac{e^2/r}{\frac{e^2}{a_o}} \qquad \{\text{III-30}\}$$

which rearranges to

$$\hat{H}^{a.u.} = -\frac{1}{2} a_o^2 \nabla^2 - \frac{1}{(r/a_o)} \qquad \{\text{III-31}\}$$

Noting that

$$a_o^2 \nabla^2 = \frac{\partial^2}{\partial (x/a_o)^2} + \frac{\partial^2}{\partial (y/a_o)^2} + \frac{\partial^2}{\partial (z/a_o)^2}$$

$$\{\text{III-32}\}$$

we may see that the distance r and coordinates x, y and z have been converted to Bohr a.u.. Consequently (r/a_o) as well as (x/a_o), (y/a_o), and (z/a_o) may be replaced by r, x, y, and z measured in bohrs (see Figure III-1).

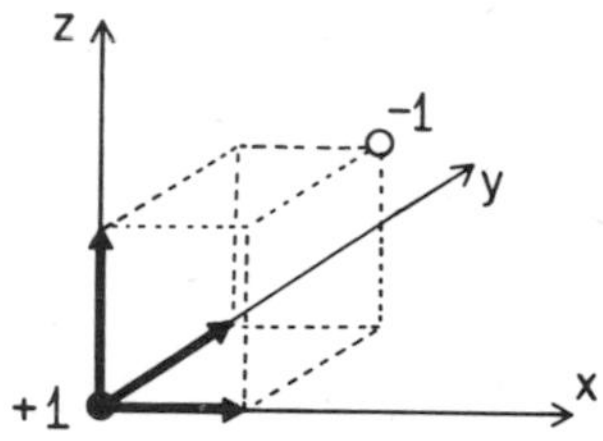

Figure III-1 Hydrogen atom in Cartesian coordinate system with charges and distances in atomic units

Thus the conversion leads to a very simple expression for the Hamiltonian operator of the hydrogen atom in a.u.

$$\hat{H}^{a.u.} = -\frac{1}{2}\nabla^2 - \frac{1}{r} \qquad \{III-33\}$$

The first term represents the kinetic energy (KE) of the

electron (in a fixed nucleus approximation the KE of the nucleus is not considered). The well known solutions to the problem, the hydrogen-like atomic orbitals are given in Table III-2. The Hamiltonian operator for other atomic systems (e.g. He, Li) can also be written in a correspondingly simple form in a.u. as illustrated in Figure III-2. A new type of interaction, the electron-electron repulsion represented by $1/r_{ij}$ (and the cause of the previously discussed difficulties in exactly solving many electron problems) has entered into the Hamiltonian which of course was not present in the one electron case.

(ii) Numeric Atomic Orbitals

Hartree and Fock showed that in order to obtain the radial part R of the atomic orbitals η

$$\eta = R \underbrace{\theta \quad \Phi}_{\text{spherical harmonic}} \qquad \{III\text{-}34\}$$

one must solve a differential equation analogous to that of the hydrogen problem

$$\frac{d^2P_{n\ell}(r)}{dr^2} + 2(E-V) - \frac{\ell(\ell+1)}{r^2} P_{n\ell}(r) = 0 \qquad \{III\text{-}35\}$$

where

$$P_{n,\ell}(r) = rR_{n\ell}(r) \qquad \{III\text{-}36\}$$

However due to the complexity of the potential V(r), (the presence of the electron-electron repulsion term), no direct analytic solution is possible. Numerical integration by the method of finite differences [see J. C. Slater, Quantum Theory of Matter, 2nd Ed., McGraw-Hill (1968), pp. 134-152] for a given r is possible and yields a wavefunction (see Table III-3

TABLE III-2 ANALYTIC ATOMIC ORBITALS FOR HYDROGEN

Orbital quantum numbers n	ℓ	m	Orbital designation	Radial function,[b] $R_{n\ell}(r)$	Angular function,[c] $\phi_{\ell m}$ $\frac{x}{r}, \frac{y}{r}, \frac{z}{r}$
1	0	0	1s	$2e^{-r}$	$\frac{1}{2\sqrt{\pi}}$
2	0	0	2s	$\frac{1}{2\sqrt{2}}(2-r)e^{-r/2}$	$\frac{1}{2\sqrt{\pi}}$
2	1	$(1)^d$	$2p_x$	$\frac{1}{2\sqrt{6}}re^{-r/2}$	$\frac{\sqrt{3}(x/r)}{2\sqrt{\pi}}$
2	1	0	$2p_z$	$\frac{1}{2\sqrt{6}}re^{-r/2}$	$\frac{\sqrt{3}(z/r)}{2\sqrt{\pi}}$
2	1	$(-1)^d$	$2p_y$	$\frac{1}{2\sqrt{6}}re^{-r/2}$	$\frac{\sqrt{3}(y/r)}{2\sqrt{\pi}}$
3	0	0	3s	$\frac{2}{81\sqrt{3}}(27-18r+2r^2)e^{-r/3}$	$\frac{1}{2\sqrt{\pi}}$
3	1	$(1)^d$	$3p_x$	$\frac{4}{81\sqrt{6}}(6r-r^2)e^{-r/3}$	$\frac{\sqrt{3}(x/r)}{2\sqrt{\pi}}$
3	1	0	$3p_z$	$\frac{4}{81\sqrt{6}}(6r-r^2)e^{-r/3}$	$\frac{\sqrt{3}(z/r)}{2\sqrt{\pi}}$
3	1	$(-1)^d$	$3p_y$	$\frac{4}{81\sqrt{6}}(6r-r^2)e^{-r/3}$	$\frac{\sqrt{3}(y/r)}{2\sqrt{\pi}}$
3	2	$(2)^d$	$3d_{x^2-y^2}$	$\frac{4}{81\sqrt{30}}r^2e^{-r/3}$	$\frac{\sqrt{15}[(x^2-y^2)/r^2]}{4\sqrt{\pi}}$
3	2	$(1)^d$	$3d_{xz}$	$\frac{4}{81\sqrt{30}}r^2e^{-r/3}$	$\frac{\sqrt{30}(xz/r^2)}{2\sqrt{2\pi}}$
3	2	0	$3d_{z^2}$	$\frac{4}{81\sqrt{30}}r^2e^{-r/3}$	$\frac{\sqrt{5}[(3z^2-r^2)/r^2]}{4\sqrt{\pi}}$
3	2	$(-1)^d$	$3d_{yz}$	$\frac{4}{81\sqrt{30}}r^2e^{-r/3}$	$\frac{\sqrt{30}(yz/r^2)}{2\sqrt{2\pi}}$
3	2	$(-2)^d$	$3d_{xy}$	$\frac{4}{81\sqrt{30}}r^2e^{-r/3}$	$\frac{\sqrt{15}(xy/r^2)}{2\sqrt{\pi}}$

[a]Both the radial and the angular functions are normalized to one.

[b]To convert to a general radial function for a one-electron atom with any nuclear charge Z, replace r by Zr/a_o and multiply each function by $(Z/a_o)^{3/2}$.

[c]Often expressed in the spherical coordinates θ and φ by replacing x with r sin θ x cos φ, y with r sin θ sin φ, and z with r cos θ.

[d]It is not correct to assign m_ℓ values to the real functions x, y, x^2-y^2, xz, yz, and xy. However, this fiction will help us find the correct terms for linear molecules.

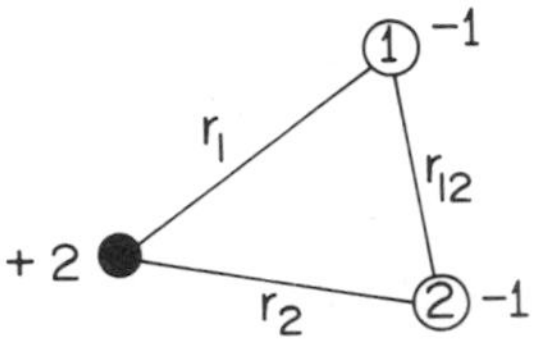

$$H^{au}_{He} = \underbrace{-\frac{\nabla_1^2}{2} - \frac{\nabla_2^2}{2}}_{K.E.} \underbrace{- \frac{2}{r_1} - \frac{2}{r_2} + \frac{1}{r_{12}}}_{P.E.}$$

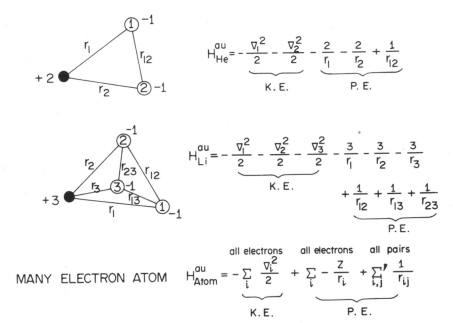

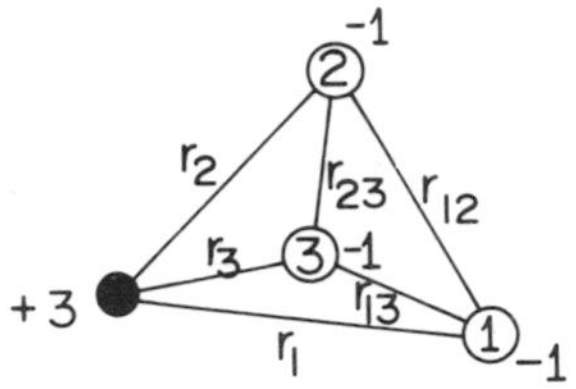

$$H^{au}_{Li} = \underbrace{-\frac{\nabla_1^2}{2} - \frac{\nabla_2^2}{2} - \frac{\nabla_3^2}{2}}_{K.E.} - \frac{3}{r_1} - \frac{3}{r_2} - \frac{3}{r_3} \underbrace{+ \frac{1}{r_{12}} + \frac{1}{r_{13}} + \frac{1}{r_{23}}}_{P.E.}$$

MANY ELECTRON ATOM

$$H^{au}_{Atom} = \underbrace{-\sum_i^{\text{all electrons}} \frac{\nabla_i^2}{2}}_{K.E.} + \underbrace{\sum_i^{\text{all electrons}} -\frac{Z}{r_i} + \sum_{i,j}^{\prime\ \text{all pairs}} \frac{1}{r_{ij}}}_{P.E.}$$

Figure III-2 Hamiltonian operators for atomic systems

TABLE III-3 NUMERICAL HARTREE-FOCK ATOMIC ORBITALS FOR CARBON

Shell	(1s)	(2s)	(2p)
ε	-21.378	-1.2895	-0.6603
x[a]	$P(x)$[b]	$P(x)$	$P(x)$
0	0	0	0
0.01	0.1329	0.0307	0.0002
0.02	0.2582	0.0596	0.0006
0.03	0.3762	0.0867	0.0014
0.04	0.4871	0.1123	0.0025
0.05	0.5914	0.1362	0.0039
0.06	0.6892	0.1586	0.0055
0.07	0.7810	0.1795	0.0074
0.08	0.8670	0.1990	0.0095
0.09	0.9473	0.2171	0.0119
0.10	1.0224	0.2339	0.0145
0.12	1.1576	0.2637	0.0203
0.14	1.2745	0.2888	0.0268
0.16	1.3746	0.3096	0.0340
0.18	1.4596	0.3265	0.0419
0.20	1.5309	0.3396	0.0503
0.22	1.5899	0.3493	0.0593
0.24	1.6377	0.3559	0.0686
0.26	1.6754	0.3597	0.0784
0.28	1.7040	0.3608	-
0.30	1.7246	0.3595	0.0989
0.34	1.7445	0.3505	0.1206
0.38	1.7410	0.3341	0.1430
0.42	1.7191	0.3116	0.1659
0.46	1.6828	0.2841	0.1892
0.50	1.6355	0.2526	0.2126
0.54	1.5800	0.2180	0.2360
0.58	1.5186	0.1808	0.2592
0.62	1.4532	0.1417	0.2822
0.66	1.3853	0.1014	0.3049
0.70	1.3162	0.0602	0.3271
0.78	1.1782	-0.0234	0.3700
0.86	1.0449	-0.1064	0.4107
0.94	0.9198	-0.1870	0.4488
1.02	0.8048	-0.2640	0.4843
1.10	0.7007	-0.3366	0.5170
1.18	0.6075	-0.4042	0.5471
1.26	0.5249	-0.4665	0.5745
1.34	0.4521	-0.5234	0.5993
1.42	0.3884	-0.5748	0.6215
1.50	0.3329	-0.6208	0.6412
1.66	0.2432	-0.6973	0.6736
1.82	0.1765	-0.7545	0.6973
1.98	0.1275	-0.7946	0.7132
2.14	0.0917	-0.8196	0.7223
2.30	0.0657	-0.8319	0.7254
2.46	0.0469	-0.8335	0.7235
2.62	0.0334	-0.8264	0.7173
2.78	0.0238	-0.8122	0.7076
2.94	0.0169	-0.7925	0.6951
3.10	0.0119	-0.7685	0.6804
3.42	0.0060	-0.7121	0.6460
3.74	0.0030	-0.6498	0.6079
4.06	0.0015	-0.5863	0.5684
4.38	0.0007	-0.5248	0.5292
5.98	0	-0.2833	0.3600
8.86	0	-0.0810	0.1654
12.06	0	-0.0180	0.0637
17.82	0	-0.0010	0.0099

[a] $x = \frac{r}{\mu}$ where

$$\mu = \frac{1}{2}\left(\frac{3\pi}{4}\right)^{2/3} Z^{-1/3} = 0.8853\ Z^{-1/3}.$$

For carbon $Z = 6$ and $\mu \approx 0.4872$.

[b] $P(x)$ is given by {III-35}.

for the carbon atom) as a table of numbers rather than in analytical form.

Although the concept of orbitals is an arbitrary and approximate one a brief consideration of the numerical method suggests that a straightforward numerical approach (if it could be made to work) would not be very satisfactory. The chemical information would be present but in an inconvenient, inaccessible form. It might well be impossible to discover the concepts and ideas of chemistry in such results.

(iii) Analytic Atomic Orbitals

From our point of view analytic functions which may be used to approximate AO are of particular value. Such atomic functions (perhaps with additional functions added) provide a basis set for MO calculations within the LCAO-MO method. As might be imagined analytic functions are easier to handle computationally than the numerical Hartree-Fock atomic orbitals just discussed.

Historically the first such approximations used were hydrogen-like orbitals followed by nodeless exponential-type functions (Slater type orbitals, STO) and then Gaussian-type functions (GTF). Today the most popular large program packages for MO calculations (e.g. IBMOL,POLYATOM, GAUSSIAN 70) all employ Gaussian type functions. Pure STO present extreme computational difficulty for multi-centre electron-electron repulsion integrals in molecules of general geometry. However, STO give a more accurate approximation in the sense that for the same number of STO as GTF a lower energy is obtained with STO. However, the computational ease of GTF allows the use of more such functions in a calculation. Further details concerning GTF will be given in Chapter XIII.

7. Introduction to Molecules

In this section, we shall briefly consider a number of points relating to molecules and their theoretical treatment.

(i) The Molecular Hamiltonian

The nonrelativistic Hamiltonian operator for a molecule of N nuclei and n electrons is

$$\hat{H} = -\frac{1}{2}\sum_{k=1}^{N}\frac{1}{M_k}\nabla^2_k - \frac{1}{2}\sum_{\mu=1}^{n}\nabla^2_\mu - \sum_{\mu=1}^{n}\sum_{k=1}^{N}\frac{Z_k}{r_{\mu k}}$$

1. 2. 3. {III-37}

$$+ \sum_{k<\ell}^{N}\frac{Z_k Z_\ell}{r_{k\ell}} + \sum_{\mu<\nu}^{N}\frac{1}{r_{\mu\nu}}$$

4. 5.

All nuclear and electronic coordinates are referred to the centre of mass of the system. The various terms are (1.) the kinetic energy of the nuclei where M_k is the mass in atomic units of the kth nucleus (2.) the electrons' kinetic energy (3.) the electron-nucleus-attractive potential energy (4.) the nuclear-nuclear repulsive potential energy and (5.) the electron-electron-repulsive potential energy. The exact Schrödinger equation which this Hamiltonian satisfies can be written

$$\hat{H}\Psi(r,R) = E\Psi(r,R) \qquad \{III\text{-}38\}$$

where E is the total internal energy of the molecule. The wavefunction $\Psi(r,R)$ depends upon both the electronic coordinates r and the nuclear coordinates R. Whereas the entire energy of an atom can be viewed as electronic (relative to a stationary centre of mass) molecules have electronic, vibrational, and rotational motions all coupled in a rather subtle fashion. However, certain assumptions and approximations can greatly simplify matters - the most important such approximation being due to Born and Oppenheimer.

(ii) <u>The Born-Oppenheimer Approximation</u>

According to Born and Oppenheimer the solutions of the Schrödinger equation can be expanded in a power series in $M^{-1/4}$ (where M is the total mass of the system). If this function is much less than unity then approximate solutions to {III-38} may be obtained by solving the wave equation for a series of fixed nuclear positions. This electronic energy can then be used as the potential energy for the wavefunction of the nuclei alone. This approximation rests upon the physical picture of massive nuclei moving so slowly relative to the electronic motions that the electrons can be thought of as being in quasi-stationary states during the course of the nuclear vibrations.

According to the Born-Oppenheimer approximation the total wavefunction of the molecule can be written

$$\Psi(r,R) = \psi_R(r)\phi(R) \qquad \{III-39\}$$

where $\psi_R(r)$ is the electronic wavefunction for fixed nuclear positions and depends only parametrically upon the nuclear coordinates. The function $\phi(R)$ is called the nuclear wavefunction.

Rewriting the Hamiltonian {III-37} as

$$\hat{H} = \hat{T}_R + \hat{h} + \hat{V} \qquad \{III-40\}$$

where $\hat{T}_R$ is the kinetic energy operator of the nuclei, $\hat{h}$ is a sum of operators each of the form

$$\hat{h}_\mu = -\frac{\nabla^2_\mu}{2} - \sum_k \frac{Z_k}{r_{\mu k}} \qquad \{III-41\}$$

and $\hat{V}$ is the potential-energy operator for nuclear-nuclear repulsions and electron-electron repulsions. The operator $(\hat{h} + \hat{V})$ is referred to as the electronic hamiltonian.

$$(\hat{h}+\hat{V})\psi_R(r) = E(R)\psi_R(r) \qquad \{III-42\}$$

E(R) is usually referred to as the molecular energy in the fixed-nuclei approximation. For a diatomic molecule a plot of E(R) versus the internuclear distance R leads to the well-known potential-energy curve the existence of which, of course, depends on the validity of the Born-Oppenheimer approximation.

Of course, more complicated molecules lead to a more complicated functional dependence of the energy on geometry leading to surfaces and hypersurfaces (of which more later).

(iii) Analytic Molecular Orbitals

At the very heart of this course lies the problem of finding analytic functions to describe the electronic structure of molecules. Such functions or molecular orbitals are in general multicentred and delocalized over the entire molecule. In the methods we will consider in much detail later, the molecular orbitals will be approximated by a linear combination of atomic orbitals (LCAO-MO).

(iv) The Virial and Hellmann-Feynman Theorems

Two theorems that aid in the interpretation of chemical bonding - i.e. in analysis of wavefunctions - will be mentioned briefly in this section.

(a) The Virial Theorem

Consider a system in a stationary state ψ and let $\hat{H}$ be its time-independent Hamiltonian then for any linear operator $\hat{A}$ which does not involve the time it is easy to show that

$$\int \psi^* \, [\hat{H},\hat{A}]\psi d\tau = 0 \qquad \{III-43\}$$

which is termed the hypervirial theorem. It states that the expectation values of time independent operators do not vary with time in stationary states.

Choosing $\hat{A}$ to be

$$\sum_i \hat{q}_i\hat{p}_i = -ih \sum_i q_i \frac{\partial}{\partial q_i} \qquad \{III-44\}$$

with the sum over 3n Cartesian coordinates of n particles, then

$$[\hat{H}, \sum_i \hat{q}_i\hat{p}_i] = ih \sum_i q_i \frac{\partial \hat{V}}{\partial q_i} - 2ih\hat{T} \qquad \{III-45\}$$

where $\hat{T}$ and $\hat{V}$ are the kinetic- and potential-energy operators for the system. Application of the hypervirial theorem gives

$$<\psi|\sum_i \hat{q}_i \frac{\partial \hat{V}}{\partial q_i}|\psi> \quad = \quad 2<\psi|\hat{T}|\psi> \qquad \{III-46\}$$

or with a notation change to denote the quantum mechanical average value of the quantities

$$<\sum_i \hat{q}_i \frac{\partial \hat{V}}{\partial q_i}> \quad = \quad 2<\hat{T}> \qquad \{III-47\}$$

Equation {III-43} is the quantum mechanical virial theorem. Its validity is restricted to stationary states. The form of $\hat{V}$ can cause the virial theorem to take on a particularly simple form. If $\hat{V}$ is a homogeneous function of degree n (in Cartesian coordinates) then Euler's theorem gives:

$$\sum_i \hat{q}_i \frac{\partial \hat{V}}{\partial q_i} \quad = \quad n\hat{V} \qquad \{III-48\}$$

thus

$$2<\hat{T}> \quad = \quad n<\hat{V}> \qquad \{III-49\}$$

and since

$$<\hat{T}> \ + \ <\hat{V}> \quad = \quad E \qquad \{III-50\}$$

$$<\hat{V}> \quad + \quad \frac{2E}{n+2} \qquad \{III-51a\}$$

$$<\hat{T}> \quad = \quad \frac{nE}{n+2} \qquad \{III-51b\}$$

In particular if $\hat{V}$ is a homogeneous function of degree -1 then

$$2<\hat{T}> \quad = \quad -<\hat{V}> \qquad \{III-52a\}$$

$$<\hat{V}> \quad = \quad 2E \qquad \{III-52b\}$$

$$<\hat{T}> \quad = \quad -E \qquad \{III-52c\}$$

The true wavefunctions for a system with V a homogeneous function of the coordinates must satisfy the form of the virial theorem {III-43}. By inserting a variational parameter as a multiplier of each Cartesian coordinate and choosing this parameter to minimize the variational integral, we can make any trial variation function satisfy the virial theorem. This process is called "scaling". Hartree-Fock wavefunctions should satisfy the virial theorem and calculation of the ratio of T to V has been used as one test of the quality of a wavefunction.

(b) The Hellmann-Feynman Theorem

The Born-Oppenheimer approximation states that one can view the nuclear motions of a molecule as occurring in a potential field E(R) provided by the electrons. This implies that the forces acting upon the nuclei are expressible as gradients of E(R).

The generalized Hellmann-Feynman theorem is

$$\frac{\partial E}{\partial \lambda} = \left\langle \frac{\partial \hat{H}}{\partial \lambda} \right\rangle = \int \psi_n^* \frac{\partial \hat{H}}{\partial \lambda} \psi_n d\tau \qquad \{III-53\}$$

where λ is some parameter present in $\hat{H}$ (for example λ may be a nuclear coordinate). The theorem states that the slope of $E(\lambda)$ versus λ can be calculated as the expectation value of the operator $\partial\hat{H}/\partial\lambda$. The form of the theorem is reminiscent of perturbation theory in that changes in the energy of a system are related to changes in the hamiltonian.

Hellmann and Feynman independently applied {III-53} to molecules with λ a nuclear Cartesian coordinate. The result is that the effective force acting on a nucleus in a molecule can be calculated by simple electrostatics as the sum of the Coulombic forces exerted by the other nuclei and by a hypothetical electron cloud whose charge density is found by solving the electronic Schrödinger equation. This statement is called the electrostatic theorem and follows reasonably from the Born-Oppenheimer approximation since the rapid motion of the electrons allows the electronic wavefunction and probability density to adjust immediately to changes in nuclear configuration; the rapid motion of the electrons causes the nuclei to "see" the electrons as a charge cloud rather than as discrete particles. The fact that the effective forces on the nuclei are electrostatic affirms that there are no "mysterious quantum-mechanical forces" acting in molecules.

8. Standard States

Since the Hamiltonian explicitly accounts for all possible interactions between the electrons and nuclei the energy of such interactions is zero if and only if all particles are infinitely separated. This infinite separation corresponds to the full ionization which represents the zero energy reference point of the Quantum Chemical Standard State.

$$\left.\begin{array}{lll} H^{+} & + & e^{-} \\ He^{2+} & + & 2e^{-} \\ Li^{3+} & + & 3e^{-} \\ Be^{4+} & + & 4e^{-} \\ B^{5+} & + & 5e^{-} \\ C^{6+} & + & 6e^{-} \\ N^{7+} & + & 7e^{-} \\ O^{8+} & + & 8e^{-} \\ F^{9+} & + & 9e^{-} \\ Ne^{10+} & + & 10e^{-} \\ \vdots & & \end{array}\right\} \quad E = 0.000000 \qquad \{III-54\}$$

Consequently all atomic energies are negative implying a greater degree of stability than the infinitely separated electrons and bare nuclei and their absolute value is simply equal to the sum of the ionization potentials (I_i)

$$E_{Atom} = - \sum_{i=1}^{\text{all electrons}} I_i \qquad \{III-55\}$$

The total experimental energy values for the ground states of the first ten atoms of the periodic system in the Quantum Chemical Standard State as may be calculated according to equation {II-17} are listed below:

$$
\begin{aligned}
E(H) &= -0.5000 \\
E(He) &= -2.9037 \\
E(Li) &= -7.4780 \\
E(Be) &= -14.6685 \\
E(B) &= -24.6579 \\
E(C) &= -37.8558 \\
E(N) &= -54.6122 \\
E(O) &= -75.1101 \\
E(F) &= -99.8053 \\
E(Ne) &= -129.0560
\end{aligned}
\qquad \{III\text{-}56\}
$$

The values were taken from M. Krauss, N.B.S. Technical Note 438 (1967) & P. E. Cade and W. M. Huo, J.C.P. 47, 649 (1967), where an analysis of recommended "experimental" values is given in detail. The values quoted here should be adequate for comparison with molecular energies.

Molecules are still lower on the Quantum Chemical Energy Scale corresponding to stability relative to separated atoms. One of the two components of their stability is the atomization or total dissociation energy which is the sum of all bond dissociation energies

$$E_{Atomization} = - \sum_{i}^{\text{all bonds}} D_i \qquad \{III\text{-}57\}$$

The other component of the molecular stabilization relative to the Quantum Chemical Standard State is the zero point vibrational (ZPV) energy which is related to the fundamental vibrational frequencies

$$E_{ZPV} = -\frac{1}{2} hc \sum_{i}^{\text{all normal modes}} \bar{\nu}_i \qquad \{III-58\}$$

Of course molecules cannot have energies less than their zero point vibrational energy level but due to the Born-Oppenheimer theorem in which the nuclear and electronic motions are separated molecular energies below the ZPV vibrational energy values may be computed. The total molecular energy is computable as the sum of {III-56}, {III-57} and {III-58}

$$E_{Molecule} = E_{Atoms} + E_{Atomization} + E_{ZPV} \qquad \{III-59\}$$

The components of the total molecular energies of the ten-electron hydrides (CH_4, NH_3, H_2O, HF) are summarized in Table III-4.

Equation {III-59} leads us immediately to the definition of the Chemical Standard State. Because molecules can never occupy an energy state below their zero point vibrational level, E_{ZPV} is omitted from equation {III-59} for the computation of molecular energy in the Chemical Standard State (CSS).

In addition the separated atoms are considered to be the reference point in the CSS consequently the atomic energies (E_{Atom}) are set equal to zero in equation {III-59}. Thus, the only term left in equation {III-59} is the total dissociation ($E_{Atomization}$) or total bond energy

$$E^{CSS}_{Molecule} = E_{Atomization} \qquad \{III-60\}$$

which in turn is defined by equation {III-57}.

The Thermodynamic Standard State is not only the first one, historically speaking, that was established as a reference state but it is the most arbitrary in terms of bonding. Its practicality originates from the convention that elements in their naturally occurring form at 25°C and 1 atmosphere pressure are arbitrarily regarded as the zero energy refer-

TABLE III-4 EXPERIMENTAL TOTAL ENERGIES AND ENERGY COMPONENTS FOR SELECTED TEN-ELECTRON HYDRIDES

Molecule	Reference	Energies[e] (hartree) Atomic	Atomization	Zero Point	Total
CH_4	a	-39.856	-0.625	-0.043	-40.524
NH_3	b	-56.098	-0.447	-0.033	-56.578
OH_2	c	-76.116	-0.349	-0.021	-76.486
FH	d	-99.8053	-0.235	-0.019	-100.059

[a,d] Calculated from experimental data given by G. Herzberg according to {III-59}.

[b] R. E. Kari and I. G. Csizmadia, J. Chem. Phys. 56, 4337 (1972).

[c] I. G. Csizmadia in The Chemistry of the Thiol Group, S. Patai, Ed., Wiley-Interscience, N.Y. (1975).

[e] The energies presented must be viewed with some caution. For example, for ammonia other references yield -56.5829, -56.563, -56.588 and -56.578 a.u. for the total molecular energies with similar small differences for the other energy quantities.

ence point. Molecular energies or heats of formation are
expressed with respect to this reference point. The major
disadvantage of this convention is that most of the elements
are not gaseous diatomic molecules at room temperature and
atmospheric pressure like H_2, N_2, O_2 and F_2 are. The choice
of the most stable allotropic form must also be considered.
For example, phosphorous occurs both in yellow and red forms
and carbon is known in both diamond and graphite forms. For
example, carbon the most basic element of organic compounds is
regarded to be zero in the form of graphite and the atomization
of graphite must be complete before molecules may be formed.
Clearly the heat of formation of even the simplest organic
compound methane incorporates a great many changes in bonding
therefore its heat of formation is not a simple measure of
bond strength or of molecular stability

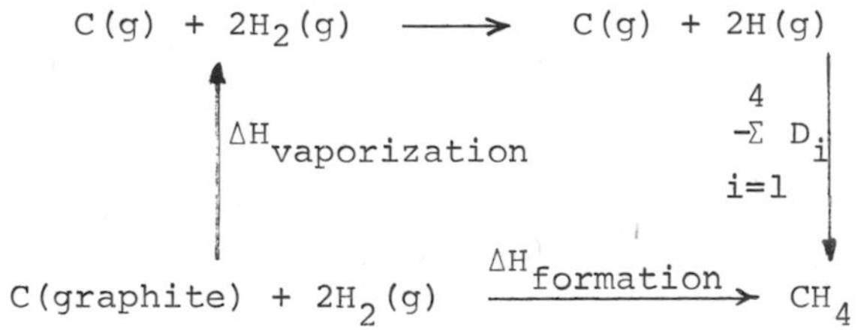

$$C(g) + 2H_2(g) \longrightarrow C(g) + 2H(g) \qquad \{III-61\}$$

with $\Delta H_{vaporization}$ and $-\sum_{i=1}^{4} D_i$ and $\Delta H_{formation}$

$$C(graphite) + 2H_2(g) \xrightarrow{\Delta H_{formation}} CH_4$$

However, none of these standard states represent any
fundamental problems or fundamental advantage simply because
the concern of the chemist is always with energy differences.
This is necessarily so since in order to observe anything in
chemistry a "transition" from an initial state ($E_{initial}$) to
a final state (E_{final}) is required (e.g. reactants $\rightarrow$ products):

$$\Delta E = E_{final} - E_{initial} \qquad \{III-62\}$$

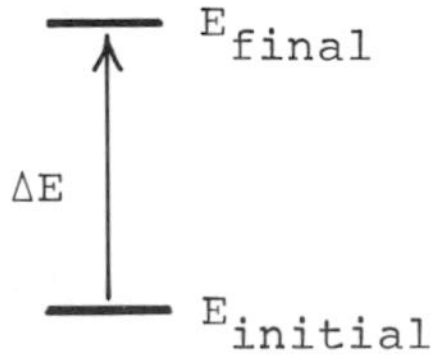

In other words in the measurement of an equilibrium constant for the final and the initial states the energy difference between two minima on the energy hyper-surface is determined, while in the measurement of a rate constant the final state corresponds to a saddle point and the initial state is a minimum. Alternatively in absorption spectroscopy the final state is an upper excited state and the initial state is the lower state while in emission spectroscopy the opposite is the case. Thus the energy difference calculated according to {III-62} is the same irrespective of the standard state used for the expression of E_{final} and $E_{initial}$.

The three standard states are shown in Figure III-3 for the simplest molecular system. H_2 and a more complicated situation is illustrated in Figure III-4 for FNO_2 and various possible fragments. Note that in the latter case the energy difference, calculated as in {III-62}, associated with any fragmentation pattern is the same regardless of the standard state used.

For computational convenience, however, as was discussed earlier, the Quantum Chemical Standard State will be used exclusively and all quantities including energy will be expressed in atomic units, although occasionally they will be converted to units more often used in experimental chemistry.

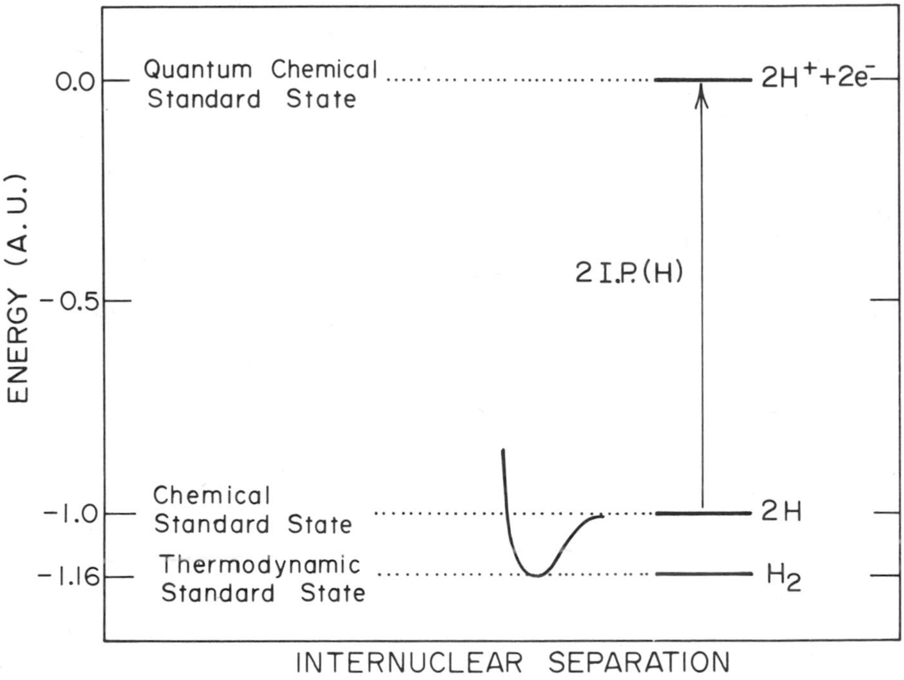

Figure III-3 Standard states for hydrogen

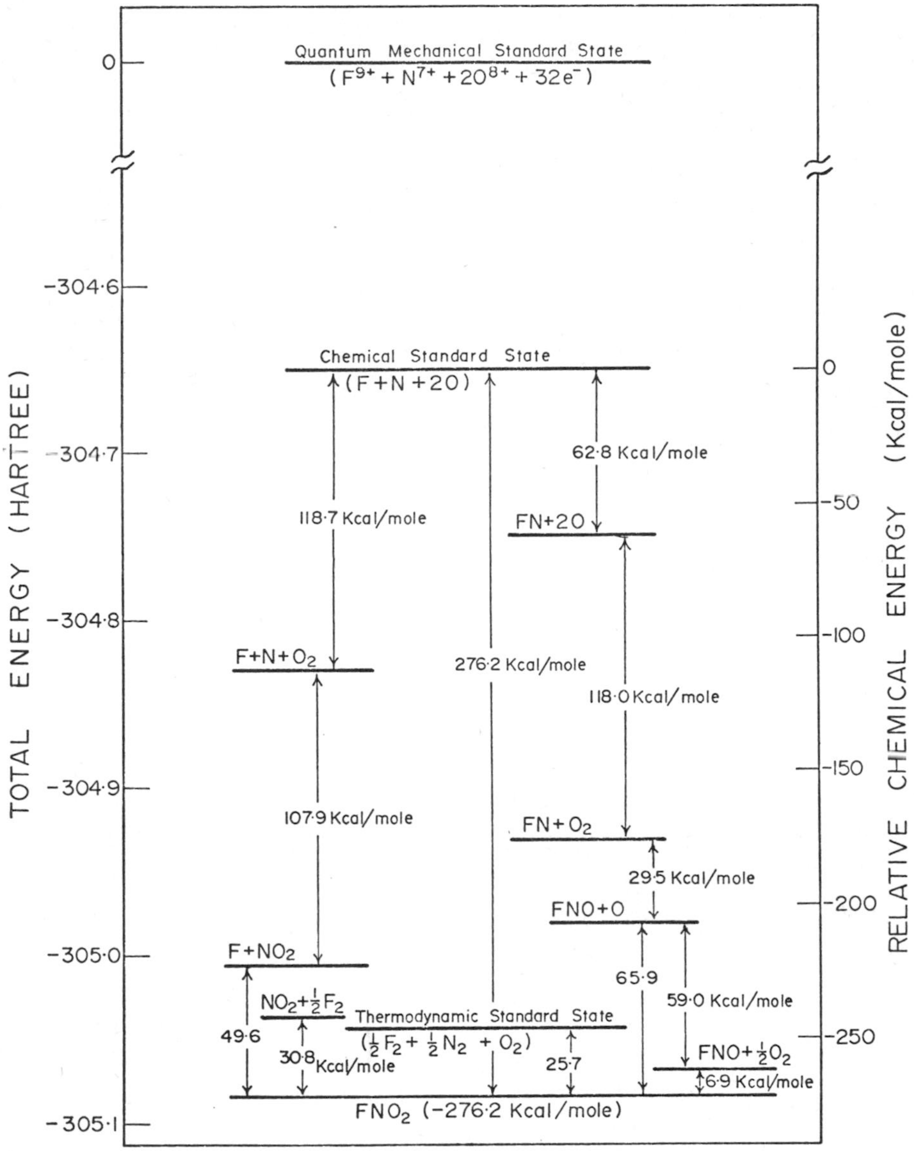

Figure III-4 Standard states for FNO_2 and relative energies for its fragments

SECTION B

THEORY OF CLOSED ELECTRONIC SHELLS

CHAPTER IV

NON-EMPIRICAL OR HARTREE-FOCK MO THEORY

1. Molecular Orbital Wavefunctions

The mathematical formulation of molecular orbital theory as a problem in infinite dimensional Hermitian vector space (Hilbert space) requires that an infinite number of AO be used to form a complete basis set in which to expand the MO. Practical considerations dictate the use of truncated basis sets, and the first stage of any calculation is the decision on the degree of truncation (i.e. the size) of the AO basis.

Before undertaking a more general treatment of MO calculations we consider H_2 as a specific example and begin by limiting ourselves to a single 1s AO for each proton. The pair of 1s AO can be referred to as a MINIMAL BASIS since it is not normally possible to undertake the computation with fewer than two AO.

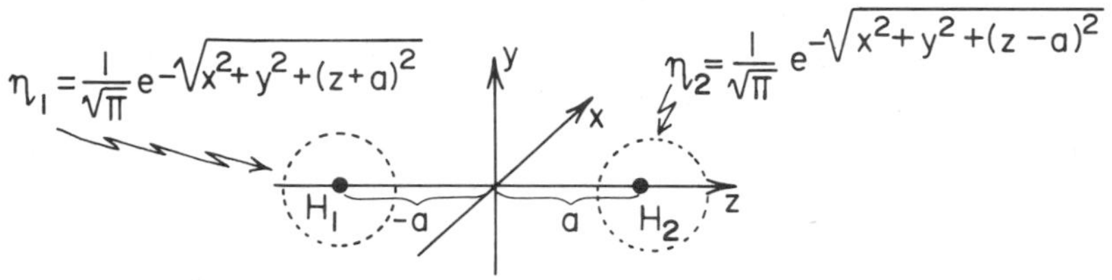

Figure IV.1 The spatial arrangement of H_2 and the pair of 1s functions

These two AO (η_1 and η_2) will yield two MO (ϕ_1 and ϕ_2), which correspond to σ_g and σ_u*. The required transformation from the AO basis to the MO basis may be written in matrix notation as

$$\phi = \eta \underline{\underline{C}} \qquad \{IV-1\}$$

or, in detail,

$$(\phi_1 \ \phi_2) = (\eta_1 \ \eta_2) \begin{pmatrix} c_{11} & c_{12} \\ c_{21} & c_{22} \end{pmatrix} \qquad \{IV-2\}$$

As discussed elsewhere the coefficient matrix $\underline{\underline{C}}$ may be obtained from the variation principle which is equivalent in practice to the process of solving a set of iterative eigenproblem equations (cf. the SCF procedure described in the next section). Since in this method the orthonormality condition is always imposed (i.e. the solutions, the MO, form an orthonormal set) this condition should be required here as well. The simplicity of the present problem permits the elements of the coefficient matrix to be obtained solely from the orthonormality conditions. For the normalization of ϕ_1 we may write:

$$\langle\phi_1|\phi_1\rangle = \langle C_{11}\eta_1 + C_{21}\eta_2|C_{11}\eta_1 + C_{21}\eta_2\rangle = 1 \qquad \{IV-3\}$$

Choosing $C_{11} = C_{21}$ because of the symmetry properties of the system (one of the hydrogen AO cannot have a greater contribution to an MO than the other) we obtain

$$C_{11}C_{11}\underbrace{\langle\eta_1|\eta_1\rangle}_{1} + 2\,C_{11}C_{11}\underbrace{\langle\eta_1|\eta_2\rangle}_{S_{12}} + C_{11}C_{11}\underbrace{\langle\eta_2|\eta_2\rangle}_{1} = 1 \qquad \{IV-4\}$$

$$2\,C^2_{11} + 2\,C^2_{11}\,S_{12} = 1 \qquad \{IV-5\}$$

or

$$C_{11} = C_{21} = \frac{1}{\sqrt{2(1 + S_{12})}} \qquad \{IV-6\}$$

For orthogonality of ϕ_1 and ϕ_2, the inner product should vanish, i.e.

$$\langle\phi_1|\phi_2\rangle = \langle C_{11}\eta_1 + C_{11}\eta_2|C_{12}\eta_1 + C_{22}\eta_2\rangle = 0 \qquad \{IV-7\}$$

$$C_{11}C_{12}\underbrace{\langle\eta_1|\eta_1\rangle}_{1} + C_{11}C_{12}\underbrace{\langle\eta_2|\eta_1\rangle}_{S_{12}} + C_{11}C_{22}\underbrace{\langle\eta_1|\eta_2\rangle}_{S_{12}} + C_{11}C_{22}\underbrace{\langle\eta_2|\eta_2\rangle}_{1} = 0 \qquad \{IV-8\}$$

$$(C_{12} + C_{22}) + (C_{12} + C_{22})\,S_{12} = 0 \qquad \{IV-9\}$$

or

$$(C_{12} + C_{22})\,(1 + S_{12}) = 0 \qquad \{IV-10\}$$

since $(1 + S_{12})$ cannot be zero,

$$C_{12} + C_{22} = 0 \qquad \{IV-11\}$$

or

$$C_{12} = -C_{22} \qquad \{IV-12\}$$

From the normalization of ϕ_2,

$$<\phi_2|\phi_2> = <C_{12}\eta_1 - C_{12}\eta_2|C_{12}\eta_1 - C_{12}\eta_2> = 1 \qquad \{IV-13\}$$

$$C^2_{12} \underbrace{<\eta_1|\eta_1>}_{1} - 2\,C^2_{12} \underbrace{<\eta_1|\eta_2>}_{S_{12}} + C^2_{12} \underbrace{<\eta_2|\eta_2>}_{1} = 0 \qquad \{IV-14\}$$

$$2\,C^2_{12} - 2\,C^2_{12}\,S_{12} = 1 \qquad \{IV-15\}$$

or

$$C_{12} = \frac{1}{\sqrt{2(1 - S_{12})}} \qquad \{IV-16\}$$

The coefficient matrix thus has the following elements

$$\underline{\underline{C}} = \begin{pmatrix} \frac{1}{\sqrt{2(1 + S_{12})}} & \frac{1}{\sqrt{2(1 - S_{12})}} \\ \frac{1}{\sqrt{2(1 + S_{12})}} & \frac{1}{\sqrt{2(1 - S_{12})}} \end{pmatrix} \qquad \{IV-17\}$$

It should be noted that the elements of $\underline{\underline{C}}$ depend on the value of the overlap integral S_{12} which may vary between 0 and 1 depending on the internuclear distance (cf. Figure IV-2).

Explicit knowledge of the two MO permits calculation of the two MO energies (ε_1 and ε_2) but a detailed knowledge of the total wavefunction $\Psi(1,2)$, which describes the complete two-electron system is necessary to calculate the total energy (E) of the system.

The electronic configuration of H_2 may be represented in the form of an orbital energy diagram (cf. Figure IV-3). Several important concepts, needed for the construction of a

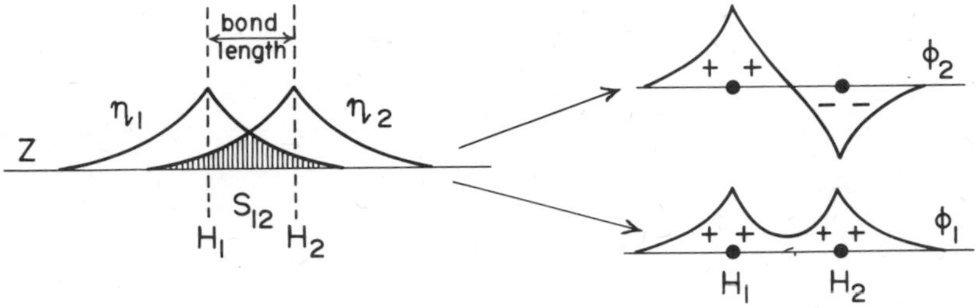

Figure IV.2 A schematic illustration of MO (ϕ_1 and ϕ_2) formation from AO (η_1 and η_2) for the H_2 molecule

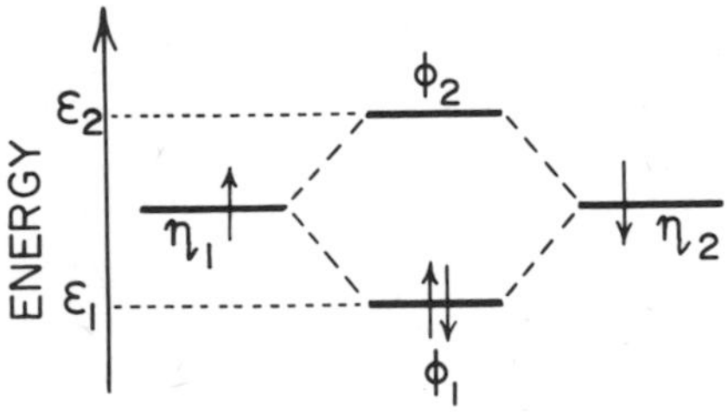

Figure IV.3 Orbital energy level diagram for H_2

two electron wavefunction, are illustrated in such a diagram.

(i) ϕ_1 is doubly occupied and ϕ_2 is unoccupied (i.e. ϕ_2 is a "virtual orbital"), (ii) the directional character of the arrows implies the explicit incorporation of electron spin; (iii) double occupancy of ϕ_1 is permitted for electrons having opposite spin only (Pauli principle) as indicated by the opposite directions of the arrows.

To account mathematically for electron spin it may be recalled that an orbital is a one-electron space-function. To describe spin it is necessary to introduce an additional co-ordinate over the three dimensions (x, y, z) of the physical space and it is convenient to employ two functions, α and β.

These are combined with each orbital to give two <u>spin-orbitals</u>:

$$\phi_1(\mu)\alpha(\mu) = \phi_1(x_\mu, y_\mu, z_\mu)\alpha(s_\mu)$$
$$\phi_1(\nu)\beta(\nu) = \phi_1(x_\nu, y_\nu, z_\nu)\beta(s_\nu) \qquad \{IV\text{-}18a,b\}$$

where μ and ν are electron labels symbolizing explicit use of the coordinates of the electrons μ and ν. Integration involving spin orbitals is performed over the four dimensional space:

$$d\tau_\mu = dx_\mu \,.\, dy_\mu \,.\, dz_\mu \,.\, ds_\mu$$
$$d\tau_\nu = dx_\nu \,.\, dy_\nu \,.\, dz_\nu \,.\, ds_\nu \qquad \{IV\text{-}19a,b\}$$

The spin functions α and β are orthonormal, i.e.

$$<\phi_1(\mu)\alpha(\mu)|\phi_1(\mu)\alpha(\mu)> \; = \; \underbrace{<\phi_1(\mu)|\phi_1(\mu)>}_{1} \; \underbrace{<\alpha(\mu)|\alpha(\mu)>}_{1} \; \equiv 1 \qquad \{IV\text{-}20\}$$

$$<\phi_1(\mu)\alpha(\mu)|\phi_1(\mu)\beta(\mu)> \; = \; \underbrace{<\phi_1(\mu)|\phi_1(\mu)>}_{1} \; \underbrace{<\alpha(\mu)|\beta(\mu)>}_{0} \; \equiv 0 \qquad \{IV\text{-}21\}$$

The total wavefunction $\psi(1,2)$ must incorporate both $\phi_1(1)\alpha(1)$ and $\phi_1(2)\beta(2)$. Further, since $\psi(1,2)$ represents a concurrent situation (i.e. when electron 1 "occupies" $\phi_1\alpha$ and electron 2 "occupies" $\phi_1\beta$), the total wavefunction must be expressed in the form of a product,

$$\phi_1(1)\alpha(1)\phi_1(2)\beta(2) \quad \text{or} \quad \phi_1(2)\alpha(2)\phi_1(1)\beta(1). \qquad \{\text{IV-22a,b}\}$$

These two expressions are equally probable since they differ only in the exchange of the electron labels and the total wavefunction must include them with equal weight. This can be achieved via a linear combination of the two having either a positive or a negative sign. Since the electronic wavefunction must be antisymmetric with respect to permutation of the particle labels (i.e. change sign when the particle labels are interchanged) the required expression is

$$\phi_1(1)\alpha(1)\phi_1(2)\beta(2) - \phi_1(2)\alpha(2)\phi_1(1)\beta(1) \qquad \{\text{IV-23}\}$$

which must be normalized. Since permutation of the electron labels, used to generate the total wavefunction, produced two terms for a two electron system, it can be expected that a total of N! terms are required for an N electron system, i.e. that the normalization factor is $1/\sqrt{N!}$. Applying the usual rule of normalization permits us to write $\Psi(1,2)$ in the following form:

$$\Psi(1,2) = \frac{1}{\sqrt{2!}}\left[\phi_1(1)\alpha(1)\phi_1(2)\beta(2) - \phi_1(2)\alpha(2)\phi_1(1)\beta(1)\right]$$

$$\{\text{IV-24}\}$$

As has been seen previously a determinant has the antisymmetric property required of a many-electron wavefunction, i.e. that the above expression may conveniently be written in the form of a determinant (Slater determinant),

$$\psi(1,2) = \frac{1}{\sqrt{2!}} \cdot \begin{vmatrix} \phi_1(1)\alpha(1) & \phi_1(1)\beta(1) \\ \phi_1(2)\alpha(2) & \phi_1(2)\beta(2) \end{vmatrix} \qquad \{IV-25\}$$

We now note that $\psi(1,2)$ is an electronic wavefunction only and contains no information about the motion of the nuclei. Our calculated $\psi(1,2)$ therefore refers to a fixed nuclear arrangement (the Born-Oppenheimer approximation).

The electronic Hamiltonian for H_2(a specific example of the more general Hamiltonian given in (III-7) is:

$$\hat{H}^e_{1,2} = -\frac{1}{2}\nabla^2_1 - \frac{1}{r_{I1}} - \frac{1}{r_{II1}} - \frac{1}{2}\nabla^2_2 - \frac{1}{r_{I2}} - \frac{1}{r_{II2}} + \frac{1}{r_{12}} \qquad \{IV-26\}$$

Kinetic energy

nuclear-electron attraction energy

electron-electron repulsion energy

The various distances that explicitly occur in the Hamiltonian are illustrated in the following Figure.

The first three terms in this expression belong to the electron labelled 1 and the second three terms are associated with the electron labelled 2. These composite operators may be denoted h_1 and h_2 respectively. The last term is the electron-electron interaction and is frequently denoted g_{12}, in this term r_{12} depends on the coordinates of both electrons because it is the vector component of $(x_2 - x_1)$, $(y_2 - y_1)$ and $(z_2 - z_1)$. The two-electron Hamiltonian may then be written as

$$\hat{H}^e_{1,2} = \hat{h}_1 + \hat{h}_2 + \hat{g}_{1,2} \qquad \{IV-27\}$$

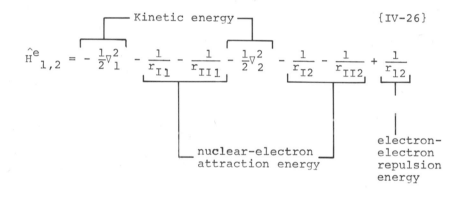

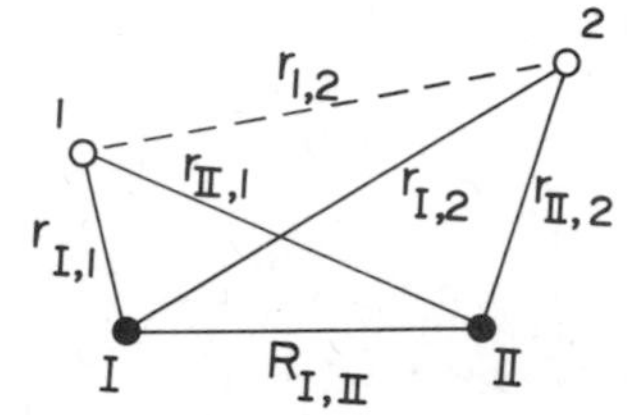

Figure IV-4 Distances in H_2 that determine the Hamiltonian operator

The electronic energy (E^e) is the expectation value of $\hat{H}^e_{1,2}$ over $\Psi(1,2)$, i.e. as the matrix representative of $\hat{H}^e_{1,2}$ in a one-dimensional vector space with the basis vector $\Psi(1,2)$ (cf. Chapter II, Section 6).

$$E^e = \langle \psi(1,2) | \hat{H}^e_{1,2} | \psi(1,2) \rangle \qquad \{IV\text{-}28\}$$

Expanding this equation leads to

$$E^e = \left\langle \frac{1}{\sqrt{2!}} \begin{bmatrix} \phi_1(1)\alpha(1) & \phi_1(1)\beta(1) \\ \phi_1(2)\alpha(2) & \phi_1(2)\beta(2) \end{bmatrix} \middle| \hat{H}^e_{1,2} \middle| \frac{1}{\sqrt{2!}} \begin{bmatrix} \phi_1(1)\alpha(1) & \phi_1(1)\beta(1) \\ \phi_1(2)\alpha(2) & \phi_1(2)\beta(2) \end{bmatrix} \right\rangle \qquad \{IV\text{-}29\}$$

$$= \frac{1}{2} \langle \phi_1(1)\phi_1(2)\{\alpha(1)\beta(2)-\alpha(2)\beta(1)\} | \hat{H}^e_{1,2} | \phi_1(1)\phi_1(2)\{\alpha(1)\beta(2)-\alpha(2)\beta(1)\} \rangle$$

The Hamiltonian operator we are using contains no spin operators and thus operates on the space functions only so that integration of the spin functions may be performed separately

$$E^e = \frac{1}{2} \langle \phi_1(1)\phi_1(2)|\hat{H}^e{}_{1,2}|\phi_1(1)\phi_1(2)\rangle \;.\; \langle \alpha(1)\beta(2)-\alpha(2)\beta(1)|\alpha(1)\beta(2)-\alpha(2)\beta(1)\rangle$$

$$= \frac{1}{2} I_{spin} \langle \phi_1(1)\phi_1(2)|\hat{H}^e{}_{1,2}|\phi_1(1)\phi_1(2)\rangle \qquad \{IV-30\}$$

We calculate I_{spin} first since it consists of simple overlap integrals over spin functions

$$\begin{aligned} I_{spin} = & \langle \alpha(1)\beta(2)|\alpha(1)\beta(2)\rangle - \\ & - \langle \alpha(2)\beta(1)|\alpha(1)\beta(2)\rangle - \\ & - \langle \alpha(1)\beta(2)|\alpha(2)\beta(1)\rangle + \\ & + \langle \alpha(2)\beta(1)|\alpha(2)\beta(1)\rangle \end{aligned} \qquad \{IV-31\}$$

Following separation of the spin variables according to the electron labels 1 and 2 the spin integral has the following form:

$$I_{spin} = \overbrace{\langle \alpha(1)|\alpha(1)\rangle}^{1}\overbrace{\langle \beta(2)|\beta(2)\rangle}^{1} - \overbrace{\langle \beta(1)|\alpha(1)\rangle}^{0}\overbrace{\langle \alpha(2)|\beta(2)\rangle}^{0} - \underbrace{\langle \alpha(1)|\beta(1)\rangle}_{0}\underbrace{\langle \beta(2)|\alpha(2)\rangle}_{0} + \underbrace{\langle \beta(1)|\beta(1)\rangle}_{1}\underbrace{\langle \alpha(2)|\alpha(2)\rangle}_{1} = 2$$

$$\{IV-32\}$$

The consequences of the orthonormality condition are indicated in this equation above or below the individual terms and lead, finally, to a value of 2 for the spin integral.

$$E^e = \frac{1}{2} \;.\; 2 \;.\; \langle \phi_1(1)\phi_1(2)|\hat{H}^e{}_{1,2}|\phi_1(1)\phi_1(2)\rangle = \langle \phi_1(1)\phi_1(2)|\hat{h}_1 + \hat{h}_2 + \hat{g}_{12}|\phi_1(1) \;_1(2)\rangle$$

$$\{IV-33\}$$

Now, since h_1 operates on electron 1 and h_2 operates on electron 2 the above expression may be simplified slightly

$$E^e = \langle \phi_1(1)|\hat{h}_1|\phi_1(1)\rangle \overbrace{\langle \phi_1(2)|\phi_1(2)\rangle}^{1} + \overbrace{\langle \phi_1(1)|\phi_1(1)\rangle}^{1} \langle \phi_1(2)|\hat{h}_2|\phi_1(2)\rangle + \langle \phi_1(1)\phi_1(2)|\hat{g}_{12}|\phi_1(1)\phi_1(2)\rangle \qquad \{IV-34\}$$

Further, since $\hat{h}_1$ and $\hat{h}_2$ differ only in terms of the running index of the electron label the first two terms on the right hand side of the above equation are equal.

$$E^e = 2\langle \phi_1(1)|\hat{h}_1|\phi_1(1)\rangle + \langle \phi_1(1)\phi_1(2)|\hat{g}_{12}|\phi_1(1)\phi_1(2)\rangle \qquad \{IV-35\}$$

The first term in this expression is usually referred to as a one-electron integral since it involves integration over the spatial coordinates of only one electron, i.e. over three dimensions; the second term is referred to as a two-electron integral since it involves integration over the spatial coordinates of two electrons, i.e. over six dimensions.

These are molecular integrals over the MO basis, i.e. they are elements of the matrix representatives of the operators $\hat{h}_1$ and $\hat{g}_{12}$ in the MO basis. Since the coefficient matrix is known, we may substitute the expression

$$\phi_1 = \frac{1}{\sqrt{2(1 + S_{12})}} (\eta_1 + \eta_2) \qquad \{IV-36\}$$

into the molecular integrals $\langle \phi_1|\hat{h}_1|\phi_1\rangle$ and $\langle \phi_1\phi_1|\hat{g}_{12}|\phi_1\phi_1\rangle$ to obtain an expression for E^e in terms of the AO basis.

Although it is instructive to carry out the substitution to discover how many molecular integrals over the AO basis are associated with this problem, it will not be performed here as it represents a simple arithmetical exercise. In any event, it is convenient at this point to introduce the similarity transformation in computational quantum chemistry. According to the previous discussion (cf. Chapter II, Section 7) and remembering the vector notation of ϕ_1

$$\phi_1 = (\eta_1 \ \eta_2) \begin{pmatrix} C_{11} \\ C_{11} \end{pmatrix} \qquad \{IV-37\}$$

we may write the following relationship:

$$<\phi_1|\hat{h}|\phi_1> = \sum_{i=1}^{2} \sum_{j=1}^{2} C_{11} <\eta_i|\hat{h}|\eta_j> C_{11} \qquad \{IV-38\}$$

or, in more detail:

$$<\phi_1|\hat{h}|\phi_1> = (C_{11}C_{11}) \begin{pmatrix} <\eta_1|\hat{h}|\eta_1> & <\eta_1|\hat{h}|\eta_2> \\ <\eta_2|\hat{h}|\eta_1> & <\eta_2|\hat{h}|\eta_2> \end{pmatrix} \begin{pmatrix} C_{11} \\ C_{11} \end{pmatrix} \qquad \{IV-39\}$$

Note, that in this particular case $C_{1i} = C_{1j} = C_{11}$.

If for some reason the integrals

$$<\phi_1|\hat{h}|\phi_2> \quad \text{and} \quad <\phi_2|\hat{h}|\phi_2>$$

are also required the full coefficient matrix must then be used

$$\underline{\underline{h}}^{\phi} = \underline{\underline{C}}^{\dagger} \, \underline{\underline{h}}^{\eta} \, \underline{\underline{C}} \qquad \{IV-40\}$$

Both the direct substitution and the similarity transformation lead to the same numerical result but the latter is more suitable for computation. When the multiplications are performed, each integral term has a coefficient which incorporates the square of coefficients: C^2_{11}. This is a measure of electron density and C^2_{11} is referred to as the 1,1 element of the density matrix $\underline{\underline{\rho}}$. For the hydrogen molecule the density matrix may be obtained in the following way:

$$\underline{\underline{\rho}} \equiv \begin{pmatrix} \rho_{11} & \rho_{12} \\ & \\ \rho_{21} & \rho_{22} \end{pmatrix} = \begin{pmatrix} c_{11} \\ \\ c_{11} \end{pmatrix} \begin{pmatrix} c_{11} & c_{11} \end{pmatrix} \qquad \{IV-41\}$$

Note that only vectors associated with occupied MO are used to generate the density matrix.

Transformation of the two-electron integrals will not be discussed in detail here but will be considered in Section 3. The process is somewhat more involved because the two electron integrals form a 4 dimensional array rather than a 2 dimensional matrix as was the case for $\underline{\underline{h}}^{\eta}$. The following general expression is illustrative:

$$\langle \phi_p(1)\phi_r(2)|\hat{g}_{12}|\phi_q(1)\phi_s(2)\rangle = \sum_{i=1}^{N}\sum_{k=1}^{N}\sum_{j=1}^{N}\sum_{\ell=1}^{N} C_{pi}C_{rk}\langle \eta_i(1)\eta_k(2)|\hat{g}_{12}|\eta_j(1)\eta_\ell(2)\rangle C_{jq}C_{\ell s} \qquad \{IV-42\}$$

For the case of H_2, $p = r = q = s = 1$ and $N = 2$.

In computations on more complex chemical systems, where the number of two-electron integrals may be $>10^6$, a different convention is used to label the two-electron integrals. The efficient transformation of two electron integrals presents a very difficult computational problem of some importance.

The notation $\langle \phi_1(1)\phi_1(2)|\hat{g}_{12}|\phi_1(1)\phi_1(2)\rangle$ is a standard one and is valid for both real and complex functions. When the one-electron function ϕ_1 is complex, the product $\phi_1(1)\phi_1(2)$ preceding the operator $\hat{g}_{12}$ is the complex conjugate. This is usually understood and not written explicitly. In practice, because the molecular orbitals (such as ϕ_1) are almost always real quantities the order of the functions may be changed. The convention, in computational quantum chemistry, is to employ the charge density notation, in which functions with electron label 1 are collected on one side and functions associated with electron label 2 are collected on the other side of the operator:

$$\{\phi_1(1)\phi_1(1)|\phi_1(2)\phi_1(2)\} \equiv <\phi_1(1)\phi_1(2)|g_{12}|\phi_1(1)\phi_1(2)> \quad \{IV\text{-}43\}$$

To distinguish this notation from the previous one, the operator g_{12} is omitted from the expression.

In the charge density notation the general expression for the two electron integral transformation from the AO basis to the MO basis has the form;

$$\{\phi_p(1)\phi_q(1)|\phi_r(2)\phi_s(2)\} = \sum_{i=1}^{N}\sum_{j=1}^{N}\sum_{k=1}^{N}\sum_{\ell=1}^{N} C_{pi}C_{qj}\{\eta_i(1)\eta_j(1)|\eta_k(2)\eta_\ell(2)\} C_{kr}C_{\ell s} \quad \{IV\text{-}44\}$$

which is considerably simplified for the case of the 2-orbital treatment of H_2.

$$\{\phi_1\phi_i|\phi_1\phi_1\} = \sum_{i=1}^{2}\sum_{j=1}^{2}\sum_{k=1}^{2}\sum_{\ell=1}^{2} C^2_{11}\{\eta_i\eta_j|\eta_k\eta_1\} C^2_{11} = \sum_{i=1}^{2}\sum_{j=1}^{2}\sum_{k=1}^{2}\sum_{\ell=1}^{2} \rho_{11}\{\eta_i\eta_j|\eta_k\eta_\ell\}\rho_{11} \quad \{IV\text{-}45\}$$

In closing this section it is convenient to summarize the expansion method that has been employed for H_2 (formation of molecular orbitals by linear combination of atomic orbitals: LCAO-MO). We begin with a non-orthogonal (but normalized) set of AO $\{\eta\}$ which are transformed directly to an orthonormal set of MO $\{\phi\}$. This is performed in two steps. The first involves orthogonalization; this transformation introduces "kinks" in the orbitals at the positions of the other nuclei so that the product of any two functions (e.g. $\chi_1\chi_2$) of the orthogonalized orbitals (OO) contains equal "negative and positive contributions" which cancel upon integration.

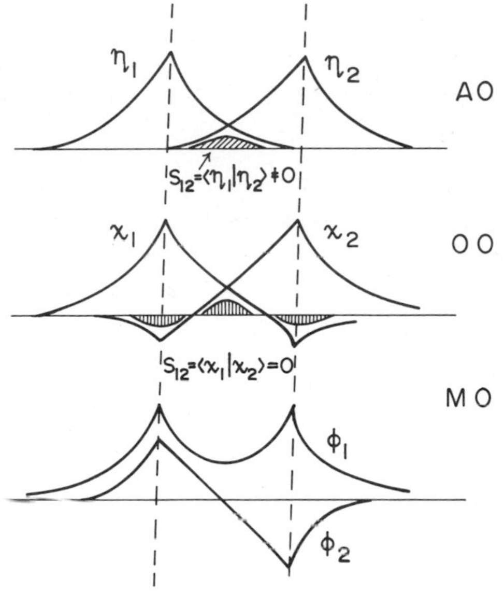

Figure IV-5 A schematic illustration for the sequence transformation of AO to MO

Transformation from the OO basis set (χ) to the MO basis set (ϕ) involves an orthogonal transformation which is mathematically equivalent to the rotation of one N-dimensional vector space into another without changing the orthonormality:

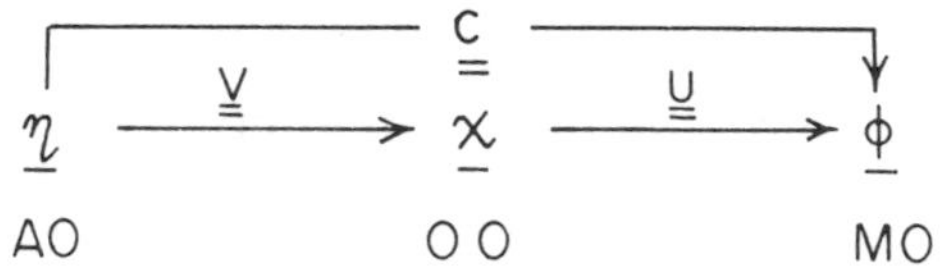

$$\underset{AO}{\underline{\eta}} \xrightarrow{\underline{\underline{V}}} \underset{OO}{\underline{\chi}} \xrightarrow{\underline{\underline{U}}} \underset{MO}{\underline{\phi}} \qquad (\underline{\eta} \xrightarrow{\underline{\underline{C}}} \underline{\phi}) \qquad \{IV\text{-}46a\}$$

This two-step process requires a separation of the $\underline{\underline{C}}$ matrix into two separate matrices: $\underline{\underline{V}}$ and $\underline{\underline{U}}$. Matrix $\underline{\underline{V}}$ performs the orthogonalization of η and $\underline{\underline{U}}$ rotates χ into ϕ. (For real orbitals $\underline{\underline{U}}$ is an orthogonal matrix, otherwise it is unitary.)

In matrix notation the process is written as

$$\phi = \eta\,\underline{\underline{C}} = \eta\,\underline{\underline{V}}\,\underline{\underline{U}} = \chi\,\underline{\underline{U}} \qquad \{IV\text{-}46b\}$$

and in detail:

$$(\phi_1\phi_2 \cdots\cdots \phi_n) = (\eta_1\eta_2 \cdots\cdots \eta_n) \begin{pmatrix} C_{11} & C_{12} & \cdots\cdots & C_{1n} \\ C_{21} & C_{22} & \cdots\cdots & C_{2n} \\ \vdots & \vdots & & \vdots \\ C_{n1} & C_{n2} & \cdots\cdots & C_{nn} \end{pmatrix} =$$

$$= (\eta_1\eta_2 \cdots\cdots \eta_n) \begin{pmatrix} V_{11} & V_{12} & \cdots\cdots & V_{1n} \\ V_{21} & V_{22} & \cdots\cdots & V_{2n} \\ \vdots & \vdots & & \vdots \\ V_{n1} & V_{n2} & & V_{nn} \end{pmatrix} \begin{pmatrix} U_{11} & U_{12} & \cdots\cdots & U_{1n} \\ U_{21} & U_{22} & \cdots\cdots & U_{2n} \\ \vdots & \vdots & & \vdots \\ U_{n1} & U_{n2} & & U_{nn} \end{pmatrix} =$$

$$= (\chi_1\chi_2 \cdots\cdots \chi_n) \begin{pmatrix} U_{11} & U_{12} & \cdots\cdots & U_{1n} \\ U_{21} & U_{22} & \cdots\cdots & U_{2n} \\ \vdots & \vdots & & \vdots \\ U_{n1} & U_{n2} & \cdots\cdots & U_{nn} \end{pmatrix} \qquad \{IV\text{-}47\}$$

It is easy to illustrate the vector analogy of orbital transformations for the case of H_2 because of the two dimensional nature of the problem. The vector analogy is valid for larger molecules even though a pictorial description of the process is impossible.

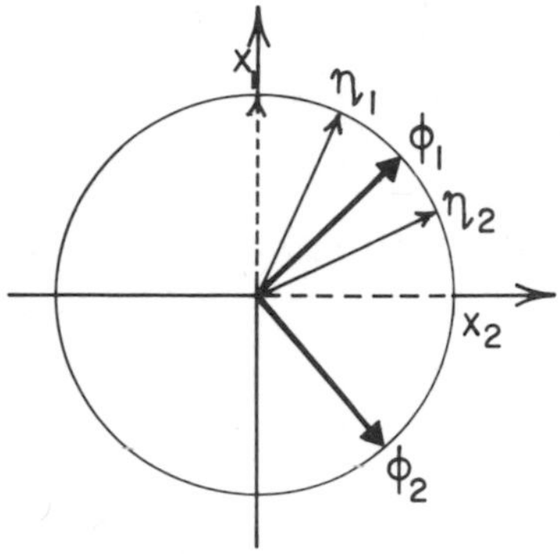

Figure IV-6 A vector model of AO and MO orbitals associated with H_2

2. <u>The Molecular Hartree-Fock Problem (Roothaan's Method)</u>

Generalisation of the concepts discussed in the previous section for H_2 leads to the following expressions:

(i) the wavefunction which describes the electronic ground state of a 2 M electron system has the form of a 2 M x 2 M Slater determinant

$$\Phi_o \equiv \Phi_o(1,2,\ \ldots,\ 2M) = \frac{1}{\sqrt{(2M)!}} \begin{vmatrix} \phi_1(1)\alpha(1) & \phi_1(1)\beta(1) & \ldots & \phi_M(1)\alpha(1) & \phi_M(1)\beta(1) \\ \phi_1(2)\alpha(2) & \phi_1(2)\beta(2) & & \phi_M(1)\alpha(2) & \phi_M(1)\beta(1) \\ \vdots & \vdots & & \vdots & \vdots \\ \phi_1(2M)\alpha(2M) & \phi_1(2M)\beta(2M) & ,\ldots & \phi_M(2M)\alpha(2M) & \phi_M(2M)\beta(2M) \end{vmatrix} \quad \{IV-48\}$$

(ii) the electronic Hamiltonian for a 2 M electron system may be written as

$$\hat{H} \equiv \hat{H}(1,2, \ldots, 2\,M) = \sum_{\mu=1}^{2\,M} \hat{h}_{\mu} + \sum_{\mu>\nu}^{M(2\,M-1)} \hat{g}_{\mu\nu} \qquad \{IV\text{-}49\}$$

(iii) the energy value after substitution of Ψ_o and $\hat{H}$ into the expectation value equation is

$$E = \langle \Psi_o | \hat{H} | \Psi_o \rangle, \qquad \{IV\text{-}50\}$$

and integrating out the spin variables, has the form:

$$E = 2\sum_{p}^{M} \langle \phi_p(1)|\hat{h}_1|\phi_p(1)\rangle + \sum_{p}^{M}\sum_{q}^{M} [2\langle \phi_p(1)\phi_q(2)|\hat{g}_{12}|\phi_p(1)\phi_q(2)\rangle - \langle \phi_p(1)\phi_p(2)|\hat{g}_{12}|\phi_q(1)\phi_q(2)\rangle\} \qquad \{IV\text{-}51\}$$

where the two electron integrals (the last two terms in the above equation) are the Coulomb and Exchange integrals respectively. Note that in the Coulomb integrals electron No. 1 is associated with orbital ϕ_p and electron No. 2 is associated with orbital ϕ_q. This distinction between Coulomb and Exchange terms becomes clearer in the electron density formalism where orbitals associated with electron No. 1 are collected to be in front of the operator while those associated with particle No. 2 are written behind the operator.

$$E = 2\sum_{p}^{M} \langle \phi_p(1)|\hat{h}_1|\phi_p(1)\rangle + \sum_{p}^{M}\sum_{p}^{M} [2\{\phi_p(1)\phi_p(1)|\phi_q(2)\phi_q(2)\} - \{\phi_p(1)\phi_q(1)|\phi_p(2)\phi_q(2)\}] \qquad \{IV\text{-}52\}$$

Note, that in order to distinguish the electron density

formalism ({IV-52}) from the traditional notation ({IV-51}) the brackets were changed from II to I and the electron-electron repulsion operator $g_{\mu\nu}$ has been omitted.

The above expression: {IV-52} can conveniently be presented in abbreviated form,

$$E = 2 \sum_{p}^{M} h^{\phi}_{pp} + \sum_{p}^{M} \sum_{q}^{M} \left(2 J^{\phi}_{pq} - K^{\phi}_{pq}\right) \qquad \{IV-53\}$$

where J_{pq} and K_{pq} symbolize the Coulomb and Exchange integrals respectively. The superscript ϕ indicates that these matrix representatives are over the MO basis.

At this stage it should be pointed out that the diagonal elements of the Coulomb and Exchange integrals are identical

$$J_{pp} = K_{pp} \qquad \{IV-54\}$$

It may also be noted that both J_{pq} and K_{pq} are positive quantities however K_{pq} enters with a negative sign in the expression of energy therefore this exchange term is frequently considered to be attractive. This is a matter of opinion and there is no way to justify this point of view (or its opposite). The exchange term K_{pq} has no classical analogue while the Coulombic term J_{pq} is analogous to the electrostatic Coulomb repulsion.

The following inequalities indicate the relative magnitudes of these quantities

$$0 \leqq K_{pq} \leqq J_{pq} \leqq \frac{1}{2} (J_{pp} + J_{qq}) \qquad \{IV-55\}$$

The J and K integrals are conveniently expressed as pseudo one-electron integrals by defining pseudo-one-electron hermitian operators $\hat{J}_p$ and $\hat{K}_p$ such that:

$$J_{pq} = \langle \phi_p | \hat{J}_q | \phi_p \rangle = \langle \phi_q | \hat{J}_p | \phi_q \rangle \qquad \{IV-56\}$$

$$K^{\Phi}_{pq} = <\phi_p|\hat{K}_q|\phi_p> = <\phi_q|\hat{K}_p|\phi_q> \qquad \{IV-57\}$$

The energy expression may then be written as:

$$E = 2\sum_{p}^{M}<\phi_p|\hat{h} + \sum_{q}^{M}{}'(2\hat{J}_q-\hat{K}_q)|\phi_p> \equiv 2\sum_{p}\int\phi_p^*\{\hat{h} + \sum_{q}{}'(2\hat{J}_q-\hat{K}_q)\}\phi_p d\tau$$

$$\{IV-58\}$$

According to the Variation Theorem the energy may be optimized by variation of ϕ. We therefore wish to minimize E with respect to ϕ_p. To do this we need δE which will be set equal to zero. To obtain δE each ϕ_p is varied by an infinitesimal amount $\delta\phi_p$, the variation in energy becomes:

$$\delta E = 2\sum_{p}^{M}\int(\delta\phi^*)\{h + \sum_{q}^{M}{}'(2\,J_q - K_q)\}\phi_p\,d\tau + 2\sum_{p}^{M}\int(\delta\phi_p)\{h^* + \sum_{q}^{M}{}'(2\,J^*_q - K^*_q)\}\phi_p^*\,d\tau = 0$$

$$\{IV-59\}$$

Introducing the orthonormality condition for the MO basis: $\{\phi\}$

$$S_{pq} \equiv <\phi_p|\phi_q> \equiv \int\phi_p^*\phi_q d\tau = \delta_{pq} \qquad \{IV-60\}$$

permits us to write the orthonormality restriction on $\delta\phi_p$:

$$\delta S_{pq} = \int(\delta\phi_p^*)\phi_q d\tau + \int(\delta\phi_q)\phi_p^* d\tau = 0 \qquad \{IV-61\}$$

The appropriate combination (cf. Appendix) of these two equations ({IV-59}) and ({IV-61}) leads to the following expression:

$$\{\hat{h} + \sum_{q}^{M}(2\,\hat{J}_q - \hat{K}_q)\}\phi_p = \phi_p\,\varepsilon_{pp} \qquad \{IV-62\}$$

where the operator, involving the one-electron operator (h) and the two-electron operators (J_q and K_q), is frequently called the Fock operator and abbreviated $\hat{F}$.

$$\hat{F}\phi_p = \phi_p \varepsilon_{pp} \qquad \{IV-63\}$$

The elements of the diagonal matrix ε are in a sense the molecular orbital energies. This expression represents the Hartree-Fock integro-differential equation in matrix notation. To carry out the actual computation the matrix representative of the Fock operator must be generated over the MO basis. The elements of the Fock matrix over the MO basis ($\underline{\underline{F}}^{\phi}$) may be written as

$$F^{\phi}_{st} \equiv \langle \phi_s | F | \phi_t \rangle \qquad \{IV-64\}$$

and, converting the Hartree-Fock operator equation to a matrix equation, we obtain

$$\underline{\underline{F}}^{\phi} = \underline{\underline{\varepsilon}} \qquad \{IV-65\}$$

since $S^{\phi}_{st} = \langle \phi_s | \phi_t \rangle = \delta_{st}$.
Because ϕ is unknown we substitute

$$\phi = \eta \, \underline{\underline{C}} \qquad \{IV-66\}$$

and

$$\phi^{\dagger} = \underline{\underline{C}}^{\dagger} \, \eta^{\dagger} \qquad \{IV-67\}$$

to obtain the Hartree-Fock matrix equation over the AO basis

$$\underline{\underline{C}} \, \underline{\underline{F}}^{\eta} \, \underline{\underline{C}} = \underline{\underline{C}} \, \underline{\underline{S}}^{\eta} \, \underline{\underline{C}} \, \underline{\underline{\varepsilon}} \qquad \{IV-68\}$$

where

$$\underline{\underline{F}}^{\eta} = \underline{\underline{h}}^{\eta} + 2 \, \underline{\underline{J}}^{\eta} - \underline{\underline{K}}^{\eta} \qquad \{IV-69\}$$

and

$$S^{\eta}_{ij} = \langle \eta_i | \eta_j \rangle \qquad \{IV-70\}$$

These molecular integrals (in the pseudo one-electron form) have the usual form:

$$h^{\eta}_{ij}(1) = \langle \eta_i(1)|\hat{h}|\eta_j(1)\rangle \qquad \{IV-71\}$$

$$J^{\eta}_{ij}(1) = \sum_{k=1}^{N} \sum_{\ell=1}^{N} \{\eta_i(1)\eta_j(1)|\eta_k(2)\eta_\ell(2)\}\ \rho_{k\ell} \qquad \{IV-72\}$$

$$K^{\eta}_{ij}(1) = \sum_{k=1} \sum_{\ell=1} \{\eta_i(1)\eta_k(1)|\eta_j(2)\eta_\ell(2)\}\ \rho_{k\ell} \qquad \{IV-73\}$$

where $\rho_{k\ell}$ is the k, ℓ-th element of the density matrix: $\underline{\underline{\rho}} = \underline{\underline{C}}_M \underline{\underline{C}}_M^{\dagger}$

$$\underline{\underline{\rho}} = \begin{pmatrix} c_{11} & c_{12} & \cdots & c_{1M} \\ c_{21} & c_{22} & \cdots & c_{2M} \\ \vdots & \vdots & & \vdots \\ c_{N1} & c_{N2} & \cdots & c_{NM} \end{pmatrix} \begin{pmatrix} c_{11} & c_{21} & \cdots & c_{N1} \\ c_{12} & c_{22} & \cdots & c_{N2} \\ \vdots & \vdots & & \vdots \\ c_{1M} & c_{2M} & \cdots & c_{NM} \end{pmatrix}$$

N x N $\qquad$ N X M $\qquad$ M x N $\qquad$ {IV-74}

Thus

$$F^{\eta}_{ij} = h^{\eta}_{ij} + \sum_{k}^{N} \sum_{\ell}^{N} [2\{\eta_i\eta_j|\eta_k\eta_\ell\} - \{\eta_i\eta_k|\eta_j\eta_\ell\}]\rho_{k\ell} \qquad \{IV-75\}$$

or briefly

$$F^{\eta}_{ij} = h^{\eta}_{ij} + 2\ J^{\eta}_{ij} - K^{\eta}_{ij} \qquad \{IV-76\}$$

If we wish to write E in terms of the AO integrals we get the following expression

$$E = 2 \sum_{i=1}^{N} \sum_{j=1}^{N} \rho_{ij}h^{\eta}_{ji} + 2 \sum_{i=1}^{N} \sum_{j=1}^{N} \rho_{ij}J^{\eta}_{ji} - \sum_{i=1}^{N} \sum_{j=1}^{N} \rho_{ij}K^{\eta}_{ji} \qquad \{IV-77\}$$

Note that there is a factor of 2 in the first term (this one includes h_{ji}). One of these may be combined with the last two terms so that the Fock matrix (F_{ji}) may be explicitly included in the above expression:

$$E = \sum_{i=1}^{N} \sum_{j=1}^{N} \rho_{ij} h^{\eta}_{ji} + \sum_{i=1}^{N} \sum_{j=1}^{N} \rho_{ij} (h^{\eta}_{ji} + 2 J^{\eta}_{ji} - K^{\eta}_{ji})$$

$$= \sum_{i=1}^{N} \sum_{j=1}^{N} \rho_{ij} h^{\eta}_{ji} + \sum_{i=1}^{N} \sum_{j=1}^{N} \rho_{ij} F^{\eta}_{ji}$$

$$= \sum_{i=1}^{N} \sum_{j=1}^{N} \rho_{ij} (h^{\eta}_{ji} + F^{\eta}_{ji}) \qquad \{IV\text{-}78\}$$

If one notes that the innermost summation (over j) will eliminate the running index j and one has only i,i elements then it is clear that the outermost summation (over i) is simply the summation of the diagonal elements of a <u>diagonal matrix</u> or in other words, it is the trace of the matrix

$$E = \sum_{i=1}^{N} \left\{ \sum_{j=1}^{N} \rho_{ij} (h^{\eta}_{ji} + F^{\eta}_{ji}) \right\}$$

$$= \text{trace} \left\{ \sum_{j=1}^{N} \rho_{ij} (h^{\eta}_{ji} + F^{\eta}_{ji}) \right\} \qquad \{IV\text{-}79\}$$

$$= \text{trace} \{ \underline{\underline{\rho}} (\underline{\underline{h}}^{\eta} + \underline{\underline{F}}^{\eta}) \}$$

Another point of interest is to relate the orbital energies to total energy. We saw in equation {IV-65} that the matrix representative of the Fock operator over the MO basis is simply the MO energy matrix. It should have also been clear from the discussion between equations {IV-65} and {IV-79} that when this quantity F^{ϕ}_{pp} is transformed to the AO basis (cf. similarity transformation in Chapter II, Section 7) one has the following equivalence:

$$\varepsilon_p \equiv F^{\phi}_{pp} \equiv \langle \phi_p|\hat{F}|\phi_p\rangle = \sum_i^N \sum_j^N C_{ip}\langle \eta_i|\hat{F}|\eta_j\rangle C_{jp} \qquad \{IV-80\}$$

If one sums up all the occupied MO energies then the result is the following

$$\sum_{p=1}^{M} \varepsilon_p = \sum_{i=1}^{N} \sum_{j=1}^{N} \rho_{ij} \langle \eta_j|\hat{F}|\eta_i\rangle = \sum_i^N \sum_j^N \rho_{ij} F^{\eta}_{ji} \qquad \{IV-80a\}$$

Now we would be ready to combine equations {IV-80} and {IV-78} but the latter expression also contains h^{η}_{ji} so let us express the h^{η}_{ji} element from equation {IV-76} first

$$h^{\eta}_{ji} = F^{\eta}_{ji} - (2\,J^{\eta}_{ji} - K^{\eta}_{ji}) \qquad \{IV-81\}$$

and substitute this into {IV-78} to get the following expression

$$E = \sum_{i=1}^{N} \sum_{j=1}^{N} \rho_{ij}(h^{\eta}_{ji} + F^{\eta}_{ji}) = \sum_{i=1}^{N} \sum_{j=1}^{N} \rho_{ij}\{2F^{\eta}_{ji} - (2J^{\eta}_{ji} - K^{\eta}_{ji})\}$$

$$= 2 \sum_{i=1}^{N} \sum_{j=1}^{N} \rho_{ij}F^{\eta}_{ji} - \sum_{i=1}^{N} \sum_{j=1}^{N} \rho_{ij}(2J^{\eta}_{ji} - K^{\eta}_{ji}) \qquad \{IV-82\}$$

Now we may use equations {IV-80} and {IV-80a} to see how to relate E to $\Sigma\varepsilon_p$.

$$E = 2\sum_{i=1}^{N}\sum_{j=1}^{N} \rho_{ij}F^{\eta}_{ji} - \sum_{i=1}^{N}\sum_{j=1}^{N} \rho_{ij}(2\,J^{\eta}_{ji} - K^{\eta}_{ji})$$

$$= 2\sum_{i=1}^{M} \varepsilon_i - \sum_{i=1}^{N}\sum_{j=1}^{N} \rho_{ij}(2\,J_{ji} - K_{ji}) \qquad \{IV-83\}$$

This equation is quite important in the sense that it indicates that the total energy is twice the sum of the doubly occupied MO minus the electron-electron repulsion. This will be useful to evaluate the nature of approximation of the semi-empirical MO theories (cf. Chapter V).

Let us return to our main line of thought that we left off at equations {IV-75} - {IV-77}.

The basic problem is that even when all the integrals are known we cannot calculate $\underline{\underline{F}}^{\eta}$ (required for the Hartree-Fock equation) and E because both depend on $\underline{\underline{\rho}}$ which, in turn, is constructed from $\underline{\underline{C}}$ and it is this quantity which we want to obtain from the Hartree-Fock equation. This requires that the eigen-problem equation be solved in an iterative manner by a process referred to as the Self Consistent Field (SCF) method.

A discussion of the SCF method requires first a consideration of the orthogonality problem. The basis set $\{\eta\}$ is not orthogonal so that the coefficient matrix $\underline{\underline{C}}$ which diagonalizes $\underline{\underline{F}}$ cannot be found by an orthogonal transformation. Consequently, it is necessary to transform the non-orthogonal set $\{\eta\}$ into an orthogonal set $\{\chi\}$ and, correspondingly, $\underline{\underline{F}}^{\eta}$ into $\underline{\underline{F}}^{\chi}$. Now $\underline{\underline{F}}^{\chi}$ may be diagonalized by an orthogonal transformation (such as Jacobi's diagonalization) and this will permit calculation of the coefficient matrix.

Since orthogonalization is a special type of transformation (cf. end of Chapter IV, Section 1) we separate the $\underline{\underline{C}}$ matrix into two matrices $\underline{\underline{V}}$ and $\underline{\underline{U}}$

$$\phi = \underline{\eta}\,\underline{\underline{C}} = \underline{\eta}\,\underline{\underline{V}}\,\underline{\underline{U}} = \underline{\chi}\,\underline{\underline{U}} \qquad \{IV-84\}$$

also for column notation:

$$\phi^\dagger = \underline{\underline{U}}^\dagger \, \underline{\underline{V}}^\dagger \, \eta^\dagger = \underline{\underline{U}}^\dagger \, \chi^\dagger \qquad \{IV\text{-}85\}$$

$$(\phi_1 \phi_2 \;\cdots\; \phi_N) = (\eta_1 \eta_2 \;\cdots\; \eta_N) \begin{pmatrix} v_{11} & v_{12} & \cdots & v_{1N} \\ v_{21} & v_{22} & \cdots & v_{2N} \\ \vdots & \vdots & & \vdots \\ v_{N1} & v_{N2} & \cdots & v_{NN} \end{pmatrix} \begin{pmatrix} u_{11} & u_{12} & \cdots & u_{1N} \\ u_{21} & u_{22} & \cdots & u_{2N} \\ \vdots & \vdots & & \vdots \\ u_{N1} & u_{N2} & \cdots & u_{NN} \end{pmatrix}$$

{IV-86}

Substituting into the Hartree-Fock equation, we have

$$\underline{\underline{C}}^\dagger \, \underline{\underline{F}}^\eta \, \underline{\underline{C}} = \underline{\underline{C}}^\dagger \, \underline{\underline{S}}^\eta \, \underline{\underline{C}} \, \underline{\underline{\varepsilon}} \qquad \{IV\text{-}87\}$$

$$\underline{\underline{U}}^\dagger \underbrace{\underline{\underline{V}}^\dagger \, \underline{\underline{F}}^\eta \, \underline{\underline{V}}}_{\underline{\underline{F}}^\chi} \underline{\underline{U}} = \underline{\underline{U}}^\dagger \underbrace{\underline{\underline{V}}^\dagger \, \underline{\underline{S}}^\eta \, \underline{\underline{V}}}_{\underline{\underline{S}}^\chi = \underline{\underline{1}}} \underline{\underline{U}} \, \underline{\underline{\varepsilon}} \qquad \{IV\text{-}88\}$$

$$\underline{\underline{U}}^\dagger \, \underline{\underline{F}}^\chi \, \underline{\underline{U}} = \underline{\underline{U}}^\dagger \, \underline{\underline{U}} \, \underline{\underline{\varepsilon}} \qquad \{IV\text{-}89\}$$

$$\underline{\underline{U}}^\dagger \, \underline{\underline{F}}^\chi \underline{\underline{U}} = \underline{\underline{\varepsilon}} \qquad \{IV\text{-}90\}$$

$\underline{\underline{\varepsilon}}$ is a diagonal matrix so that $\underline{\underline{F}}^\chi$ is to be diagonalized by $\underline{\underline{U}}$. Löwdin showed that the orthogonalizing matrix may be chosen to be $\underline{\underline{S}}^{-1/2}$, i.e. $\underline{\underline{V}} = \underline{\underline{S}}^{-1/2}$. Accepting this choice leads to the symmetric or Löwdin orthogonalization (cf. Figure II-5).

The process in which $\underline{\underline{S}}^{-1/2}$ is evaluated is based on the special properties of diagonal matrices, according to which any operation on the diagonal element is equivalent to the same operation on the whole matrix*. We therefore convert $\underline{\underline{S}}$ into its diagonal form $\underline{\underline{D}}$, take the -1/2 power of the diagonal elements, and then convert this diagonal matrix back to its non-diagonal form.

$$\underline{\underline{\theta}}^{\dagger}\ \underline{\underline{S}}\ \underline{\underline{\theta}} = \underline{\underline{D}} \qquad \{IV\text{-}91\}$$

where $\underline{\underline{\theta}}$ is an orthogonal matrix. The back conversion proceeds in the opposite sense

$$\underline{\underline{\theta}}(\underline{\underline{\theta}}^{\dagger}\ \underline{\underline{S}}\ \underline{\underline{\theta}})\underline{\underline{\theta}}^{\dagger} = \underline{\underline{\theta}}\ \underline{\underline{D}}\ \underline{\underline{\theta}}^{\dagger} \qquad \{IV\text{-}92\}$$

$$(\underline{\underline{\theta}}\underline{\underline{\theta}}^{\dagger})\underline{\underline{S}}(\underline{\underline{\theta}}\underline{\underline{\theta}}^{\dagger}) = \underline{\underline{\theta}}\ \underline{\underline{D}}\ \underline{\underline{\theta}}^{\dagger} \qquad \{IV\text{-}93\}$$

Since

$$\underline{\underline{\theta}}\underline{\underline{\theta}}^{\dagger} = \underline{\underline{1}}$$

$$\underline{\underline{S}} = \underline{\underline{\theta}}\ \underline{\underline{D}}\ \underline{\underline{\theta}}^{\dagger} \qquad \{IV\text{-}94\}$$

The process proceeds as follows:

1st step: Diagonalization

$$\underline{\underline{\theta}}^{\dagger}\ \underline{\underline{S}}\ \underline{\underline{\theta}} = \underline{\underline{D}} \equiv \begin{pmatrix} d_{11} & & 0 \\ & d_{22} & \\ 0 & & d_{nn} \end{pmatrix} \qquad \{IV\text{-}95\}$$

---

*The following procedure is based on the general feature of matrices:

$$\underline{\underline{U}}^{\dagger}\ \underline{\underline{A}}\ \underline{\underline{U}} = \underline{\underline{D}} \quad \text{and} \quad \underline{\underline{A}} = \underline{\underline{U}}\ \underline{\underline{D}}\ \underline{\underline{U}}^{\dagger}$$

however

$$f(\underline{\underline{A}}) = \underline{\underline{U}}[f(\underline{\underline{D}})]\underline{\underline{U}}^{\dagger}$$

2nd step: Raising D to the -1/2 power

$$D^{-1/2} = \begin{pmatrix} 1/\sqrt{d}_{11} & & 0 \\ & 1/\sqrt{d}_{22} & \\ 0 & & 1/\sqrt{d}_{nn} \end{pmatrix} \qquad \{IV-96\}$$

3rd step: Back transformation

$$\mathbf{S}^{-1/2} = \mathbf{\Theta}\,\mathbf{D}^{-1/2}\,\mathbf{\Theta}^{\dagger} \qquad \{IV-97\}$$

We may now check the value of $\mathbf{V}^{\dagger}\,\mathbf{S}\,\mathbf{V}$ if $\mathbf{V} = \mathbf{S}^{-1/2}$

$$\mathbf{V}^{\dagger} = (\,\mathbf{\Theta}\,\mathbf{D}^{-1/2}\,\mathbf{\Theta}^{\dagger})^{\dagger} = (\,\mathbf{\Theta}^{\dagger})^{\dagger}(\mathbf{D}^{-1/2})^{\dagger}\mathbf{\Theta}^{\dagger} = \mathbf{\Theta}\,\mathbf{D}^{-1/2}\,\mathbf{\Theta}^{\dagger} \qquad \{IV-98\}$$

since $(\mathbf{D}^{-1/2})^{\dagger} = \mathbf{D}^{-1/2}$ is diagonal.

$$\mathbf{V}^{\dagger}\mathbf{S}\,\mathbf{V} = (\mathbf{\Theta}\,\mathbf{D}^{-1/2}\,\mathbf{\Theta}^{\dagger})\mathbf{S}(\mathbf{\Theta}\,\mathbf{D}^{-1/2}\,\mathbf{\Theta}^{\dagger}) = \mathbf{\Theta}\,\mathbf{D}^{-1/2}(\mathbf{\Theta}^{\dagger}\mathbf{S}\,\mathbf{\Theta})\mathbf{D}^{-1/2}\,\mathbf{\Theta}^{\dagger}$$

$$= \mathbf{\Theta}\,\mathbf{D}^{-1/2}\,\mathbf{D}\,\mathbf{D}^{-1/2}\,\mathbf{\Theta}^{\dagger} \qquad \{IV-99\}$$

Since diagonal matrices commute

$$\mathbf{V}^{\dagger}\mathbf{S}\,\mathbf{V} = \mathbf{\Theta}\,\mathbf{D}^{-1/2}\,\mathbf{D}^{-1/2}\,\mathbf{D}\,\mathbf{\Theta}^{\dagger} = \mathbf{\Theta}\,\mathbf{D}^{-1}\,\mathbf{D}\,\mathbf{\Theta}^{\dagger}$$

$$= \mathbf{\Theta}\,\mathbf{\Theta}^{\dagger} = \mathbf{1} \qquad \{IV-100\}$$

Now with $\mathbf{S}^{-1/2}$ in hand

$$\mathbf{U}^{\dagger}\,\mathbf{F}^{\chi}\,\mathbf{U} = \boldsymbol{\varepsilon} \qquad \text{and} \qquad \mathbf{F}^{\chi} = \mathbf{S}^{-1/2}\,\mathbf{F}^{\eta}\,\mathbf{S}^{-1/2} \qquad \{IV-101\}$$

The iterative SCF procedure can now be presented, most conveniently, in the form of a self-explanatory flow-chart.

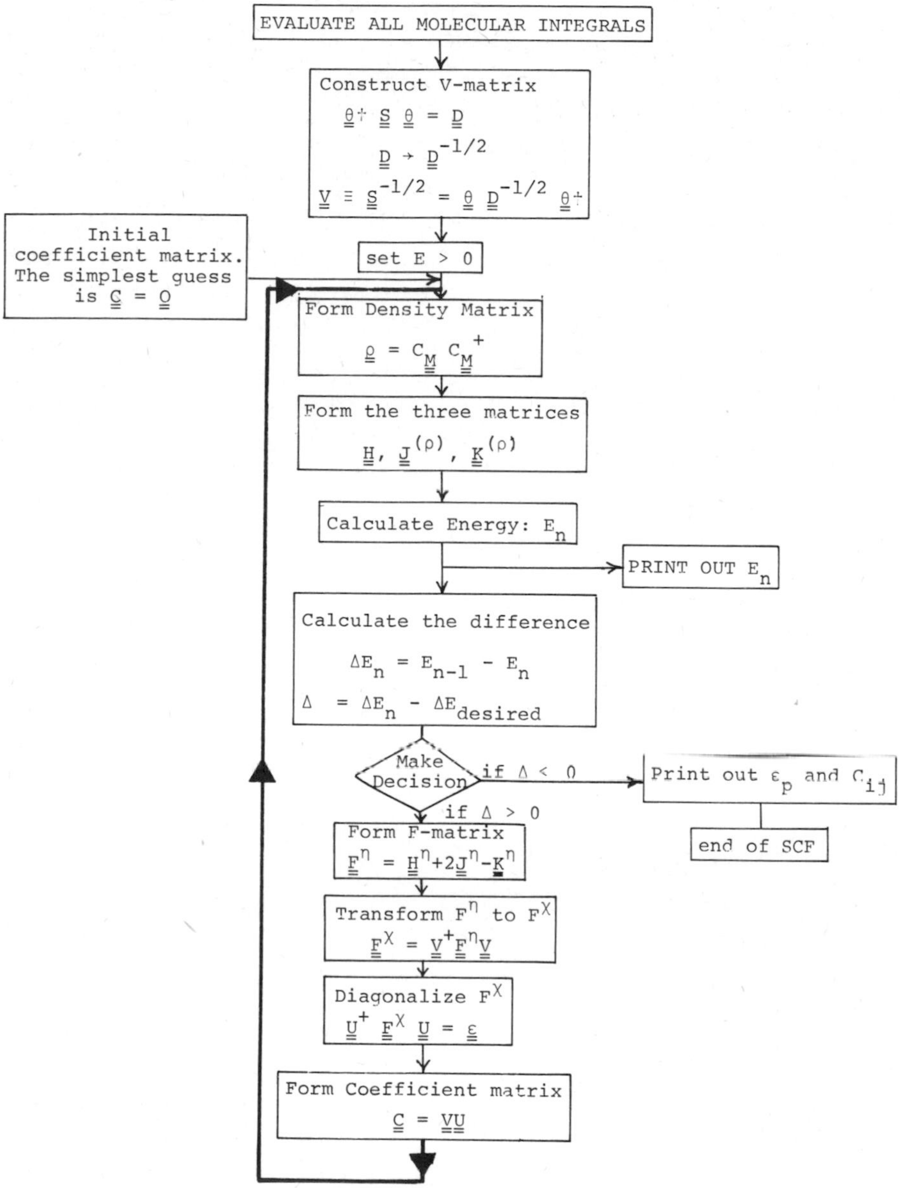

Figure IV-7. Flowchart for the iterative Self-Consistent-Field (SCF) method.

In the course of the SCF procedure the total energy (E) is lowered in each iteration cycle (i.e. E_n gradually becomes a larger negative quantity as n increases). Convergence may be measured by the difference between the energy values associated with two successive iterations ($\Delta E_n = E_{n-1} - E_n$). Since we do not know the energy to be expected after the first iteration (which depends on the initial coefficient matrix) it is best to set the initial value of E greater than zero. It is also possible to judge convergence in terms of the density matrix $\underline{\underline{\rho}}$ ($\Delta \underline{\underline{\rho}}_n = \underline{\underline{\rho}}_{n-1} - \underline{\underline{\rho}}_n$).

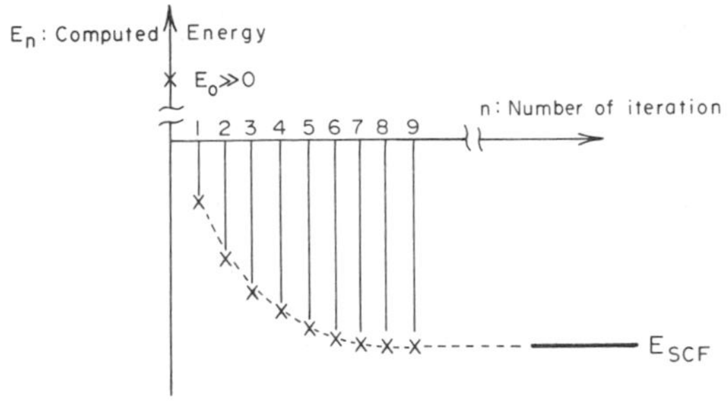

Figure IV-8 A schematic illustration for the convergency of an SCF calculation

Another point of interest is the consequence of choosing the simplest possible initial coefficient matrix, i.e. $\underline{\underline{C}} = \underline{\underline{0}}$. In this event the density matrix becomes zero ($\underline{\underline{\rho}} = \underline{\underline{0}}$) and terms ($\underline{\underline{J}}$, $\underline{\underline{K}}$ and $\underline{\underline{E}}$) that incorporate it will vanish. However, the $\underline{\underline{F}}$-matrix will not since only the last two terms in the expression $\underline{\underline{F}} = \underline{\underline{H}} + 2\,\underline{\underline{J}} - \underline{\underline{K}}$ depend on ρ. The approximation $\underline{\underline{F}}^{\eta} = \underline{\underline{H}}^{\eta}$ will then be operative. This approximation will be discussed in the next chapter in connection with the electron-independent approximations.

3. <u>Integral Transformation</u>

As was mentioned briefly in Chapter IV, Section 1, the transformation of integrals from one basis to another (e.g. AO to CMO) is important for example in the generation of localized molecular orbitals (LMO) or as a preliminary step in the configuration interaction (CI) method. The one-electron integrals may be handled by a straightforward similarity transformation as illustrated for the 3 orbital case in equations {IV-102-104}.

$$\underline{\underline{I}}^{\eta} \xrightarrow[\text{transformation}]{\text{similarity}} \underline{\underline{I}}^{\phi} = \begin{pmatrix} \langle\phi_1|h|\phi_1\rangle & \langle\phi_1|h|\phi_2\rangle & \langle\phi_1|h|\phi_3\rangle \\ & \langle\phi_2|h|\phi_2\rangle & \langle\phi_2|h|\phi_3\rangle \\ & & \langle\phi_3|h|\phi_3\rangle \end{pmatrix}$$

{IV-102}

$$I^{\phi}(p,r) = \langle\phi_p|\hat{h}|\phi_r\rangle = \sum_i \sum_j C_{pi}\langle\eta_i|\hat{h}|\eta_j\rangle C_{jr} \qquad \text{\{IV-103\}}$$

$$\underline{\underline{I}}^{\phi} = \underline{\underline{C}}^{\dagger}\, \underline{\underline{I}}^{\eta}\, \underline{\underline{C}} \qquad \text{\{IV-104\}}$$

In dealing with the two-electron integrals the 4 dimensional matrix $\{\eta_i\eta_j|\eta_k\eta_\ell\}$ or $\{ij|k\ell\}$ is converted to a pseudo 2-dimensional matrix by defining the following compound indices:

$$ij = \frac{i(i-1)}{2} + j \quad \text{and} \quad k\ell = \frac{k(k-1)}{2} + \ell \qquad \text{\{IV-105\}}$$

(See Table IV-1 for a 3-orbital example.)

It should be noted that although for the 3-orbital case there are $3^2 \times 3^2 = 81$ elements, only 21 elements are distinctly different because of the following identities:

$$\{ij|k\ell\} = \{ji|k\ell\} = \{ij|\ell k\} = \{ji|\ell k\} \qquad \text{\{IV-106a\}}$$

$$= \{k\ell|ij\} = \{\ell k|ij\} = \{k\ell|ji\} = \{\ell k|ji\} \qquad \text{\{IV-106b\}}$$

TABLE IV-2 GENERAL PSEUDO TWO DIMENSIONAL TWO ELECTRON INTEGRAL MATRIX

	kl	1	2	3	4	5	6	7	8	9
ij	i,j k,l	1,1	1,2	1,3	2,1	2,2	2,3	3,1	3,2	3,3
1	1,1	{11|11}	{11|12}	{11|13}	{11|21}	{11|22}	{11|23}	{11|31}	{11|32}	{11|33}
2	1,2		{12|12}	{12|13}	{12|21}	{12|22}	{12|23}	{12|31}	{12|32}	{12|33}
3	1,3			{13|13}	{13|21}	{13|22}	{13|23}	{13|31}	{13|32}	{13|33}
4	2,1				{21|21}	{21|22}	{21|23}	{21|31}	{21|32}	{21|33}
5	2,2					{22|22}	{22|23}	{22|31}	{22|32}	{22|33}
6	2,3						{23|23}	{23|31}	{23|32}	{23|33}
7	3,1							{31|31}	{31|32}	{31|33}
8	3,2								{32|32}	{32|33}
9	3,3									{33|33}

Note that the formulae in {IV-105} are only valid for $i \geq j$ and $k \geq \ell$. However, with recognition of the identities in {IV-106} these formulae are sufficient to handle all distinctly different integrals.

TABLE IV-1 REDUCED PSEUDO TWO DIMENSIONAL TWO ELECTRON INTEGRAL MATRIX

ij \ kℓ	1	2	3	4	5	6
1	{11|11}	{11|21}	{11|22}	{11|31}	{11|32}	{11|33}
2		{21|21}	{21|22}	{21|31}	{21|32}	{21|33}
3			{22|22}	{22|31}	{22|32}	{22|33}
4				{31|31}	{31|32}	{31|33}
5					{32|32}	{32|33}
6						{33|33}

The transformation of the two-electron integrals may be written as specified by equation {IV-107}.

$$II^{\phi}(p,r,s,t) \equiv \{\phi_p\phi_r|\phi_s\phi_t\} = \sum_i \sum_j \sum_k \sum_\ell C^*_{pi}C^*_{rj}\{\eta_i\eta_j|\eta_k\eta_\ell\}C_{ks}C_{\ell t}$$

$$= \sum_{ij}\sum_{k\ell} u_{pr,ij}\, II^{\eta}(ij,k\ell)\, u_{k\ell,st} \qquad \{IV-107\}$$

Where: $C_{pi}C_{rj} = u_{pr,ij}$; $C_{ks}C_{\ell t} = u_{k\ell,st}$ are elements of a $\underline{\underline{U}}$ matrix and the two-electron integrals are labelled as elements of a pseudo-two-dimensional matrix $\underline{\underline{II}}^{\eta}(ij,k\ell)$

$$\underline{\underline{U}} = \underline{\underline{C}} \times \underline{\underline{C}} \qquad \{IV-108\}$$

$$\underline{\underline{II}}^{\phi} = \underline{\underline{U}}^{\dagger}\, \underline{\underline{II}}^{\eta}\underline{\underline{U}} \qquad \{IV-109\}$$

However, for the two-electron integrals because of the very large number of integrals involved, care must be taken to make the transformation as efficient as possible. In the charge density notation the general expression for the two electron integral transformation from the AO basis to the MO basis was given by {IV-107}. The transformation could be performed by a straightforward 8-fold DO loop but such a procedure becomes extremely inefficient as the number of integrals becomes large. A more efficient transformation following Sutcliffe's suggestion carries out the multiplication by the coefficients in a stepwise fashion:

$$\{\phi_p\eta_j|\eta_k\eta_\ell\} = \sum_i C_{pi}\{\eta_i\eta_j|\eta_k\eta_\ell\} \equiv \beta^{p}{}_{jk\ell} \qquad \{IV-110a\}$$

$$\{\phi_p\phi_r|\eta_k\eta_\ell\} = \sum_j C_{rj}\beta^{p}{}_{jk\ell} \equiv \Gamma^{pr}{}_{k\ell} \qquad \{IV-110b\}$$

$$\{\phi_p\phi_r|\phi_s\eta_\ell\} = \sum_k C_{ks}\Gamma^{pr}{}_{k\ell} \equiv \Delta^{prs}{}_{\ell} \qquad \{IV-110c\}$$

$$\{\phi_p\phi_r|\phi_s\phi_t\} = \sum_1 C_{1t}\ \Delta_1^{prs} \qquad \{IV\text{-}110d\}$$

4. Mulliken's Population Analysis or Charge Distribution in Terms of AO

Molecular orbitals are always doubly occupied in a closed shell system but other orbitals may have fractional occupancy. For this reason it is more appropriate to refer to the "population" rather than the occupancy of a set of orbitals.

Returning to the diatomic example considered in section 1 of this chapter, any normalized MO may be written as a linear combination of normalized AOs η_1 and η_2

$$(\phi_1\phi_2) = (\eta_1\eta_2)\begin{pmatrix} C_{11} & C_{12} \\ C_{21} & C_{22} \end{pmatrix} \qquad \{IV\text{-}111\}$$

The population of the i-th MO will be N_i (here $N_1=2$ and $N_2=0$).

$$\phi_1 = \eta_1 C_{11} + \eta_2 C_{21} \qquad \{IV\text{-}112\}$$

Squaring and integrating over all space gives, with a normalization to the total number of electrons N_1 rather than unity:

$$\begin{aligned} N_1 &= N_1\int\phi_1^2 d\tau = N_1\int(\eta_1 C_{11} + \eta_2 C_{21})^2 d\tau \\ &= N_1 C_{11}^2\int\eta_1^2 d\tau + 2N_1 C_{11}C_{21}\int\eta_1\eta_2 d\tau + N_1 C_{21}^2\int\eta_2^2 d\tau \\ &= N_1 C_{11}^2 + 2N_1 C_{11}C_{21}S_{12} + N_1 C_{21}^2 \qquad \{IV\text{-}113\} \end{aligned}$$

where S_{12} is the overlap integral $\int\eta_1\eta_2 d\tau$.

Rewriting IV-113 making use of the definition of the density matrix (eq. IV-41) gives

$$N_1 = N_1\rho_{11} + 2N_1\rho_{12}S_{12} + N_1\rho_{22} \qquad \{IV\text{-}114\}$$

The populations $N_1\rho_{11}$ and $N_1\rho_{22}$ are called the atomic populations, and $2N_1\rho_{12}S_{12}$ is called the overlap population. A population matrix $\underline{\underline{P}}$ may then be defined as

$$\underline{\underline{P}} = N_1 \begin{pmatrix} \rho_{11}S_{11} & \rho_{12}S_{12} \\ \rho_{21}S_{21} & \rho_{22}S_{22} \end{pmatrix} \qquad \{IV\text{-}115\}$$

Note that $S_{11}=S_{22}=1$, $\rho_{12}=\rho_{21}$ and $S_{12}=S_{21}$ and hence $\underline{\underline{P}}$ is symmetric. The general matrix element may be defined as

$$P_{ij} = N_1\rho_{ij}S_{ij} \qquad \{IV\text{-}116\}$$

and from IV-114

$$\sum_i \sum_j P_{ij}=N_1 \qquad \{IV\text{-}117\}$$

In general, the MOs are expressed as a linear combination of many AOs:

$$\phi_i = \sum_j \eta_j C_{ji} \qquad \{IV\text{-}118\}$$

$$\phi_i^2 = (\sum_j \eta_j C_{ji})^2 = (\sum_j \eta_j C_{ji})(\sum_k \eta_k C_{ki}) \qquad \{IV\text{-}119\}$$

$$= \sum_{jk}\sum \eta_j \eta_k C_{ji} C_{ki}$$

Integrating over all space, the orbital population N_i is

$$N_i = N_i \int \phi_1^2 d\tau = N_i \sum_{jk}\sum C_{ji}C_{ki} \int \eta_j \eta_k d\tau \qquad \{IV\text{-}120\}$$

$$= \sum_{jk}\sum N_i C_{ji} C_{ki} S_{jk}$$

where S_{jk} is the overlap integral $\int \eta_j \eta_k d\tau$. Summing over all occupied MOs gives

$$N = \sum_i N_i = \sum_i N_i \sum_{jk}\sum C_{ji} C_{ki} S_{jk} \qquad \{IV\text{-}121\}$$

$$= \sum_{jk}\sum S_{jk} \ (\sum_i N_i C_{ji} C_{ki})$$

where N is the total number of electrons. In the general case, the orbital occupancy N_i would be incorporated into the density matrix (eq. IV-74 would be modified), but in a closed shell system N_i = 2 for all the occupied MOs, so a factor 2 may be brought outside the summation. From equation IV-74

$$\rho_{jk} = \sum_{i}^{\text{Occ MO}} C_{ji}C_{ik}^{\dagger} = \sum_{i}^{\text{Occ Mo}} C_{ji}C_{ki}^{*} = \sum_{i} C_{ji}C_{ki} \qquad \{IV\text{-}122\}$$

($C_{ki}^{*} = C_{ki}$ when MO are real). Then equation IV-121 becomes

$$N = 2 \sum_{jk}\sum \rho_{jk}S_{jk} \qquad \{IV\text{-}123\}$$

A population matrix $\underline{\underline{P}}$ may be defined by

$$P_{ij} = 2\rho_{ij}S_{ij} \qquad \{IV\text{-}124\}$$

$$\underline{\underline{P}} = 2 \begin{pmatrix} \rho_{11}S_{11} & \rho_{12}S_{12} & \cdots & \rho_{1n}S_{1n} \\ \rho_{21}S_{21} & \rho_{22}S_{22} & \cdots & \rho_{2n}S_{2n} \\ \vdots & \vdots & \cdots & \vdots \\ \rho_{n1}S_{n1} & \rho_{n2}S_{n2} & \cdots & \rho_{nn}S_{nn} \end{pmatrix} \qquad \{IV\text{-}125\}$$

The diagonal elements P_{ii}(=$2\rho_{ii}$, since S_{ii}=1) represent the "atomic charge" (in units of number of electrons) and the off-diagonal elements P_{ij} are related to the simple idea of "bond-order". Note that $P_{ij} = P_{ji}$ since both $\underline{\underline{\rho}}$ and $\underline{\underline{S}}$ are symmetric, and that from IV-123

$$\sum_{ij}\sum P_{ij} = N \qquad \{IV\text{-}126\}$$

The final result of the Mulliken Population Analysis is

$$P_{ij} = 2\rho_{ij}S_{ij} \qquad \{IV\text{-}127\}$$

The $\underline{\underline{P}}$ matrix provides an orbital population analysis. The summation of all P_{ij} elements associated with atomic orbital i gives the gross orbital charge of AO_i, GOC_i:

$$GOC_i = P_{ii} + (1/2) \sum_{j \neq i} P_{ij} + (1/2) \sum_{j \neq i} P_{ji}$$

$$= P_{ii} + (1/2) \sum_{j \neq i} (P_{ij} + P_{ji}) \qquad \{IV\text{-}128\}$$

$$= P_{ii} + \sum_{j \neq i} P_{ij} = \sum_j P_{ij}$$

since $P_{ij} = P_{ji}$. The factor 1/2 is introduced because only half the overlap density P_{ij} is counted with orbital i.

The summation of all the P_{ij} elements associated with a given pair of atoms A and B reduces the "orbital by orbital" population matrix $\underline{\underline{P}}$ (see Figure IV-9a) to an "atom by atom" population matrix $\underline{\underline{R}}$ (see Figure IV-9b)

$$R_{AB} = \sum_{i}^{\text{AO on A}} \sum_{j}^{\text{AO on B}} P_{ij} \qquad \{IV\text{-}129\}$$

The total number of electrons associated with atom A, N_A, is given by

$$N_A = R_{AA} + (1/2) \sum_{\substack{B \\ B \neq A}}^{\text{atoms}} R_{AB} + (1/2) \sum_{\substack{B \\ B \neq A}}^{\text{atoms}} R_{BA}$$

$$= R_{AA} + (1/2) \sum_{\substack{B \\ B \neq A}}^{\text{atoms}} (R_{AB} + R_{BA}) \qquad \{IV\text{-}130\}$$

$$= R_{AA} + \sum_{\substack{B \\ B \neq A}}^{\text{atoms}} R_{AB} = \sum_{B}^{\text{all atoms}} R_{AB}$$

since $R_{AB} = R_{BA}$. Once again a factor of 1/2 is needed since only half the electrons shared between A and B (measured by R_{AB}) are counted with A. The net charge Q_A associated with A is given by:

$$Q_A = Z_A - N_A \qquad \{IV\text{-}131\}$$

where Z_A is the nuclear charge of atom A. The net charges for CH_3SH are presented in Figure IV-9c.

Note that this definition of the populations P_{ij}, R_{AB} and R_A has the property

$$N = \sum_{i,j} P_{ij} = \sum_{A,B} R_{AB} = \sum_{A} R_A \qquad \{IV\text{-}132\}$$

Although population analysis is very popular since it preserves many chemical concepts, it should be noted that it is an arbitrary procedure. Any non-arbitrary method of dividing a molecule into atomic fragments is much more involved.

Mulliken first proposed his analysis in R.S. Mulliken, J. Chem. Phys. 23, 1833 (1955).

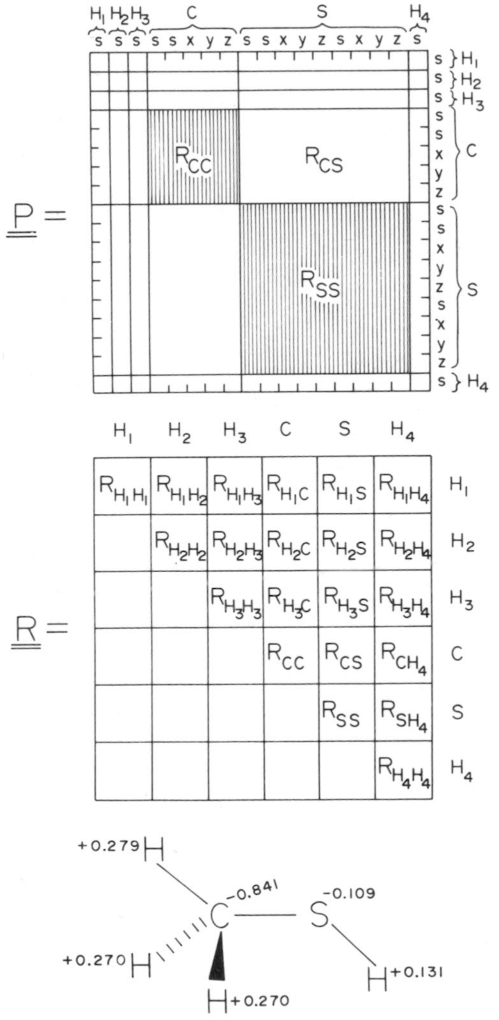

a. Orbital by orbital Population Matrix, $\underline{\underline{P}}$

b. Atom by atom Population Matrix, $\underline{\underline{R}}$

c. Net Atomic Charges

Figure IV-9 Mulliken Population Analysis for Methane Thiol

CHAPTER V

SEMI-EMPIRICAL MOLECULAR ORBITAL THEORIES

1. The Nature of the Approximation

The word "approximation" can have various meanings. If we adopt the definition

$$\text{Approximation} \equiv \text{Mathematical Neglect} \qquad \{V\text{-}1\}$$

the Hartree-Fock method (non-empirical SCF-MO theory) discussed in Chapter IV is not an approximate method because it includes no mathematical neglect. It may not be the most general method (cf. Chapter VIII) but it is nevertheless rigorous. The various semi-empirical MO theories, which may formally be derived from the exact SCF formalism, include approximations and the use of empirical parameters to different degrees. In order to appreciate the significance of these simplifications the essence of the SCF formalism is reviewed briefly below.

The MO are obtained from the AO by a linear transformation:

$$(\phi_1 \phi_2 \; .... \; \phi_N) = (\eta_1 \eta_2 \; .... \; \eta_N) \begin{pmatrix} c_{11} & c_{12} & .... & c_{1N} \\ c_{21} & c_{22} & .... & c_{2N} \\ \vdots & \vdots & & \vdots \\ c_{N1} & c_{N2} & & c_{NN} \end{pmatrix} \qquad \{V\text{-}2\}$$

These MO $\{\phi_i\}$ are eigenvectors of an effective molecular hamiltonian, the Fock operator. The solution $\{\phi_i\}$ is to be obtained, in the form of the coefficient matrix $\underline{\underline{C}}$, from the Hartree-Fock eigenproblem equation

$$\underline{\underline{C}}^{\dagger} \, \underline{\underline{F}}^{\eta} \, \underline{\underline{C}} = \underline{\underline{C}}^{\dagger} \, \underline{\underline{S}}^{\eta} \, \underline{\underline{C}} \, \underline{\underline{\varepsilon}} \qquad \{V\text{-}3\}$$

or, simply:

$$\underline{\underline{F}}^{\eta} \, \underline{\underline{C}} = \underline{\underline{S}}^{\eta} \, \underline{\underline{C}} \, \underline{\underline{\varepsilon}} \qquad \{V\text{-}4\}$$

where $\underline{\underline{S}}^{\eta}$ is the overlap matrix and $\underline{\underline{F}}^{\eta}$ is the matrix representative of the pseudo-one-electron-operator, the Fock operator, over the AO basis $\{\eta_i\}$. The F-matrix is assembled from the one-electron and two-electron molecular integrals as follows:

$$\underline{\underline{F}}^{\eta} = \underline{\underline{H}}^{\eta} + 2\,\underline{\underline{J}}^{\eta} - \underline{\underline{K}}^{\eta} \qquad \{V-5\}$$

where

$$H^{\eta}_{ij} = \langle \eta_i | \hat{h} | \eta_j \rangle \qquad \{V-6\}$$

$$J^{\eta}_{ij} = \sum_{k}^{N} \sum_{\ell}^{N} \rho_{k\ell} \{ \eta_i \eta_j | \eta_k \eta_\ell \} \qquad \{V-7\}$$

$$K^{\eta}_{ij} = \sum_{k}^{N} \sum_{\ell}^{N} \rho_{k\ell} \{ \eta_i \eta_k | \eta_j \eta_\ell \} \qquad \{V-8\}$$

where $\rho_{k\ell}$ represents the $k\ell^{th}$ element of the density matrix

$$\rho_{k\ell} = \sum_{r=1}^{M} C_{kr} C_{r\ell} = (C_{k1} C_{k2} \cdots C_{kM}) \begin{pmatrix} C_{1\ell} \\ C_{2\ell} \\ \vdots \\ C_{M\ell} \end{pmatrix} \qquad \{V-9\}$$

It should be noted that $\underline{\underline{F}}^{\eta}$ <u>depends on the number of electrons</u> since it includes $\underline{\underline{J}}^{\eta}$ and $\underline{\underline{K}}^{\eta}$ which, in turn, include $\underline{\underline{\rho}}$ that has been calculated from the M eigenvectors associated with the M doubly occupied MO. There are two ways to simplify the above formalism.

(i) We may approximate $\underline{\underline{F}}$ by regarding it as <u>electron independent</u>. This corresponds to the situation after the first SCF iteration if the initial coefficient matrix is assumed to be a null matrix, i.e.

$$\text{if } \underline{\underline{C}} = \underline{\underline{O}} \text{ (null matrix) then } \underline{\underline{\rho}} = \underline{\underline{O}}, \qquad \{V-10\}$$

$$\underline{\underline{J}} = \underline{\underline{K}} = \underline{\underline{O}} \qquad \{V\text{-}11\}$$

and

$$\underline{\underline{F}}^{\eta} = \underline{\underline{H}}^{\eta} \qquad \{V\text{-}12\}$$

so that

$$\underline{\underline{U}}^{\dagger}\, \underline{\underline{H}}^{\chi}\, \underline{\underline{U}} = \underline{\underline{\varepsilon}} \qquad \{V\text{-}13\}$$

This leads to a very crude estimate of the MO. In practice the value of $\underline{\underline{F}}$ has to be estimated from experimental data such as ionization potentials (IP). If the estimated $\underline{\underline{F}}$ matrix is permitted to vary by incorporation of the density matrix (i.e. $\underline{\underline{F}} = \underline{\underline{F}}(\underline{\underline{\rho}})$) the method is then electron dependent. This can be achieved within the framework of semi-empirical MO theories by parametrizing $\underline{\underline{J}}$ and $\underline{\underline{K}}$.

(ii) We may reduce the size of the basis set. Exclusion of all 1s type AO (i.e. neglect of all core electrons) leads to a method that may be called an all-valence-electron (AVE) theory. Neglect of all but one p-π orbital on each atom is referred to as π-electron theory. Combination of these two general methods of approximation leads to four different groups of theories summarized in the table below.

Table V-1 A Summary of the Four Different Types of Approximations in Semi-Empirical MO Theory

Type of orbital system \ Type of approximation	Electron independent	Electron dependent
π	SHMO	PPP
$\sigma + \pi$	EHMO	CNDO

Although the AVE approximation ($(\sigma + \pi)$ theory) may be applied to any chemical system, the very restrictive approximation, in which only one orbital per atom is included in the basis set may only generally be applied to planar conjugated molecules.

The abbreviations in the above table refer to the names of the following semi-empirical methods:

SHMO: Simple Hückel MO Theory
EHMO: Extended Hückel MO Theory
PPP: Pariser-Parr-Pople Theory
CNDO: Complete Neglect of Differential Overlap Theory

2. The Simple Hückel Molecular Orbital (SHMO) Theory

The SHMO theory is the oldest and simplest theory of all. It is the simplest because it contains the largest number of approximations as follows:

1. Only $2p_z$ AO are included in the basis set and these are assumed to be orthonormal, i.e. one starts with an assumed orthonormal basis $\{\chi_i\}$ rather than the usual non-orthonormal basis: $\{\eta_i\}$

$$\langle \chi_i | \chi_j \rangle = \delta_{ij} \qquad \{V\text{-}14a,b\}$$

or

$$\underline{\underline{S}}^{\chi} = \underline{\underline{1}}$$

Although S_{ij} can be shown to be as high as 0.25 for neighbouring atoms, the SHMO approximation assumes this quantity to be zero. However, because orbitals are not used explicitly in the calculation, this is a tolerable assumption. Moreover, the inclusion of overlap presents no great difficulty but also provides little further information for the ground states of molecules.

2. The second assumption concerns the matrix elements of F_{ij}. The diagonal elements are assumed to be the same when only carbon atoms are considered

$$F_{11} = F_{22} = \;\ldots\; = F_{ii} = F_{jj} = \;\ldots\; = \alpha \qquad \{V\text{-}15\}$$

Estimation of off-diagonal matrix elements is more tenuous because it is assumed that all bond lengths are equal. (It is possible to produce a self consistent Hückel method with β proportional to some measure of bond length but again at the qualitative level at which Hückel theory is most useful little is gained.)

$F_{ij} = \beta$ if i and j are adjacent (bonded atoms)

$F_{ij} = 0$ if i and j are not adjacent (non-bonded atoms)

The simplified Hartree-Fock matrix equation which results can be solved by diagonalization of the assumed $\underline{\underline{F}}^{\chi}$ matrix since the AO are assumed orthonormal (i.e. $\underline{\phi} = \underline{\chi}\ \underline{\underline{U}}$)

$$\underline{\underline{U}}^{\dagger}\ \underline{\underline{F}}^{\chi}\ \underline{\underline{U}} = \underline{\underline{\varepsilon}} \qquad \{V\text{-}16\}$$

It would be preferable to have numbers as matrix elements of $\underline{\underline{F}}$ rather than the symbols α and β. To bypass this problem it is customary to consider only the change in energy (relative to the atomic state) which occurs when a π-system is formed from a number of carbon atoms. This requires a <u>relative</u> Fock matrix $\underline{\underline{F}}^{rel}$, defined as follows:

$$\underline{\underline{F}}^{rel} = \underline{\underline{F}}^{Molecule} - \underline{\underline{F}}^{Atoms} \qquad \{V\text{-}17\}$$

By definition $\underline{\underline{F}}^{Atoms}$ is a diagonal matrix of α's.

$$\underline{\underline{F}}^{Atom} = \begin{pmatrix} \alpha & 0 & 0 \\ 0 & \alpha & 0 \\ 0 & 0 & \alpha \end{pmatrix} \qquad \{V\text{-}18\}$$

$\underline{\underline{F}}^{rel}$ will then be a matrix containing only zero and β elements.

To perform the computation it is now necessary to substitute some energy value for β. The various empirical estimates of this number vary from -13 Kcal/mole (-0.564 electron volt) to -112 Kcal/mole (-4.857 electron volt) depending on the experimental property examined. It is therefore more practical to carry out the calculation so that the results are obtained in units of β (relative to α). These results will then be equally valid irrespective of the numerical value given to β. If β is set equal to one in the matrix of $\underline{\underline{F}}^{rel}$ the highest root obtained by solving {V-16} will correspond to the lowest energy level because β is a negative quantity. It is more convenient therefore to carry out the computation in such a way that the results are obtained (relative to α) in units of |β|. This corresponds to setting β = -1. The lowest root will then correspond to the lowest energy level. This may be illustrated by an example, the allyl anion.

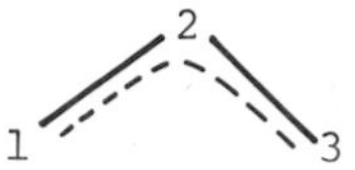

$$\underline{\underline{F}} = \begin{array}{c} \\ 1 \\ 2 \\ 3 \end{array}\begin{array}{c} \begin{array}{ccc} 1 & 2 & 3 \end{array} \\ \begin{pmatrix} \alpha & \beta & 0 \\ \beta & \alpha & \beta \\ 0 & \beta & \alpha \end{pmatrix} \end{array} \quad \text{and} \quad \underline{\underline{F}}^{rel} = \begin{pmatrix} 0 & -1 & 0 \\ -1 & 0 & -1 \\ 0 & -1 & 0 \end{pmatrix}$$

{V-19a-c}

Solving the Fock matrix equation by diagonalizing $\underline{\underline{F}}^{rel}$

$$\underline{\underline{U}}\dagger \; \underline{\underline{F}}^{rel} \; \underline{\underline{U}} = \underline{\underline{\varepsilon}}^{rel} \qquad \text{\{V-20\}}$$

we find $\underline{\underline{U}}$ and $\underline{\underline{\varepsilon}}$

$$\underline{\underline{U}} = \begin{pmatrix} \frac{1}{2} & \frac{-1}{\sqrt{2}} & \frac{1}{2} \\ \frac{1}{2}\sqrt{2} & 0 & -\frac{1}{2}\sqrt{2} \\ \frac{1}{2} & \frac{1}{\sqrt{2}} & \frac{1}{2} \end{pmatrix} \qquad \text{\{V-21\}}$$

$$\varepsilon_{11}^{rel} = -\sqrt{2}, \quad \varepsilon_{22}^{rel} = 0, \quad \varepsilon_{33}^{rel} = +\sqrt{2} \qquad \text{\{V-22a-c\}}$$

$$\begin{pmatrix} \frac{1}{2} & \frac{1}{2}\sqrt{2} & \frac{1}{2} \\ \frac{-1}{\sqrt{2}} & 0 & \frac{1}{\sqrt{2}} \\ \frac{1}{2} & -\frac{1}{2}\sqrt{2} & \frac{1}{2} \end{pmatrix} \begin{pmatrix} 0 & -1 & 0 \\ -1 & 0 & -1 \\ 0 & -1 & 0 \end{pmatrix} \begin{pmatrix} \frac{1}{2} & \frac{-1}{\sqrt{2}} & \frac{1}{2} \\ \frac{1}{2}\sqrt{2} & 0 & -\frac{1}{2}\sqrt{2} \\ \frac{1}{2} & \frac{1}{\sqrt{2}} & \frac{1}{2} \end{pmatrix} = \begin{pmatrix} -\sqrt{2} & 0 & 0 \\ 0 & 0 & 0 \\ 0 & 0 & +\sqrt{2} \end{pmatrix}$$

{V-23}

These roots or eigenvalues may now be plotted. The method is electron independent thus the eigenvalues are the same for the cation, the radical and the anion.

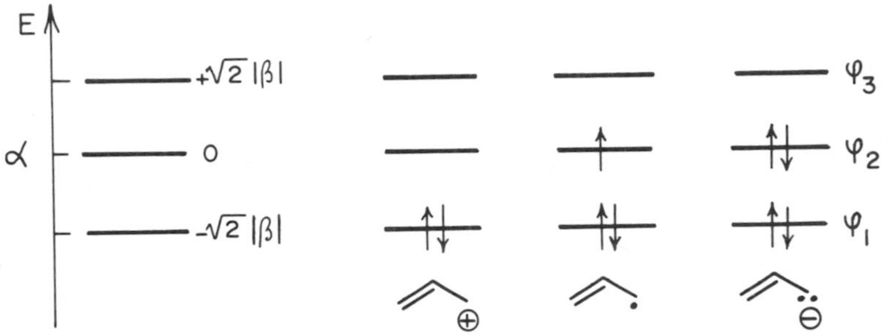

Figure V-1 Molecular Orbital energy levels for the allyl system

Note that in this result we have the opposite sign for ε from that obtained in more simplified treatments of SHMO theory. The results would have been the same if we had set $\beta = 1$ rather than -1. It should be noted that the total energy in the SHMO method is twice the sum of the energies of the occupied MO since no electron-electron repulsion is included.

$$E = \sum_i \sum_j \rho_{ij}(H_{ji} + F_{ji}) \approx \sum_i \sum_j 2\rho_{ij}F_{ji} = 2 \sum_i \varepsilon_i \qquad \{V\text{-}24\}$$

<u>Population Analysis</u>

$$\underline{\underline{P}} = 2\ \underline{\underline{\rho}}\ \underline{\underline{S}} = 2\underline{\underline{\rho}} \qquad \{V\text{-}25a,b\}$$

because in SHMO $\underline{\underline{S}} = \underline{\underline{1}}$

For the anion there are two occupied MO so that

$$\underline{\underline{\rho}} = \begin{pmatrix} \frac{1}{2} & \frac{1}{\sqrt{2}} \\ \frac{1}{2}\sqrt{2} & 0 \\ \frac{1}{2} & \frac{1}{\sqrt{2}} \end{pmatrix} \begin{pmatrix} \frac{1}{2} & \frac{1}{2}\sqrt{2} & \frac{1}{2} \\ \frac{1}{\sqrt{2}} & 0 & \frac{1}{\sqrt{2}} \end{pmatrix} = \begin{pmatrix} 0.75 & 0.35 & -0.25 \\ 0.35 & 0.50 & 0.35 \\ -0.25 & 0.35 & 0.75 \end{pmatrix}$$

{V-26}

$$\underline{\underline{P}} = 2\underline{\underline{\rho}} = \begin{pmatrix} \boxed{1.5} & 0.7 & -0.5 \\ 0.7 & \boxed{1.0} & 0.7 \\ -0.5 & 0.7 & \boxed{1.5} \end{pmatrix}$$

p_{12}, p_{23} (bond order) overlap population

$p_{11} = q_1$

$p_{22} = q_2$

$p_{33} = q_3$ (charge) atomic population

{V-27}

{V-28a,b}

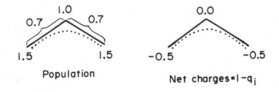

Calculations Involving Heteroatoms

SHMO calculations can be performed without difficulty for fully conjugated hydrocarbons. However the situation is more complicated when the molecule contains one or more heteroatoms (O, N, S ...) because α and β will no longer be the same for each atom or bond (α_C and β_{CC}). A new set of values (α_X and β_{C-X}) is needed for each heteroatom X.

Consider first the value of α. This is the binding energy (a negative quantity) of an electron to an atom. As will be seen this quantity may be related to the ionization potential or electronegativity of the atom in question, i.e. $|\alpha_C| < |\alpha_N| < |\alpha_O| < |\alpha_F|$. The description of C=C and C=O is thus analogous but the energy values are different as is indicated in the following diagram:

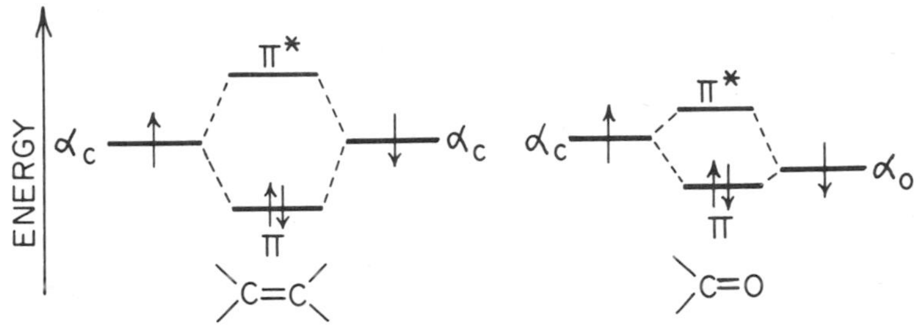

Figure V-2 A schematic comparison of orbital energy levels for a system with and without a heteroatom

However, because a numerical value is not given to α in the calculation, the heteroatom values (α_N, α_O, α_F, etc.) must be related to α_C. It is customary to express the lowering of α in units of β_{CC} (which is again a negative quantity). Thus

$$\alpha_X = \alpha_C + h_x \beta_{CC} \qquad \{V-29\}$$

where h_x is a proportionality constant. Optimization of h_x is necessary because it is a variable parameter in the above equation. The following table summarizes the h_x values used most commonly for N, O and F.

Table V-2

A summary of selected values of h_x for N, O and F

Number of Electrons in the π system	N	O	F
2	1.5	2.0	3.0
1	0.5	1.0	-

In structures such as:

$$\begin{array}{ccc} \diagdown C - N\diagup & \diagdown C - \ddot{O} & \diagdown C - F: \end{array} \qquad \{V-30\}$$

the heteroatoms contribute two electrons to the π-electron system. However, in structures such as

$$\begin{array}{cc} \diagdown C = N & \diagdown C = O \end{array} \qquad \{V-31\}$$

the heteroatom contributes only one electron to the π system so that the two situations require different h_x values.

The resonance integrals β_{CX} associated with a C-X bond are also expressed in units of β_{CC},

$$\beta_{CX} = k_{CX}\, \beta_{CC} \qquad \{V-32\}$$

the proportionality constant being labelled k_{CX}. Again it is necessary to distinguish between singly and doubly bonded systems:

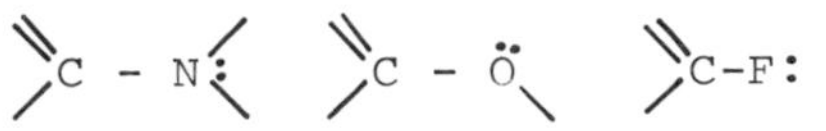

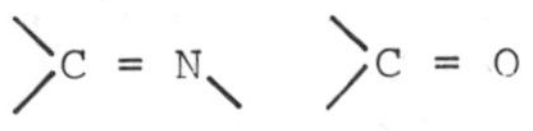

$$\diagup\kern-0.3em\diagdown \!\!\!\!C - \ddot{X} \qquad \diagdown\kern-0.3em C = X \qquad\qquad \{V\text{-}33\}$$

The most commonly used values of k_{CX} are summarized in the following table.

Table V-3

A summary of selected values of k_{CX} for various bonds

Bond	C - C	C - N	C - O	C - F
Single	0.9	0.8	0.8	0.7
Double	1.0	1.0	1.0	-

With α_X and β_{CX} expressed in units of β_{CC} the numerical calculation may proceed as before, but if the calculations are carried out in units of $|\beta_{CC}|$ both the h_X and the k_{CX} values must be taken with the opposite sign from that given in the tables.

<u>ω-Technique</u>

Because the SHMO method over-emphasizes charge separation the ω-technique was invented to compensate for this shortcoming. The method involves the following modifications of SHMO theory:

(i)   leave the off-diagonal elements unchanged (β or 0)

(ii) make the diagonal elements electron (i.e. charge) dependent by including the electron density ρ explicitly. Since ρ is part of the solution the calculation becomes an iterative one as in the SCF method.

For the rth iteration

$$\alpha^r_i = \alpha^o_i + \omega\,(1 - q^r_i)\,\beta \qquad\qquad \{V\text{-}34\}$$

where ω is a scaling constant: $1.0 \leq \omega \leq 1.4$ and the last term $\omega(1 - q^r_i)\beta$ is expected to simulate $(2J_{ii} - K_{ii} = J_{ii})$ because $q_i = 2\rho_{ii}$.

The results obtained for allyl anion are tabulated below.

Table V-4

SHMO charge distribution for allyl anion computed with and without ω-technique

	SHMO	ω-technique ($\omega = 1.0$)
$q_1 = q_3$	1.50	1.41
q_2	1.00	1.18
$1 - q_1$	0.00 -0.50 -0.50	-0.18 -0.41 -0.41

3. The Extended Hückel Molecular Orbital (EHMO) Theory

1. The EHMO method represents an all-valence electron (AVE) calculation. A subminimal basis set is chosen but it is more extensive than that used for the SHMO method. The choice of the basis is as follows:

for H,	1s
for C, N, O, F	2s, $2p_z$, $2p_x$, $2p_y$
for Si, P, S, Cl	3s, $3p_z$, $3p_x$, $3p_y$ (and possibly 3d)

Further, the basis $\{\eta_i\}$ is not assumed to be orthogonal as in SHMO theory

$$S_{ij} = \langle \eta_i | \eta_j \rangle \neq \delta_{ij} \qquad \{V-35\}$$

Therefore the overlap matrix must be computed explicitly.

2. The EHMO method is still an electron independent model requiring an assumed $\underline{\underline{F}}^{\eta}$ rather than an iterative process.

Thus: $\underline{\underline{F}}^{\eta}$ is guessed first and then the following steps are performed.

Compute $\underline{\underline{S}}^{-1/2}$ from $\underline{\underline{S}}$ as usual and then apply the necessary similarity transformation to $\underline{\underline{F}}^{\chi}$

$$\underline{\underline{F}}^{\chi} = \underline{\underline{S}}^{-1/2}\ \underline{\underline{F}}^{\eta}\ \underline{\underline{S}}^{-1/2} \qquad \{V-36\}$$

Diagonalize $\underline{\underline{F}}^{\chi}$ to obtain the eigenvalues

$$\underline{\underline{\varepsilon}} = \underline{\underline{U}}^{\dagger}\ \underline{\underline{F}}^{\chi}\ \underline{\underline{U}} \qquad \{V-37\}$$

and finally compute the coefficient matrix

$$\underline{\underline{C}} = \underline{\underline{S}}^{-1/2}\ \underline{\underline{U}} \qquad \{V-38\}$$

with $\underline{\underline{C}}_M$ being used to calculate $\underline{\underline{\rho}}$ and $\underline{\underline{P}}$.
Note that the total energy again has a simplified expression as in the SHMO method

$$E = \sum_i \sum_j \rho_{ij}(h^{\eta}_{ij} + F^{\eta}_{ij}) = \sum_i \sum_j 2\rho_{ij}F^{\eta}_{ij} = \sum_i 2\varepsilon_{ii} = 2\sum_i \varepsilon_i \qquad \{V-39\}$$

The choice of $\underline{\underline{F}}^{\eta}$ is not as simple as in SHMO theory. For the diagonal elements F_{ii} the negative of the orbital ionization potential is taken as the measure of the binding energy of the electron to the atom. For the off-diagonal elements several formulae have been proposed but a particularly common choice employs the average of the two corresponding diagonal elements scaled to the overlap integral.
F_{ii} = -IP of the atom associated with the removal of an electron from the ith AO

$$F_{ij} = K \left[\frac{F_{ii} + F_{jj}}{2}\right] S_{ij} \qquad \{V-40\}$$

where K is a scaling constant: $(1.5 < K < 2.0)$ usually taken to be 1.75. Actual evaluation of the overlap integrals S_{ij} requires explicit knowledge of the molecular geometry and a functional form for the basis functions (usually chosen to be STO).

Population Analysis and the Iterative EHMO (IEHMO) Method

In the SHMO method the ω-technique $(1.0 < \omega < 1.4)$ was used to partially compensate for the drastic approximations, using the following expression for the nth iteration

$$\alpha^{(n)} = \alpha^{(0)} + \omega(1-q^{(n-1)})\beta = \alpha^{o} + \omega Q^{(n-1)}\beta \qquad \{V-41\}$$

where (1-q) is the net charge, Q. The diagonal element associated with the rth atom may then be written as

$$\alpha_r^{(n)} = \alpha_r^{(0)} + \omega Q_r^{(n-1)}\beta \qquad \{V-42\}$$

In EHMO theory, the orbital labels do not correspond to the atomic labels $(i \neq r)$ and the effective nuclear charge differs from that of the carbon system $(Z^{eff} \neq 1)$. Consequently, a complete Mulliken population analysis must be performed to determine Q_r.

The electronic charge shared between atoms r and s is obtained by the partial summation

$$R_{rs} = \sum_{i}^{\text{AO on } r} \sum_{j}^{\text{AO on } s} P_{ij} \qquad \{V-43\}$$

where P_{ij} is the overlap population between the ith and j^{th} AO. The diagonal elements of this matrix (R_{rr}) represent the charge (in terms of number of electrons) associated with a given atom r. The net charge associated with a given atom r may be calculated by

$$Q_r = Z^{eff}_r - R_{rr} \qquad \{V-44\}$$

where Z^{eff} is the effective nuclear charge (1 for H, 4 for C, 5 for N, 6 for O, etc.) and R_{rr} is computed as before.

In the IEHMO method (as in the ω-technqiue) a similar formula is employed for the diagonal elements in the course of the charge iteration

$$F^{(n)}_{ii} = F^{(0)}_{ii} + \omega Q_r^{(n-1)} \Delta F_{ii} \qquad \{V-45\}$$

where $\omega = 1.0$ and Q_r is the net charge on the atom r on which atomic orbital i is centered. This equation corresponds to a linear interpolation either for $0 < Q_r < 1$ or for $-1 < Q_r < 0$. Depending on the sign of Q_r, one needs a separate ΔF_{ii}, as indicated below

$$\text{If } Q_r > 0, \quad \Delta F_{ii} \equiv \Delta F_{ii}^{(+)} = F_{ii}^{(+)} - F_{ii}^{(o)} \qquad \{V-46a,b\}$$

$$Q_r < 0, \quad \Delta F_{ii} \equiv \Delta F_{ii}^{(-)} = F_{ii}^{(-)} - F_{ii}^{(o)}$$

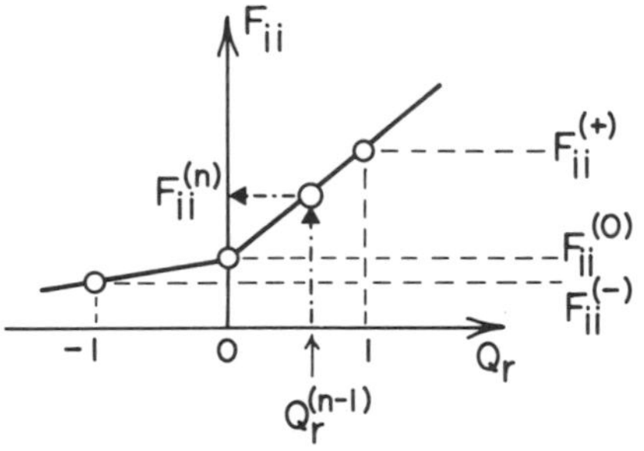

Figure V-3 Linear interpolation for diagonal matrix elements in the IEHMO theory

Note that ΔF_{ii} is the difference in two successive ionization potentials:

F_{ii} is the -IP of the ionization: $A^{(o)} \rightarrow A^{(+)} + e^{(-)}$

$F_{ii}^{(+)}$ is the -IP of the ionization: $A^{(+)} \rightarrow A^{(++)} + e^{(-)}$

$F_{ii}^{(-)}$ is the -IP of the ionization: $A^{(-)} \rightarrow A^{(o)} + e^{(-)}$

{V-47a-c}

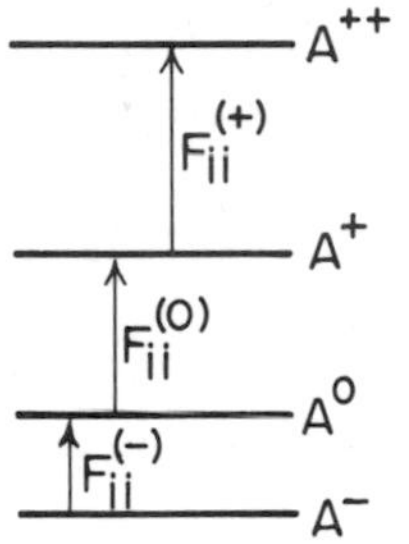

Figure V-4 Ionization parameters for the IEHMO theory

A more accurate, quadratic interpolation, treats the diagonal matrix element ($F_{ii}^{(n)}$) of each orbital as a separate parabolic function of the net atomic charge ($Q_r^{(n-1)}$) of the atom r on which the orbital is centered. The unknown parameters a and b of each parabolic function are determined by considering the "experimental", i.e. initial diagonal element (F_{ii}^{o}) of the orbital on the rth atom in the neutral ($Q_r = 0$) and either the singly positive ($Q_r = +1$) or the singly negative ($Q_r = -1$) charged atomic species, depending on the sign of the computed Q_r.

$$(F_{ii}^{+})_r = a.Q_r^2 + b.Q_r + (F_{ii}^{o})_r \qquad \{V-48\}$$

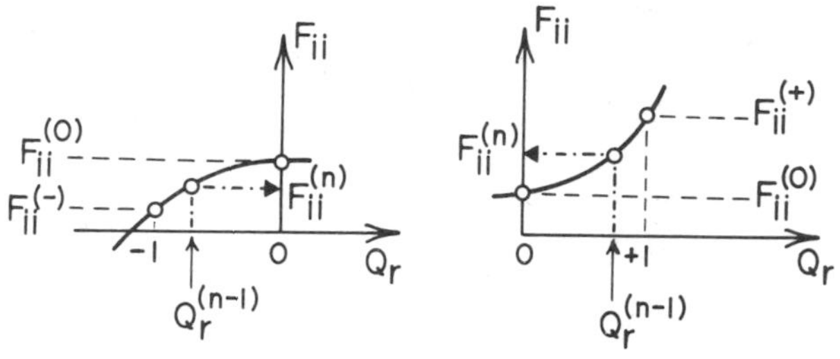

Figure V-5 Quadratic interpolation for diagonal matrix elements in the IEHMO theory

After each iteration a new F_{ii} element, $(F_{ii}^{\pm})_r$, is determined for each orbital after computation of the net charge, Q_r, of the host atom r.

Unfortunately the above formula does not converge easily to self-consistent charges. It has been shown that convergence to self-consistent net atomic charges is better if, after the nth iteration, the new F_{ii} (i.e. $F_{ii}^{(n+1)}$) is chosen such that there is a 90% to 10% linear combination with the previous element (obtained in the nth iteration).

$$(F_{ii}^{\pm})_r^{(n+1)} = 0.9\ (F_{ii}^{\pm})_r^{(n)} + 0.1\ (F_{ii}^{\pm})_r \qquad \{V\text{-}49\}$$

Somewhat more reliable charge distributions are then obtained. It can be seen that there is some similarity between the ω-technique and the IEHMO method. However there is an important difference. In the IEHMO calculation the off-diagonal elements vary during each iteration.

$$F_{ij}^{(n)} = K\left[\frac{F_{ii}^{(n)} + F_{jj}^{(n)}}{2}\right] S_{ij} \qquad \{V\text{-}50\}$$

but in the ω-technique these remain constant.

4. The Pople-Pariser-Parr (PPP) MO Theory

This is a π-electron theory but is not an electron independent theory. Consequently the non-empirical form of the F-matrix is relevant

$$F_{ij} = H_{ij} + 2\,J_{ij} - K_{ij} =$$

$$= H_{ij} + \sum_k \sum_\ell \rho_{k\ell}[2\{\eta_i\eta_j|\eta_k\eta_\ell\} - \{\eta_i\eta_k|\eta_j\eta_\ell\}] = \qquad \{V\text{-}51\}$$

$$= H_{ij} + \sum_k \sum_\ell 2\rho_{k\ell}\,[\{\eta_i\eta_j|\eta_k\eta_\ell\} - (1/2)\{\eta_i\eta_k|\eta_j\eta_\ell\}]$$

The following approximations are involved:

(1) Only π systems are considered so that the orbital labels i, j, k, ℓ may be replaced by atomic labels a, b, c, d because of the 1:1 correspondence between them (i.e. each atom has one orbital).

(2) The orbitals are assumed to form an orthonormal basis set as in SHMO theory

$$S_{ij} \equiv S_{ab} = \delta_{ab} \qquad \{V\text{-}52\}$$

This implies that $2\underline{\underline{\rho}} = \underline{\underline{P}}$.

(3) As an extension of (2) we make the following simplification for the two-electron integrals

$$\{\eta_a\eta_b|\eta_c\eta_d\} = \delta_{ab}\delta_{cd}\{\eta_a^2|\eta_c^2\} \equiv \delta_{ab}\delta_{cd}\gamma_{ac} \qquad \{V\text{-}53\}$$

where γ_{ac} should be estimated as a function of the internuclear distance R. Note that $\gamma_{ac} \to \gamma_{aa}$ as $R \to 0$.

(4) Although the method is a π-theory, the effect of the otherwise neglected σ electrons may be included in the form of the parameter H_{ij}^{core}. F_{ij} is then written as follows:

$$F_{ij} = H_{ij}^{core} + \sum_k \sum_\ell \rho_{k\ell}\{\delta_{ij}\delta_{k\ell}\gamma_{ik} - 1/2\,\delta_{ik}\delta_{j\ell}\gamma_{ij}\} \qquad \{V\text{-}54\}$$

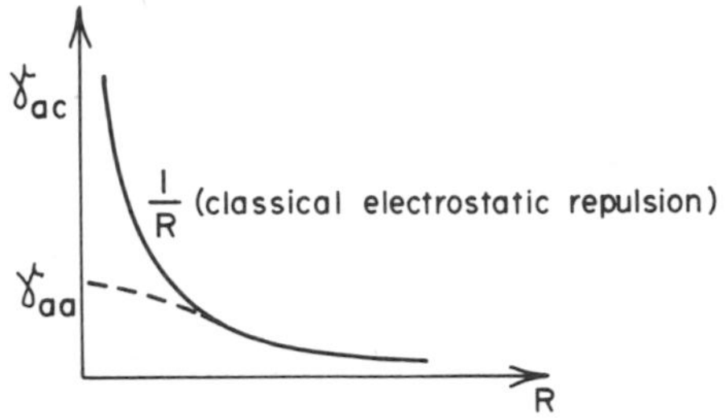

Figure V-6 A schematic representation of the variation of electron-electron repulsion with distance

We may now consider the simplification of the above expression for the off-diagonal $(i \neq j)$ and the diagonal $(i = j)$ elements.

Off-diagonal $(i \neq j)$ elements:

For $i \neq j$ the term containing δ_{ij} must vanish, i.e.

$$F_{ij} = H_{ij}^{core} + \sum_{k}\sum_{\ell}(-1/2)P_{k\ell}\delta_{ik}\delta_{j\ell}\gamma_{ij} \qquad \{V-55\}$$

Because only $k = i$ and $\ell = j$ contribute, there will be only one non-vanishing term under the double sum.

$$F_{ij} = H_{ij}^{core} - (1/2)P_{ij}\gamma_{ij} \qquad \{V-56\}$$

Two distinct cases may then be distinguished:

i and j are adjacent (bonded): $H_{ij}^{core} = \beta_{ij}^{core}$

i and j are non-adjacent (non-bonded): $H_{ij}^{core} = 0$

These approximations correspond to the SHMO formalism. This means that the matrix elements F_{ij} may be expressed in terms of bond orders (i.e. overlap-populations)

$$F_{ij} = \begin{cases} \beta_{ij}^{core} - (1/2)P_{ij}\gamma_{ij} & \text{(i and j are bonded)} \\ - (1/2)P_{ij}\gamma_{ij} & \text{(i and j are non-bonded)} \end{cases} \qquad \{V-57\}$$

where P_{ij} is the bond order between atoms i and j.

<u>Diagonal (i = j)elements</u>

Since i = j $\delta_{ij} = \delta_{ii} = 1$ Consequently

$$F_{ii} = H_{ii}^{core} + \sum_k \sum_\ell P_{k\ell} \{\delta_{\ell k}\gamma_{ki} - (1/2)\delta_{ki}\gamma_{ii}\delta_{i\ell}\} \qquad \{V-58\}$$

The first term under the double sum is zero unless $\ell = k$. Therefore (since $\delta_{k\ell} = \delta_{kk} = 1$):

$$F_{ii} = H_{ii}^{core} + \sum_k P_{kk}\{\gamma_{ik} - (1/2)\delta_{ki}\gamma_{ii}\delta_{ik}\} \qquad \{V-59\}$$

The second term under the summation sign will always be zero except for the single case of k = i. Consequently this may be removed from the summation as shown:

$$F_{ii} = H_{ii}^{core} + P_{ii}\,[\gamma_{ii} - (1/2)\delta_{ii}\gamma_{ii}] + \sum_{\substack{k \\ (k \neq i)}}{}' P_{kk}\gamma_{ik} \qquad \{V-60\}$$

$$F_{ii} = H_{ii}^{core} + (1/2)P_{ii}\gamma_{ii} + \sum_{\substack{k \\ (k \neq i)}}{}' P_{kk}\gamma_{ik} \qquad \{V-61\}$$

Introducing the atomic charge notation $P_{ii} = q_i$ and $P_{kk} = q_k$

$$F_{ii} = H_{ii}^{core} + (1/2)q_i\gamma_{ii} + \sum_{k (k\neq i)}' q_k\gamma_{ik} \qquad \{V-62\}$$

Now examine the nature of H_{ii}^{core}. It would be reasonable to relate the potential seen by an electron to the potential seen in the free atom (U_{ii}). However, this potential is variable because of interaction with other nuclei ($\Sigma_B V_{AB}$) in the molecule. This interaction may be defined, in terms of the previous quantities, as:

$$\sum_{B (B\neq A)}' V_{AB} \approx \sum_{k (k\neq i)}' \gamma_{ik} \qquad \{V-63\}$$

which leads to energy lowering because the molecule is more stable than the separated atoms; cf. {V-64}

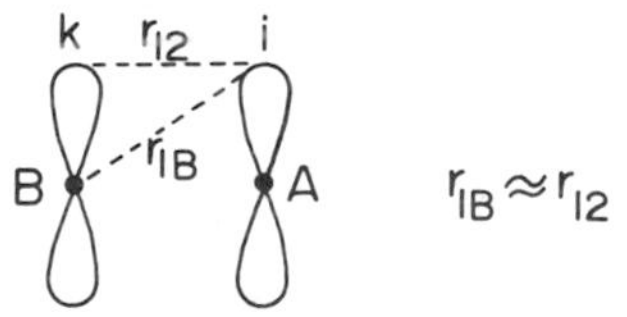

Figure V-7 A schematic illustration of the approximation of the penetration integral V_{AB}

$$H_{ii}^{core} = U_{ii} - \sum_{k (k\neq i)}' \gamma_{ik} \qquad \{V-64\}$$

$$F_{ii} = U_{ii} - \sum_{k (k \neq i)}' \gamma_{ik} + (1/2) q_i \gamma_{ii} + \sum_{k (k \neq i)}' q_k \gamma_{ik} \qquad \{V-65\}$$

therefore:

$$F_{ii} = U_{ii} + (1/2) q_i \gamma_{ii} + \sum_{k (k \neq i)}' (q_k - 1) \gamma_{ik} \qquad \{V-66\}$$

If the U_{ii} value is chosen carefully total energies comparable to those obtained in HF type calculations can be obtained, but it is more appropriate to seek relative values as in the SHMO method (ε was related to α). To do this we subtract the atomic energies from the diagonal elements of the F matrix. The atomic orbital energy is taken as

$$U_{ii} + (1/2) \gamma_{ii} \qquad \{V-67\}$$

(for a closed shell system it would be $U_{ii} + 2 J_{ii} - K_{ii} \simeq U_{ii} + J_{ii}$ but, because the $2p_z$ orbital is only half filled instead of J_{ii} half of its value $(1/2) J_{ii}$ or in the present notation $(1/2) \gamma_{ii}$ is taken)

$$F_{ii}^{rel} = F_{ii} - (U_{ii} + 1/2\ \gamma_{ii}) \qquad \{V-68\}$$

$$F_{ii}^{rel} = U_{ii} + (1/2) q_i \gamma_{ii} + \sum_{k (k \neq i)}' (q_k - 1) \gamma_{ik} - U_{ii} - (1/2) \gamma_{ii} \qquad \{V-69\}$$

$$F_{ii}^{rel} = (1/2)(q_i - 1) \gamma_{ii} + \sum_{k (k \neq i)}' (q_k - 1) \gamma_{ik} \qquad \{V-70\}$$

Taking the usual example: the allyl anion

$\underline{\underline{F}}$ is a 3x3 matrix since there are 3 orbitals in the system

$$\underline{\underline{F}}^{rel} = \begin{bmatrix} 1/2(q_1-1)\gamma_{11} + (q_2-1)\gamma_{12} + (q_3-1)\gamma_{13} & \beta_{12} - 1/2P_{12}\gamma_{12} & -1/2\, P_{13}\gamma_{13} \\ \text{same as the upper triangular since the Fock matrix is a real symmetric matrix} & 1/2(q_2-1)\gamma_{22} + (q_1-1)\gamma_{12} + (q_3-1)\gamma_{23} & \beta_{23} - 1/2P_{23}\gamma_{23} \\ & & 1/2(q_3-1)\gamma_{33} + (q_1-1)\gamma_{13} + (q_2-1)\gamma_{23} \end{bmatrix} \quad \{V\text{-}71\}$$

Because all atoms are carbon: $\gamma_{11} = \gamma_{22} = \gamma_{33}$ and also $\gamma_{12} = \gamma_{23} \neq \gamma_{13}$ as well as $\beta_{12} = \beta_{23} = \beta_{CC}$. Numerical values for β_{CC}, γ_{11}, γ_{12} and γ_{13} must now be supplied. The large number of parameters in the PPP method leads to some uncertainty as to the significance of the computed numbers. These parameters are normally selected in such a way that some physical property (frequently the u.v. spectrum) is reproduced for some molecule or series of molecules of interest.

Table V-5

A comparison of input parameters and computed net charges for allyl anion

	J. A. Pople, Trans. Farad. Soc. 49, 1375 (1953)	R. Pariser, J. Chem. Phys. 24, 250 (1956)
β_{CC}	-2.3 ev[a]	-2.37 ev
γ_{11}	17.0 ev[b]	10.96 ev
γ_{12}	9.0 ev[b]	6.90 ev
γ_{13}	5.6 ev[b]	5.68 ev
Final charge distribution	$q_1 = q_3 = 1.518$ $q_2 = 0.964$	$q_1 = q_3 = 1.495$ $q_2 = 1.010$

[a]Obtained from spectroscopy

[b]Theoretical (i.e. computed) values using STO ($2p_z$). In an actual calculation the iteration may begin with the q_i and p_{ij} values obtained from an SHMO calculation and these are: $q_1 = q_3 = 1.5$ and $q_2 = 1.0$, as well as $p_{12} = p_{23} = 1/\sqrt{2} = 0.7072$ and $p_{13} = -1/2 = -0.500$.

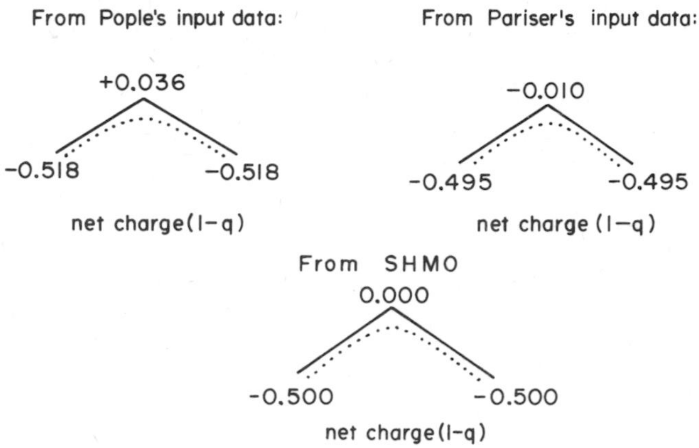

Figure V-8 A summary of PPP and SHMO results presented in Tables V-5 and V-4 respectively

Note that the population analysis in the PPP theory is the same as in SHMO theory since $\underline{\underline{P}} = 2\underline{\underline{\rho}}$ because $\underline{\underline{S}} = \underline{\underline{1}}$.

5. <u>The Complete Neglect of Differential Overlap (CNDO) Theory</u>

The CNDO method developed through an evolutionary change and the final set of approximations are labelled as CNDO/2. Comparison between CNDO/1 and CNDO/2 is given in Pople's third paper {J. Chem. Phys. <u>44</u>, 3289 (1966)} but only CNDO/2 will be discussed here. Other NDO theories employing different formulations and/or parameterizations (MINDO/1/2/3, CNDO/S, INDO, etc.) are also in common use.

1. An all valence electron (AVE) basis set such as 1s AO for H and 2s, $2p_x$, $2p_y$, $2p_z$ for C, N, O, F is chosen. Orthonormality is assumed, i.e. overlap is neglected.

$$S_{ij} \equiv \langle \eta_i | \eta_j \rangle \simeq \langle \chi_i | \chi_j \rangle = \delta_{ij} \qquad \{V\text{-}72\}$$

This means that only the following matrix equation

is solved: $\underline{\underline{U}}^{\dagger}\ \underline{\underline{F}}^{\chi}\ \underline{\underline{U}} = \underline{\underline{\varepsilon}}$.

2. To make the neglect of overlap complete, it is neglected not only in S_{ij} but in all two-electron integrals as well. This includes both

diatomic differential overlap 

and {V-73a-c}

monatomic differential overlap 

$$\{\eta_i\eta_j|\eta_k\eta_\ell\} = \delta_{ij}\delta_{k\ell}\gamma_{ik}$$

3. The rigorous SCF formalism (where all integrals are properly computed) is invariant with respect to the choice of coordinate system. However, in the CNDO approach this is not necessarily the case. The following simplification is necessary to ensure rotational invariance:

$\gamma_{ik} = \gamma_{AB}$ if γ_i is located on atom A and γ_k is located on atom B

$\gamma_{ik} = \gamma_{AA}$ if both γ_i and γ_k are located on atom A.

Since this equality is assumed regardless of the angular type of η_i and η_k (2s, 2p, 2p, $2p_z$), the approximation implies that all interactions are replaced by a spherical average.

4. Penetration integrals of the type V_{AB}, $\langle\eta_i|V_B|\eta_k\rangle$, where both η_i and η_k are on atom A are approximated on the basis of the following assumption:

$$\gamma_{AB} = \gamma_{ik} = \{\eta_i(1)\eta_i(1)|\eta_k(2)\eta_k(2)\} \equiv \{\eta_i(1)\eta_k(2)|\frac{1}{r_{12}}|\eta_i(1)\eta_k(2)\rangle$$

{V-74}

If we assume that $r_{12} \simeq r_{B1}$ (cf. Figure V-10), then we may write:

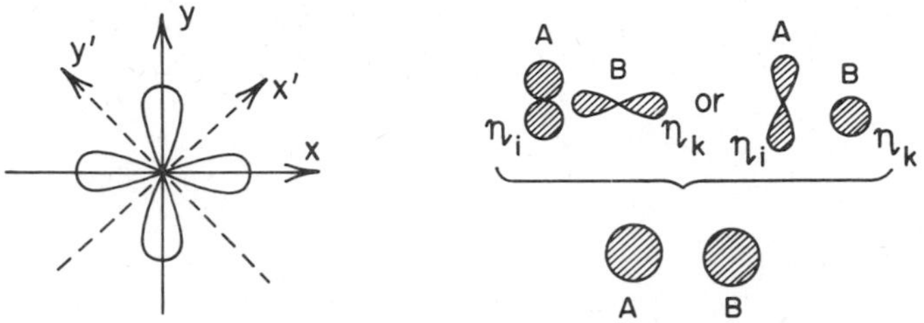

Figure V-9 A schematic representation of the problem of rotational invariance

$$\gamma_{AB} = \gamma_{ik} = \{\eta_i(1)\eta_i(1)|\eta_k(2)\eta_k(2)\} \equiv \{\eta_i(1)\eta_k(2)|\frac{1}{r_{12}}|\eta_i(1)\eta_k(2)> \qquad \{V\text{-}74\}$$

$$\gamma_{AB} = \gamma_{ik} = <\eta_i(1)\eta_k(2)|\frac{1}{r_{B1}}|\eta_i(1)\eta_k(2)> = <\eta_i(1)|\frac{1}{r_{B1}}|\eta_i(1)> <\eta_k(2)|\eta_k(2)> \qquad \{V\text{-}75\}$$

$$<\eta_i|\frac{-1}{r_{B1}}|\eta_i> \simeq -\gamma_{AB} \qquad \{V\text{-}76\}$$

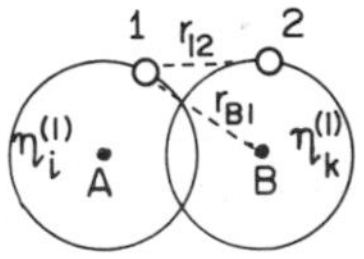

Figure V-10 A schematic illustration for the approximation of the penetration integral V_{AB}

5. Off-diagonal core elements (H_{ij}^{core}) are treated semiempirically:

$$H_{ij}^{core} = \beta_{AB} S_{ij} = \frac{\beta_A^o + \beta_B^o}{2} S_{ij} \qquad \{V\text{-}77\}$$

as in EHMO
Selection of β^o is made so that the results will agree with non-empirical calculations on H_2, C_2, N_2, etc. Alternatively a match with experimental observables is also possible.

In the SCF-MO Formalism:

$$E = \sum_i \sum_j \rho_{ji}(H_{ij} + F_{ij}) \qquad \{V\text{-}78\}$$

$$F_{ij} = H_{ij} + G_{ij} \qquad \{V\text{-}79\}$$

$$G_{ij} = 2\,J_{ij} - K_{ij} = \sum_k \sum_\ell \rho_{k\ell}\,[2\{\eta_i\eta_j|\eta_k\eta_\ell\} - \{\eta_i\eta_k|\eta_j\eta_\ell\}] \qquad \{V\text{-}80\}$$

In the CNDO/2 Formalism:

$$\underline{\underline{\rho}} = (1/2)\underline{\underline{P}} \qquad \{V\text{-}81\}$$

Also note that

$$R_{AA} = \sum_{i}^{\text{all } i \text{ on atom A}} P_{ii} \qquad \{V\text{-}82\}$$

and

$$R_{BB} = \sum_{k}^{\text{all } k \text{ on atom B}} P_{kk} \qquad \{V\text{-}83\}$$

$$E = (1/2)\sum\sum P_{ji}(H_{ij} + F_{ij}) \qquad \{V\text{-}84\}$$

$$F_{ij} = H_{ij} + G_{ij} \qquad \{V\text{-}85\}$$

$$G_{ij} = \sum_{k}\sum_{\ell} P_{k\ell}\ [\{\eta_i\eta_j|\eta_k\eta_\ell\} - (1/2)\{\eta_i\eta_k|\eta_j\eta_\ell\}] \qquad \{V\text{-}86\}$$

$$H_{ij} = \frac{\beta_A{}^{o} + \beta_B{}^{o}}{2}\ S_{ij} \equiv \beta_{AB}\ S_{ij} \qquad \{V\text{-}87\}$$

$$H_{ii} = U_{ii} - \sum_{B} Z_B\beta_{AB} \qquad \{V\text{-}88\}$$

Note that the ith orbital $\{\eta_i\}$ is on atom A and the jth orbital $\{\eta_j\}$ is on atom B.

<u>Details of H_{ii}</u>

In the SCF formalism:

$$H_{ii} = \langle\eta_i(1)|\hat{h}_i|\eta_i(1)\rangle \qquad \{V\text{-}89\}$$

where $\hat{h}$ is the one-electron operator of the Hamiltonian

$$\hat{h} \equiv \hat{h}_1 = \underbrace{\overbrace{-\tfrac{1}{2}\nabla^2}^{\text{Kinetic Energy}} \qquad - \frac{Z_A}{r_{A1}}}_{\text{Operator for atom A}} \overbrace{\ - \underbrace{\sum_{B}{}' \frac{Z_B}{r_{B1}}}_{\text{Interaction with all other nuclei}}}^{\text{Potential Energy}} \qquad \{V\text{-}90\}$$

Note again that η_i is located on atom A. Consequently we may write:

$$H_{ii} \equiv \langle\eta_i|\hat{h}|\eta_i\rangle = \langle\eta_i|-\tfrac{1}{2}\nabla^2|\eta_i\rangle - \langle\eta_i|\frac{Z_A}{r_{A1}}|\eta_i\rangle - \sum_{\substack{B \\ B\neq A}}{}' \langle\eta_i|\frac{Z_B}{r_{B1}}|\eta_i\rangle$$

$$\{V\text{-}91\}$$

$$H_{ii} = U_{ii} - \sum_{\substack{B \\ (B \neq A)}}' \langle \eta_i | \frac{Z_B}{r_B} | \eta_i \rangle = U_{ii} + \sum_{\substack{B \\ (B \neq A)}}' V_{AB} \quad \{V-92\}$$

According to approximation No. 4

$$\sum_{\substack{B \\ (B \neq A)}}' V_{AB} = \sum_{\substack{B \\ (B \neq A)}}' - Z_B \langle \eta_i | \frac{1}{r_{B1}} | \eta_i \rangle \simeq \sum_{\substack{B \\ (B \neq A)}}' - Z_B \gamma_{AB} \quad \{V-93\}$$

Thus

$$H_{ii} = U_{ii} - \sum_{\substack{B \\ (B \neq A)}}' Z_B \gamma_{AB} \quad \{V-94\}$$

Details of G_{ij}

$$G_{ij} = \sum_k \sum_\ell P_{k\ell} \left[\{\eta_i \eta_j | \eta_k \eta_\ell\} - \frac{1}{2}\{\eta_i \eta_k | \eta_j \eta_\ell\}\right] \quad \{V-95\}$$

involving approximation No. 2

$$G_{ij} = \sum_k \sum_\ell P_{k\ell} \left[\delta_{ij} \delta_{k\ell} \gamma_{ik} - \frac{1}{2} \delta_{ik} \delta_{j\ell} \gamma_{ij}\right] \quad \{V-96\}$$

Considering the possible values of the various δ functions:

$$G_{ij} = \delta_{ij} \sum_k P_{kk} \gamma_{ik} - \frac{1}{2} P_{ij} \gamma_{ij} \quad \{V-97\}$$

where

$$\gamma_{ik} = \{\eta_i^2 | \eta_k^2\} \quad \text{and} \quad \gamma_{ij} = \{\eta_i^2 | \eta_j^2\} \quad \{V-98\}$$

since δ_{ij} is unity if and only if $i = j$, the off-diagonal elements will be:

$$G_{ij} = -\frac{1}{2} P_{ij}\gamma_{ij} \qquad \text{for } i \neq j \qquad \{V\text{-}99\}$$

and the diagonal elements may be written as:

$$G_{ii} = -\frac{1}{2} P_{ii}\gamma_{ii} + \sum_{\substack{k \\ (k\neq i)}}' P_{kk}\gamma_{ik} \qquad \{V\text{-}100\}$$

Now consider the implications of approximation No. 3. There are two possibilities depending on the locations of η_i and η_k:

(i) both η_i and η_k are localized on atom A

Thus: $\gamma_{ik} \simeq \gamma_{AA}$

(ii) η_i is localized on atom A and η_k is localized on atom B

Thus: $\gamma_{ik} \simeq \gamma_{AB}$

{V-101}

Consequently:

$$G_{ii} = -\frac{1}{2} P_{ii}\gamma_{ii} + \sum_{\substack{\text{all } k \\ \text{on atoms} \\ \text{A and B}}} P_{kk}\gamma_{ik} \simeq -\frac{1}{2} P_{ii}\gamma_{AA} + \sum_{\substack{\text{all } k \\ \text{on atoms} \\ \text{A and B}}} P_{kk}\gamma_{AB}$$

$$= -\frac{1}{2} P_{ii}\gamma_{AA} + \sum_{\substack{\text{all } k \\ \text{on atom} \\ \text{A}}} P_{kk}\gamma_{AA} + \sum_{\substack{\text{all } k \\ \text{on atoms B} \\ (B\neq A)}}' P_{kk}\gamma_{AB}$$

$$= -\frac{1}{2} P_{ii}\gamma_{AA} + \gamma_{AA} \sum_{\substack{\text{all } k \\ \text{on atom A}}} P_{kk} + \sum_{\substack{\text{all } k \\ \text{on atoms B} \\ (B\neq A)}}' P_{kk}\gamma_{AB} \qquad \{V\text{-}102\}$$

Note that:

$$P_{AA} = \sum_{\substack{\text{all } k \\ \text{on atom A}}} P_{kk} \quad \text{and} \quad \sum_{\substack{\text{all } k \\ \text{on atoms B}}} P_{kk} = P_{BB} \qquad \{V\text{-}103a,b\}$$

so that

$$G_{ii} = (P_{AA} - \frac{1}{2} P_{ii})\, \gamma_{AA} + \sum_{\substack{\text{all B} \\ (B \neq A)}}' P_{BB}\gamma_{AB} \qquad \{V\text{-}104\}$$

Finally the substitution $\gamma_{ij} \simeq \gamma_{AB}$ leads to the expression for the off-diagonal matrix elements

$$G_{ij} = -\frac{1}{2} P_{ij}\gamma_{AB} \qquad \{V\text{-}105\}$$

Note that the CNDO formalism is analogous to the PPP formalism.

The $\underline{\underline{F}}$ matrix elements may now be constructed

$$F_{ii} = U_{ii} + (P_{AA} - \frac{1}{2} P_{ii})\, \gamma_{AA} + \sum_{\substack{B \\ (B \neq A)}}' (P_{BB} - Z_B)\, \gamma_{AB} \qquad \{V\text{-}106\}$$

$$F_{ij} = \beta_{AB}{}^{o}\, S_{ij} - \frac{1}{2} P_{ij}\gamma_{AB} \qquad \{V\text{-}107\}$$

<u>Parameters for U_{ii}</u>

$$-\frac{I_i + A_i}{2} = U_{ii} + (Z_A - \frac{1}{2})\, \gamma_{AA} \qquad \{V\text{-}108\}$$

This average of I_i (ionization potential) and A_i (electron affinity) associated with the ith AO seems necessary to account satisfactorily for the tendency of an AO to gain or to lose electrons. Consequently

$$U_{ii} = -\frac{1}{2} (I_i + A_i) - (Z_A - \frac{1}{2})\, \gamma_{AA} \qquad \{V\text{-}109\}$$

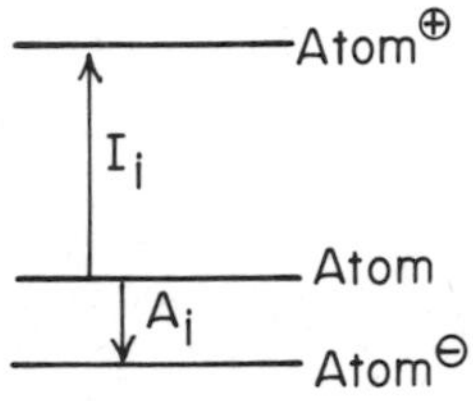

Figure V-11 Ionization parameters for diagonal matrix elements in the CNDO theory

(Note that the electron loss (I_i) or gain (A_i) involves the ith AO.) Substituting this for F_{ii} leads to the final expression for the elements of the $\underline{\underline{F}}$ matrix.

$$F_{ii} = -\frac{1}{2}(I_i + A_i) + \{(P_{AA} - Z_A) - \frac{1}{2}(P_{ii} - 1)\}\gamma_{AA} + \sum_{B (B \neq A)} (P_{BB} - Z_B)\gamma_{AB} \qquad \{V\text{-}110\}$$

and the off-diagonal elements remain the same:

$$F_{ij} = \beta_{AB}^{o} S_{ij} - \frac{1}{2} P_{ij}\gamma_{AB} \qquad \{V\text{-}111\}$$

CHAPTER VI

EXCITED AND IONIZED STATES IN THE FRAMEWORK OF CLOSED SHELL MO THEORIES

All of the previous discussion has been concerned with the ground electronic state. The HF method and the various semi-empirical theories require the solution of a matrix equation which gives the coefficient matrix ($\underline{\underline{C}}$) needed to construct the MO from the AO: $\phi = \eta\underline{C}$ {VI-1}

$$\underbrace{(\phi_1\phi_2 \; \cdots \; \phi_M}_{\text{occupied MO}} \; \cdots \; \phi_N) = (\eta_1\eta_2 \; \cdots \; \eta_N) \; \underline{\underline{C}}_M \rightarrow \left(\begin{array}{cccc|cc} c_{11} & c_{12} & \cdots & c_{1M} & \cdots & c_{1N} \\ c_{21} & c_{22} & \cdots & c_{2M} & \cdots & c_{2N} \\ \vdots & \vdots & & & & \\ c_{N1} & c_{N} & \cdots & c_{NM} & \cdots & c_{NM} \end{array} \right) \qquad \{VI\text{-}2\}$$

This matrix has different forms in the various theories. In the SHMO, PPP and CNDO theories

$$\underline{\underline{C}} = \underline{\underline{U}} \qquad \{VI\text{-}3a\}$$

and in the EHMO and Hartree-Fock (SCF)-MO theories

$$\underline{\underline{C}} = \underline{\underline{S}}^{-1/2} \; \underline{\underline{U}} \qquad \{VI\text{-}3b\}$$

The first M occupied molecular orbitals are then used to describe the 2 M electrons in their ground electronic configuration

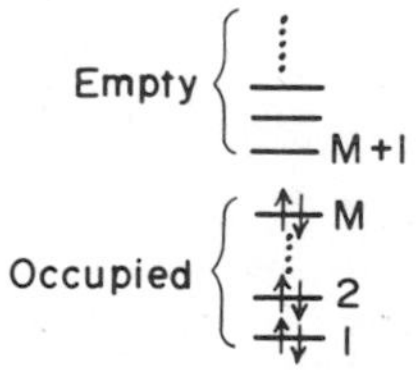

Figure VI-1 Orbital energy levels for 2 M electrons in their electronic ground state (closed shell)

$$\Phi_c(1,2,\ \ldots,\ 2M) = \frac{1}{\sqrt{(2M)!}} \begin{vmatrix} \phi_1(1)\alpha(1) & \phi_1(1)\beta(1) & \cdots\cdots & \phi_M(1)\alpha(1) & \phi_M(1)\beta(1) \\ \phi_1(2)\alpha(2) & \phi_1(2)\beta(2) & \cdots\cdots & \phi_M(2)\alpha(2) & \phi_M(2)\beta(2) \\ \vdots & \vdots & & \vdots & \vdots \\ \phi_1(2M)\alpha(2M) & \phi_1(2M)\beta(2M) & \cdots & \phi_M(2M)\alpha(2M) & \phi_M(2M)\beta(2M) \end{vmatrix}$$

{VI-4}

1. Excited Configurations

In MO theory excited states are approximated by excited configurations just as the ground state is approximated by ground configurations.

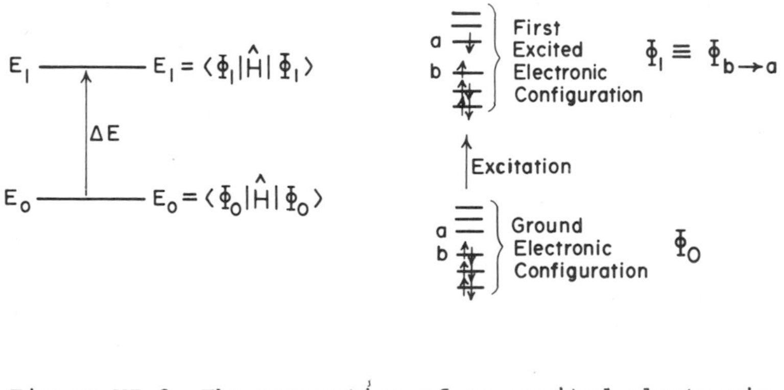

Figure VI-2 The generation of an excited electronic configuration for a closed shell system of 2M electrons

Therefore the description of one of the electrons requires that an anti-bonding (a) or virtual orbital be SUBSTITUTED into the determinantal wavefunction to replace the bonding (b) or occupied MO. Apart from this substitution $\Phi_{b\to a}$ is analogous to Φ_o.

To calculate the excitation energy (ΔE) the energies of the ground configuration ($E_o = \langle \Phi_o | \hat{H} | \Phi_o \rangle$) and the excited configuration ($E_1 = \langle \Phi_1 | \hat{H} | \Phi_1 \rangle$) must be computed.

Ground Configuration:

Total Energy:

$$E_o = \langle \Phi_o | \hat{H} | \Phi_o \rangle = \sum_{i=1}^{M} \{ 2H_{ii} + \sum_{j=1}^{M} (2J_{ii} - K_{ij}) \} \qquad \{VI\text{-}5\}$$

The HF equation has the following form:

$$\{ \hat{H}_i + \sum_{j=1}^{M} (2\hat{J}_j - \hat{K}_j) \} \phi_i \equiv \hat{F} \phi_i = \phi_i \varepsilon_i \qquad \{VI\text{-}6\}$$

or, in terms of its matrix elements over the MO basis,

$$H_{ii} + \sum_{j}^{M} (2J_{ij} - K_{ij}) = \varepsilon_{ii} \qquad \{VI\text{-}7\}$$

Excited Configurations:

It is necessary to distinguish between singlet and triplet configurations upon excitation from orbital b to orbital a:

Figure VI-3 Singlet and triplet excited configurations

Because the above wavefunctions have the same form we may denote them as $^{1,3}\Phi_{b\to a}$

$$^{1,3}E_{b\to a} = \langle {}^{1,3}\Phi_{b\to a}|H|{}^{1,3}\Phi_{b\to a}\rangle = \sum_{\substack{i \\ \text{all } b \\ \text{but not } a}}^{M} 2H_i + \sum_{\substack{i,j \\ \text{all } b \\ \text{but not } a}}^{M} (2J_{ij} - K_{ij})$$

$$\underbrace{-H_b - \sum_{i}^{M} (2J_{ib} - K_{ib})}_{-\varepsilon_b} + \underbrace{H_a + \sum (2J_{ia} - K_{ia})}_{+\varepsilon_a} - (J_{ba} - K_{ba}) \pm K_{ba} \qquad \{VI\text{-}8\}$$

where the $\pm$ signs represent the singlet and triplet states respectively. Thus

$$^{1,3}E_{b\to a} = E_o + (\varepsilon_a - \varepsilon_b) - (J_{ba} - K_{ba}) \pm K_{ba} \qquad \{VI\text{-}9\}$$

or, in detail:

$$^{1}E_{b\to a} = E_o + (\varepsilon_a - \varepsilon_b) - J_{ba} + 2K_{ba} \qquad \{VI-10\}$$

$$^{3}E_{b\to a} = E_o + (\varepsilon_a - \varepsilon_b) - J_{ba} \qquad \{VI-11\}$$

$$^{1}\Delta E_{b\to a} \equiv (^{1}E_{b\to a} - E_o) = (\varepsilon_a - \varepsilon_b) - J_{ba} + 2K_{ba} \qquad \{VI-12\}$$

$$^{3}\Delta E_{b\to a} \equiv (^{3}E_{b\to a} - E_o) = (\varepsilon_a - \varepsilon_b) - J_{ba} \qquad \{VI-13\}$$

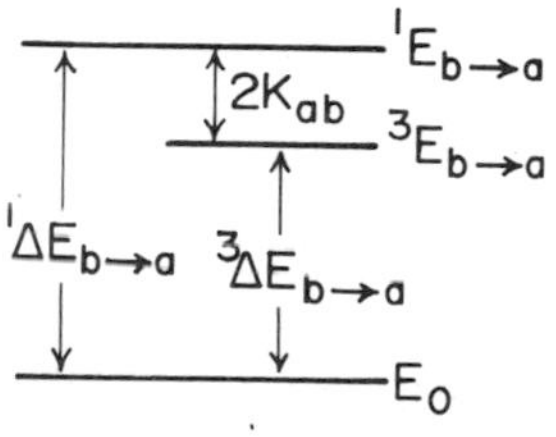

Figure VI-4 The spacing of singlet and triplet excited configuration relative to the closed shell ground electronic state

Since these results were derived by utilizing the virtual orbitals the method is frequently referred to as the "virtual orbital technique" and is valid for all electron-dependent theories, i.e. non-empirical SCF, PPP and CNDO. This computation is not possible in the SHMO and EHMO theories because there are no J and K integrals. It may be assumed that

$$^{1}\Delta E_{b\to a} \simeq {}^{3}\Delta E_{b\to a} = (\varepsilon_a - \varepsilon_b) \qquad \{VI\text{-}14\}$$

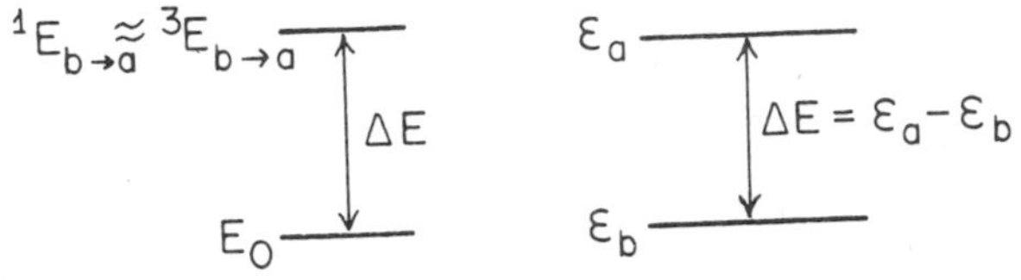

Figure IV-5 The spacing of excited configurations relative to the ground state in the electron independent models (SHMO and EHMO)

i.e. that the orbital energy difference (a single value), should fall between the singlet and triplet excitation values. Therefore, we take the arithmetic average of the two rigorous equations

$$\frac{^{1}\Delta E_{b\to a} + {}^{3}\Delta E_{b\to a}}{2} = \frac{2(\varepsilon_a - \varepsilon_b) - 2J_{ba} + 2K_{ba}}{2} \qquad \{VI\text{-}15\}$$

$$\frac{^{1}\Delta E_{b\to a} + {}^{3}\Delta E_{b\to a}}{2} = (\varepsilon_a - \varepsilon_b) - J_{ba} + K_{ba} \qquad \{VI\text{-}16\}$$

Since, in the simple LCAO theories, J and K cannot be distinguished it is assumed that these cancel:

$$(\varepsilon_a - \varepsilon_b) = \frac{^{1}\Delta E_{b\to a} + {}^{3}\Delta E_{b\to a}}{2} \qquad \{VI\text{-}17\}$$

Note that this provides a method for estimation of the numerical value of β for the simple Hückel theory (SHMO) from spectroscopic data.

2. Integral Transformations for J_{ab} and K_{ab}

$$J_{ab} = \{\phi_a\phi_a|\phi_b\phi_b\} \qquad \{VI-18\}$$

$$K_{ab} = \{\phi_a\phi_b|\phi_a\phi_b\} \qquad \{VI-19\}$$

The general expression has four running MO indices:

p, q, r, s

while the two electron integral over the AO basis has four different running indices:

i, j, k, l

Thus the transformation reads as follows after substitution of the MO - AO expansion

$$\phi_t = \sum_{m=1}^{N} \eta_m C_{mt} \qquad \{VI-20\}$$

$$\{\phi_p\phi_q|\phi_r\phi_s\} = \{\sum_i^N C_{pi}\eta_i \sum_j^N C_{qj}\eta_j|\sum \eta_k C_{kr} \sum_\ell^N \eta_\ell C_{\ell s}\}$$

$$= \sum_i^N \sum_j^N \sum_k^N \sum_\ell^N C_{pi}C_{qj}\{\eta_i\eta_j|\eta_k\eta_\ell\}C_{kr}C_{\ell s} \qquad \{VI-21\}$$

Instead of $\{\phi_p\phi_q|\phi_r\phi_s\}$ the above expressions for J_{ab} and K_{ab} occurred. In the PPP and CNDO approximations $\{\eta_i\eta_j|\eta_k\eta_\ell\} = \delta_{ij}\delta_{k\ell}\gamma_{ik}$, i.e. γ_{ik} is transformed

$$J_{ab} = \{\phi_a\phi_a|\phi_b\phi_b\} \sum_i \sum_k C^2_{ai}\, \gamma_{ik}\, C^2_{kb} \qquad \{VI\text{-}22\}$$

$$K_{ab} = \{\phi_a\phi_b|\phi_a\phi_b\} \sum_i \sum_k C_{ai}\, C_{bi}\, \gamma_{ik}\, C_{ka}\, C_{kb} \qquad \{VI\text{-}23\}$$

The above integral transformation to generate J_{ab} and K_{ab} represents a special case of the general procedure outlined in Chapter IV-3.

3. Ionized Configurations

The ionized state (a molecule-ion with loss of an electron from one of the "bonding orbitals": b) is a doublet state. Configuration $^2\Phi_b$ is formed by suppressing one of the rows and one of the columns of the ground state determinant, and assuming that the remainder is unchanged. The energy of this doublet state may be computed from this wavefunction {a(2M-1) x (2M-1) determinant} as given below:

$$^2E_b = \langle {}^2\Phi_b|\hat{H}|{}^2\Phi_b\rangle =$$

$$= \underbrace{2\sum_i^M H_i + \sum_i \sum_j (2J_{ij} - K_{ij})}_{E_o} \underbrace{- H_b - \sum_i (2J_{ib} - K_{ib})}_{-\varepsilon_b} = E_o - \varepsilon_b \qquad \{VI\text{-}24\}$$

The ionization energy $^2\Delta E_b$ then has the following simple expression

$$^2\Delta E_b = ({}^2E_b - E_o) = -\varepsilon_b \qquad \{VI\text{-}25\}$$

This means that the ionization potential $^2\Delta E_b$ associated with the removal of an electron from orbital b is the negative of that orbital energy ε_b. Since the orbital energy is, in general, a negative quantity ($\varepsilon_b < 0$) the ionization potential is a positive number.

The above result (${}^2\Delta E_b = -\varepsilon_b$) is frequently referred to as <u>Koopmans' Theorem</u>.

CHAPTER VII

HYBRID ATOMIC ORBITALS AND LOCALIZED MOLECULAR ORBITALS

1. The Concept of Hybridization

To this point we have constructed molecular orbitals (MO) from 1s, 2s, $2p_x$, $2p_y$, $2p_z$, etc. atomic orbitals (AO). Chemical problems, however, are often discussed in terms of hybrid atomic orbitals (HAO). This process of hybridization corresponds to an orthogonal transformation of a given basis set ($\chi_1 = 2s$; $\chi_2 = 2p_x$; $\chi_3 = 2p_y$; $\chi_4 = 2p_z$) to an equivalent basis set $\{\lambda_i\}$. Since the hybridization may involve the mixing (i.e. linear combination) of two (2s, $2p_x$), three (2s, $2p_x$, $2p_y$) or four (2s, $2p_x$, $2p_y$, $2p_z$) AO, it is possible to derive 3 types of hybridization referred to as sp, sp^2 and sp^3 respectively. The mathematical relationships that define these three sets of HAO and the corresponding vector models are shown in Figure VII-1.

The coefficients shown in Figure VII-1 can be derived under the condition that both the original set of AO and the set of hybrid AO are orthonormal. The transformation does not change the normalization or the orthogonality of the vectors. It is, therefore, orthogonal, and corresponds to a special case of the unitary transformation.

We may rewrite the relationships shown in Figure VII-1 in matrix notation as shown for the sp^3 hybridization in equation {VII-1}.

$$(\lambda_1\lambda_2\lambda_3\lambda_4) = (\chi_1\chi_2\chi_3\chi_4)\begin{pmatrix} \frac{1}{2} & \frac{1}{2} & \frac{1}{2} & \frac{1}{2} \\ \frac{1}{2} & \frac{1}{2} & -\frac{1}{2} & -\frac{1}{2} \\ \frac{1}{2} & -\frac{1}{2} & \frac{1}{2} & -\frac{1}{2} \\ \frac{1}{2} & -\frac{1}{2} & -\frac{1}{2} & \frac{1}{2} \end{pmatrix} \qquad \{VII-1\}$$

---

*It should be noted that the choice of direction for the first hybrid (χ_1) is arbitrary in any one of the 3 sets and by convenience the (1,1,1)point in the xyz space is chosen for the first sp^3 hybrid while the (1,1)point in the xy plane is associated with the first sp^2 hybrid.

sp^3 hybrids

$$\lambda_1 = 1/2(\chi_1 + \chi_2 + \chi_3 + \chi_4)$$

$$\lambda_2 = 1/2(\chi_1 + \chi_2 - \chi_3 - \chi_4)$$

$$\lambda_3 = 1/2(\chi_1 - \chi_2 + \chi_3 - \chi_4)$$

$$\lambda_4 = 1/2(\chi_1 - \chi_2 - \chi_3 + \chi_4)$$

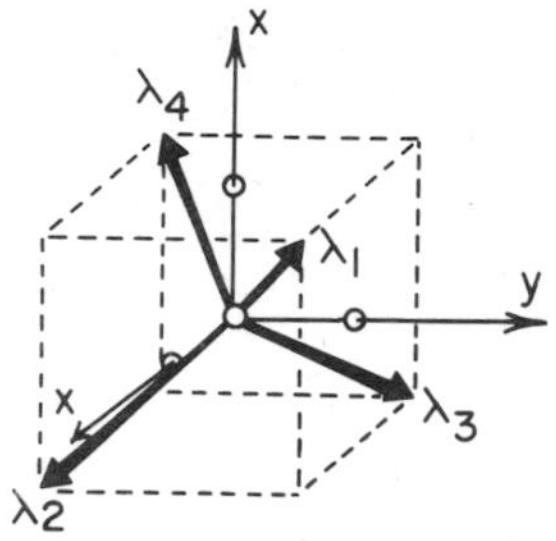

sp^2 hybrids

$$\lambda_1 = \frac{1}{\sqrt{3}}(\chi_1 + \chi_2 + \chi_3)$$

$$\lambda_2 = \frac{1}{\sqrt{3}}(\chi_1 + 0.366\ \chi_2 - 1.366\ \chi_3)$$

$$\lambda_3 = \frac{1}{\sqrt{3}}(\chi_1 - 1.366\ \chi_2 + 0.366\ \chi_3)$$

$$\lambda_4 = \chi_4$$

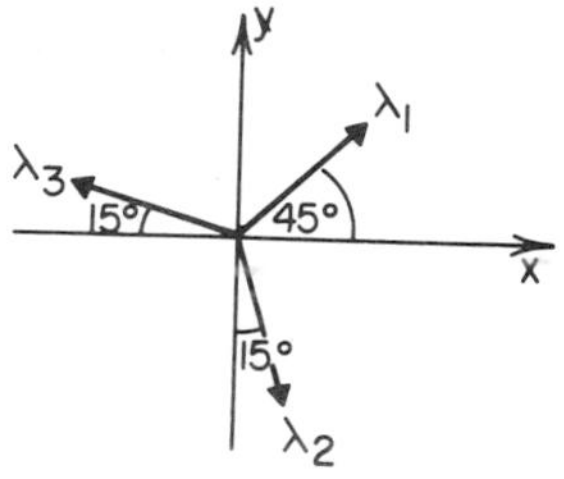

sp hybrids

$$\lambda_1 = \frac{1}{\sqrt{2}}(\chi_1 + \chi_2)$$

$$\lambda_2 = \frac{1}{\sqrt{2}}(\chi_1 - \chi_2)$$

$$\lambda_3 = \chi_3$$

$$\lambda_4 = \chi_4$$

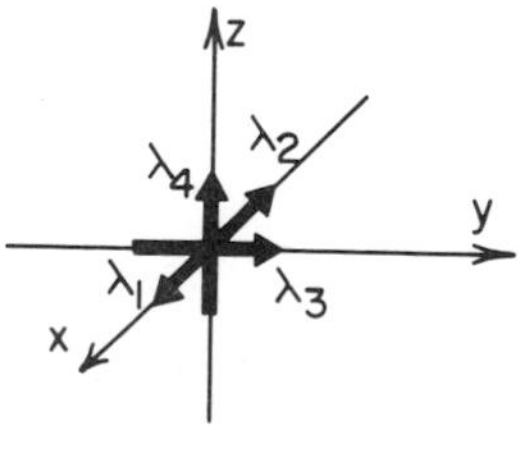

Figure VII-1 Vector model and analytical expressions of HAO

where the transforming matrix, composed of +1/2 and -1/2 may be divided into two matrices, one responsible for the hybridization and the other for renormalization.

$$\begin{pmatrix} 1 & 1 & 1 & 1 \\ 1 & 1 & -1 & -1 \\ 1 & -1 & 1 & -1 \\ 1 & -1 & -1 & 1 \end{pmatrix} \begin{pmatrix} \frac{1}{2} & 0 & 0 & 0 \\ 0 & \frac{1}{2} & 0 & 0 \\ 0 & 0 & \frac{1}{2} & 0 \\ 0 & 0 & 0 & \frac{1}{2} \end{pmatrix} = \begin{pmatrix} \frac{1}{2} & \frac{1}{2} & \frac{1}{2} & \frac{1}{2} \\ \frac{1}{2} & \frac{1}{2} & -\frac{1}{2} & -\frac{1}{2} \\ \frac{1}{2} & -\frac{1}{2} & \frac{1}{2} & -\frac{1}{2} \\ \frac{1}{2} & -\frac{1}{2} & -\frac{1}{2} & \frac{1}{2} \end{pmatrix} \qquad \{VII\text{-}2\}$$

hybridization   renormalization

The hybridization matrix specifies the orientation and the renormalization matrix the modulus of the vectors. For the sp^2 and sp hybrids the corresponding relationships are given in equations {VII-3} and {VII-4} respectively.

$$(\lambda_1\lambda_2\lambda_3\lambda_4) = (\chi_1\chi_2\chi_3\chi_4) \begin{pmatrix} 1 & 1 & 1 & 0 \\ 1 & 0.366 & -1.366 & 0 \\ 1 & -1.366 & 0.366 & 0 \\ 0 & 0 & 0 & 0 \end{pmatrix} \begin{pmatrix} \frac{1}{\sqrt{3}} & 0 & 0 & 0 \\ 0 & \frac{1}{\sqrt{3}} & 0 & 0 \\ 0 & 0 & \frac{1}{\sqrt{3}} & 0 \\ 0 & 0 & 0 & 1 \end{pmatrix} \qquad \{VII\text{-}3\}$$

$$(\lambda_1\lambda_2\lambda_3\lambda_4) = (\chi_1\chi_2\chi_3\chi_4) \begin{pmatrix} 1 & 1 & 0 & 0 \\ 1 & -1 & 0 & 0 \\ 0 & 0 & 1 & 0 \\ 0 & 0 & 0 & 1 \end{pmatrix} \begin{pmatrix} \frac{1}{\sqrt{2}} & 0 & 0 & 0 \\ 0 & \frac{1}{\sqrt{2}} & 0 & 0 \\ 0 & 0 & 1 & 0 \\ 0 & 0 & 0 & 1 \end{pmatrix} \qquad \{VII\text{-}4\}$$

Since an orthogonal transformation does not improve the basis set a hybridized basis set cannot be "better" than the original basis set. The same set of MO is obtained whether we begin with the hybridized or the original atomic basis. Only the coefficient matrix differs in the two cases.

Nevertheless it is often convenient to employ hybridized atomic orbitals, especially in Valence Bond (VB) type calculations or when a pictorial description of chemical bonding is desired.

2. The Equivalence of CMO and LMO

The molecular orbitals which produce a Fock matrix in the canonical (diagonal) form are known as the canonical molecular orbitals (CMO). These CMO are delocalized over the whole molecule whether they belong to the σ or the π representations. Consequently the CMO are symmetry adapted, i.e. they form the basis for the irreducible representation of the point group fixed by the symmetry of the molecule. On the other hand, the geometrical equivalents of localized molecular orbitals (LMO) are governed by the stereochemistry of the molecular bonding. Methane may be used to illustrate this point.

Neglecting the core electrons we have to account for 8 electrons or 4 electron pairs, which require 4 MO. In the CMO description, a totally symmetrical MO (a_1 type) and a triply degenerate MO (t-type), oriented towards the x,y and z directions (t_x, t_y, t_z) are obtained. These are illustrated in Figure VII-2.

In the CMO representation, the molecular orbital levels accordingly, have a triply degenerate energy value and a somewhat lower value associated with a_1. On the other hand the "orbital-energy" value* associated with the LMO basis must have four fold degeneracy since each of the 4 C-H bonds is equivalent (cf. Figure VII-3).

---

*The matrix representative of the Fock operator over the LMO basis ($\underline{\underline{F}}^{\Psi}$) is NOT a diagonal matrix (in contrast to $\underline{\underline{F}}^{\Phi}$ which is a diagonal matrix); therefore strictly speaking, the diagonal elements of $\underline{\underline{F}}^{\Psi}$ are not "orbital energies".

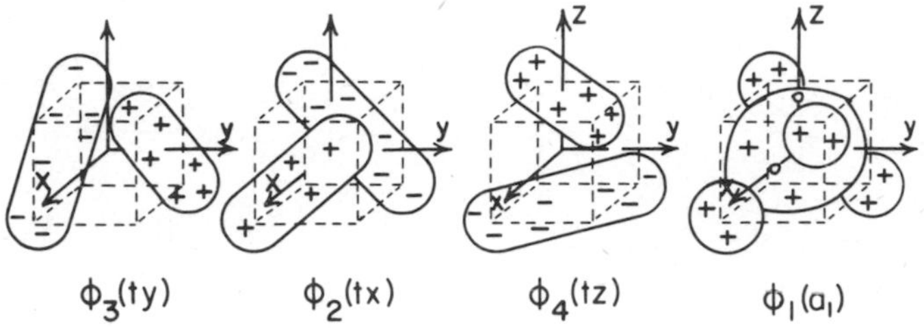

Figure VII-2 A schematic representation of the shapes and orientation of the four valence CMO of CH_4

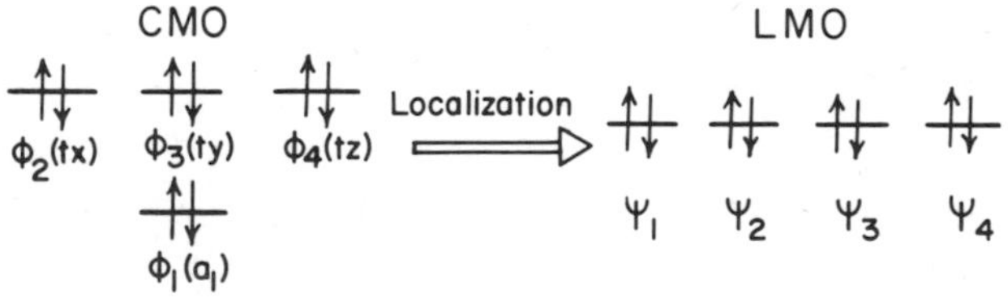

Figure VII-3 A schematic representation of the change of energy levels upon localization

The four LMO must be derived in such a way that each is directed along a C-H bond, as indicated schematically in Figure VII-4.

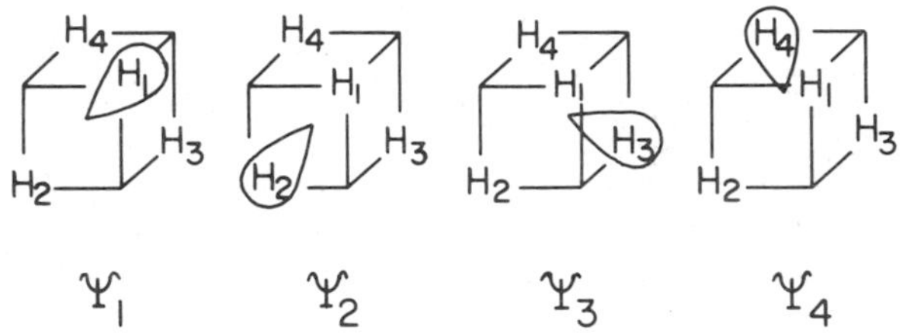

Figure VII-4 A schematic representation of the shapes and orientation of the four valence LMO of CH_4

We may then ask which set of MO is "better", the CMO or the LMO? The question can be answered objectively by evaluating the total energy of the molecule using both the CMO and LMO basis sets. The set that yields the lower total energy will be the "better" one. This requires the construction of determinantal wavefunctions from the CMO, $\{\phi_i\}$:

$$\Phi(1,2, \ldots., 2M) \qquad \{VII\text{-}5a\}$$

and from the LMO, $\{\Psi_i\}$:

$$\Phi'(1,2, \ldots., 2M) \qquad \{VII\text{-}5b\}$$

We write down the two wavefunctions in terms of their spin orbitals, employing initially a 2 electron case (one doubly occupied MO) for simplicity:

$$\Phi(1,2) = \frac{1}{\sqrt{2}} \begin{vmatrix} \phi_1(1)\alpha(1) & \phi_1(1)\beta(1) \\ \phi_1(2)\alpha(2) & \phi_1(2)\beta(2) \end{vmatrix} \quad \text{and}$$

$$\Phi'(1,2) = \frac{1}{\sqrt{2}} \begin{vmatrix} \Psi_1(1)\alpha(1) & \Psi_1(1)\beta(1) \\ \Psi_1(2)\alpha(2) & \Psi_1(2)\beta(2) \end{vmatrix} \qquad \{\text{VII-6a,b}\}$$

The Slater determinants are most conveniently written in terms of spin-orbitals, which are defined below for both the CMO and LMO bases.

$$\S_1 = \phi_1\alpha \qquad \omega_1 = \Psi_1\alpha$$

$$\S_2 = \phi_1\beta \qquad \omega_2 = \Psi_1\beta$$

$$\Phi(1,2) = \frac{1}{\sqrt{2}} \begin{vmatrix} \S_1(1) & \S_2(1) \\ \S_1(2) & \S_2(2) \end{vmatrix}$$

$$\Phi'(1,2) = \frac{1}{\sqrt{2}} \begin{vmatrix} \omega_1(1) & \omega_2(1) \\ \omega_1(2) & \omega_2(2) \end{vmatrix} \qquad \{\text{VII-7a,b}\}$$

It can be shown that $\Phi'(1,2)$ and $\Phi(1,2)$ are related to each other in the following way

$$\Phi'(1,2) = \Phi(1,2) \det |\underline{\underline{T}}| \qquad \{VII-8\}$$

Where the matrix $\underline{\underline{T}}$ is responsible for the transformation of the set of canonical spin-orbitals $\{ § \}$ into the set of localized spin-orbitals $\{ \omega \}$:

$$(\omega_1 \omega_2) = (§_1 §_2) \begin{pmatrix} T_{11} & T_{12} \\ T_{21} & T_{22} \end{pmatrix} \qquad \{VII-9\}$$

or, in matrix notation

$$\underline{\omega} = \underline{§}\ \underline{\underline{T}} \qquad \{VII-10\}$$

where the components of ω are as follows:

$$\omega_1 = §_1 T_{11} + §_2 T_{21}$$
$$\omega_2 = §_1 T_{12} + §_2 T_{22} \qquad \{VII-11a,b\}$$

To illustrate the validity of the above relationship between $\Phi(1,2)$ and $\Phi'(1,2)$ we perform the appropriate substitution:

$$\Phi'(1,2) = \frac{1}{\sqrt{2}} [\omega_1(1)\omega_2(2) - \omega_2(1)\omega_1(2)]$$

$$= \frac{1}{\sqrt{2}} [(§_1(1)T_{11} + §_2(1)T_{21})(§_1(2)T_{12} + §_2(2)T_{22}) - (§_1(1)T_{12} + §_2(1)T_{22})(§_1(2)T_{11} + §_2(2)T_{21})] \qquad \{VII-12\}$$

After the multiplications are carried out, four of the eight terms cancel the following relationship results:

$$\Phi'(1,2) = \frac{1}{\sqrt{2}} [T_{11}T_{22}\S_1(1)\S_2(2) - T_{12}T_{21}\S_1(1)\S_2(2) - T_{22}T_{11}\S_2(1)\S_1(2) + T_{21}T_{12}\S_2(1)\S_1(2)] \quad \{VII-13\}$$

By factoring out $\S_1(1)\S_2(2)$ from the first two terms and $-\S_2(1)\S_1(2)$ from the last two terms we obtain the following relationship

$$\Phi'(1,2) = \frac{1}{\sqrt{2}} [\S_1(1)\S_2(2)(T_{11}T_{22} - T_{12}T_{21}) - \S_2(1)\S_1(2)(T_{22}T_{11} - T_{21}T_{12})] \quad \{VII-14\}$$

This can be simplified further by factoring out the composite terms involving elements T_{ij}.

$$\Phi'(1,2) = \frac{1}{\sqrt{2}} [\S_1(1)\S_2(2) - \S_2(1)\S_1(2)][T_{11}T_{22} - T_{12}T_{21}] \quad \{\{VII-15\}$$

This equation may be generalized if written in its equivalent determinental form:

$$\Phi'(1,2) = \frac{1}{\sqrt{2}} \begin{vmatrix} \S_1(1) & \S_2(1) \\ \S_1(2) & \S_2(2) \end{vmatrix} \begin{vmatrix} T_{11} & T_{12} \\ T_{21} & T_{22} \end{vmatrix} \quad \{VII-16\}$$

or simply

$$\Phi'(1,2) = \Phi(1,2) \det|\underline{\underline{T}}| \quad \{VII-17\}$$

In order to use this result in a practical way we have to invoke double occupancy of the MO explicitly, i.e. we have to change the spin-orbitals to a composite expression of spatial orbital times spin function.

Starting with the 2 electron case we note the following as shown in {VII-9}:

$$(\omega_1\omega_2) = (\S_1\S_2) \begin{pmatrix} T_{11} & T_{12} \\ T_{21} & T_{22} \end{pmatrix} \quad \{VII-18\}$$

Both $\S_1$ and $\S_2$ incorporate the same CMO, with α and β spin functions respectively, so that the two-term linear combination is reduced to one term as illustrated below.

$$(\Psi_1\alpha\Psi_1\beta) = (\phi_1\alpha\phi_1\beta) \begin{pmatrix} T_{11} & 0 \\ 0 & T_{22} \end{pmatrix} \quad \{VII\text{-}19\}$$

In addition $T_{11} = T_{22}$ because both members of the spin-orbital basis include the same spatial orbital. In this particular case therefore, det $|\underline{\underline{T}}|$ has the following value,

$$|\underline{\underline{T}}| = \begin{pmatrix} T_{11} & 0 \\ 0 & T_{11} \end{pmatrix} = T_{11}^{2} - 0 = T_{11}^{2} \quad \{VII\text{-}20\}$$

We now consider a 4 electron case (two doubly occupied MO) in order to see the generality:

$$(\Psi_1\alpha,\Psi_1\beta,\Psi_2\alpha,\Psi_2\beta) = (\phi_1\alpha,\phi_1\beta,\phi_2\alpha,\phi_2\beta) \begin{pmatrix} T_{11} & \cdots & 0 & \cdots & T_{13} & 0 \\ \vdots & & & & \vdots & \\ 0 & & T_{22} & \cdots & 0 & \cdots & T_{24} \\ \vdots & & \vdots & & \vdots & \vdots \\ T_{31} & \cdots & 0 & \cdots & T_{33} & 0 \\ & & \vdots & & & \vdots \\ 0 & & T_{42} & \cdots & 0 & \cdots & T_{44} \end{pmatrix}$$

$$\{VII\text{-}21\}$$

Note that the spatial orbitals occur twice so that the $\underline{\underline{T}}$ matrix has a special form as indicated by the broken line. Furthermore, because the spatial orbitals are the same in both of the two groups (i.e. those with α spin and those with β spin), the two blocks in $\underline{\underline{T}}$ must be the same.

$$T_{11} = T_{22} = U_{11}$$

$$T_{13} = T_{24} = U_{12}$$

$$T_{31} = T_{42} = U_{21} \qquad \{VII-22a-d\}$$

$$T_{33} = T_{44} = U_{22}$$

Therefore

$$(\phi_1\alpha\phi_1\beta\Psi_2\alpha\Psi_2\beta) = (\phi_1\alpha\phi_1\beta\phi_2\alpha\phi_2\beta) \begin{pmatrix} U_{11} & \cdot\cdot & O & \cdots & U_{12} & O \\ \vdots & & & & \vdots & \\ O & & U_{11} & \cdot & O & \cdot\cdot & U_{12} \\ \vdots & & \vdots & & \vdots & & \vdots \\ U_{21} & \cdot\cdot & O & \cdots & U_{22} & O \\ & & \vdots & & & & \vdots \\ O & & U_{21} & \cdot & O & \cdot\cdot & U_{22} \end{pmatrix}$$

$$\{VII-23\}$$

In this particular case det $|\underline{\underline{T}}|$ is somewhat more complicated. However, after carrying out the cross multiplication we obtain the following expression

$$\det |\underline{\underline{T}}| = U_{11}^2\, U_{22}^2 + U_{12}^2 U_{21}^2 - 2U_{11}\, U_{12}\, U_{21}\, U_{22}$$

$$= (U_{11}U_{22} - U_{21}U_{12})^2 \qquad \{VII-24\}$$

which can be written in terms of the square of the determinant $\underline{\underline{U}}$.

$$\det |\underline{\underline{T}}| = \det |\underline{\underline{U}}|^2 \qquad \{VII-25\}$$

where $\underline{\underline{U}}$ is an M x M matrix provided $\underline{\underline{T}}$ is a 2M x 2M matrix, and this reduction in dimension is due to the double occupancy. Thus we may rewrite our relationship for the doubly occupied MO

$$\Phi'(1,2) = \Phi(1,2) \det |\underline{\underline{U}}|^2 \qquad \{VII-26\}$$

If $\underline{\underline{U}}$ is a unitary (i.e. orthogonal) matrix then $|\underline{\underline{U}}|^2$ should be written in terms of its adjoint (i.e. in terms of its transpose)

$$|\underline{\underline{U}}|^2 = |\underline{\underline{U}}^{+}\underline{\underline{U}}| = |\underline{\underline{U}}^{-1}\,\underline{\underline{U}}| = |\underline{\underline{1}}| \qquad \{VII-27\}$$

However, since

$$\underline{\underline{U}}^{+} = \underline{\underline{U}}^{-1} \qquad \{VII-28\}$$

then

$$|\underline{\underline{U}}|^2 = 1 \qquad \{VII-29\}$$

This leads us to the equivalence relationship

$$\Phi'(1,2) = \Phi(1,2) \qquad \{VII-30\}$$

provided that we choose a unitary (orthogonal) transformation to localize the CMO into a LMO basis. We then have

$$\Psi = \phi\underline{\underline{U}} \qquad \{VII-31\}$$

or, in detail,

$$(\Psi_1\Psi_2) = (\phi_1\phi_2) \begin{pmatrix} U_{11} & U_{12} \\ U_{21} & U_{22} \end{pmatrix} \qquad \{VII-32\}$$

3. Methods of Localization

The simplest way to obtain localized molecular orbitals (LMO) is to build the localization procedure into the calculation itself. This can be achieved by beginning with a hybridized atomic orbital (HAO) basis and restricting the linear combination of these HAO to form LMO to a pair of atomic orbitals only.

For CH_4, the four C-H bonds $\{\Psi_i\}$ as illustrated in Figure VII-5 are formed from the four sp^3 hybrid AO $\{\lambda_i\}$ and the 1s orbitals $\{\chi_{Hi}\}$ of the four hydrogen atoms.

$$\Psi_1 = a\lambda_1 + b\chi_{H_1}$$

$$\Psi_2 = a\lambda_2 + b\chi_{H_2} \qquad \{VII\text{-}33a\text{-}d\}$$

$$\Psi_3 = a\lambda_3 + b\chi_{H_3}$$

$$\Psi_4 = a\lambda_4 + b\chi_{H_4}$$

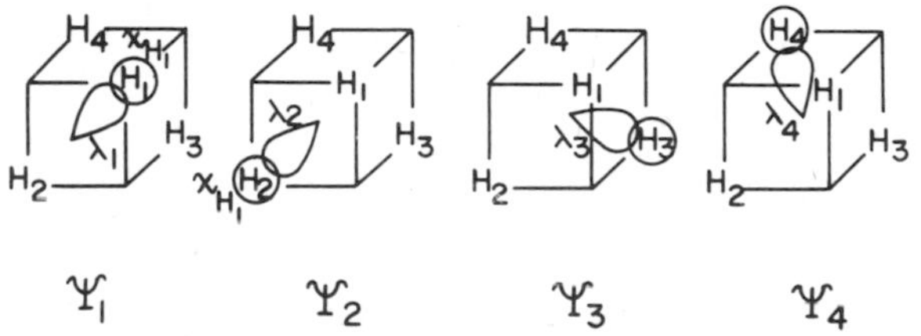

Figure VII-5 A schematic representation of the linear combination of carbon HAO and hydrogen AO to form methane LMO

However, $\{\lambda_i\}$, the HAO, also have their own expression in terms of the unhybridized AO of carbon. Take for example, the first of the series (remember that $\chi_1 = 2s$, $\chi_2 = 2p_x$, $\chi_3 = 2p_y$, $\chi_4 = 2p_z$ and $\chi_{H_1} = 1s_{H_1}$)

$$\Psi_1 = a(\chi_1 + \chi_2 + \chi_3 + \chi_4) + b(\chi_{H_1}) \qquad \{VII\text{-}34\}$$

It may be instructive to see what the equivalence of CMO and LMO (discussed in the previous section) really means in practice. Methane seems to be a good example for this exercise because from symmetry considerations we know the unitary matrix ($\underline{\underline{U}}$) that transforms the CMO $\{\Psi\}$ to the LMO $\{\Psi\}$:

$$\Psi = \phi\underline{\underline{U}} \qquad \{VII\text{-}35\}$$

For this case the $\underline{\underline{U}}$ matrix that transforms ϕ into Ψ is most conveniently the same matrix that performs the hybridization of the atomic orbitals χ.

$$(\Psi_1\Psi_2\Psi_3\Psi_4) = (\phi_1\phi_2\phi_3\phi_4)\begin{pmatrix} 1/2 & 1/2 & 1/2 & 1/2 \\ 1/2 & -1/2 & 1/2 & -1/2 \\ 1/2 & -1/2 & -1/2 & 1/2 \\ 1/2 & 1/2 & -1/2 & -1/2 \end{pmatrix} \qquad \{VII\text{-}37\}$$

In order to see the set of Ψ(LMO) expressed in terms of the AO (cf. Figure VII-1) we have to specify the expressions for the set of ϕ(CMO) the shapes and orientations of these CMO have been shown previously as given below

$$\begin{aligned} \phi_1 &= C_1\chi_1 + C_2(\chi_{H_1} + \chi_{H_2} + \chi_{H_3} + \chi_{H_4}) \\ \phi_2 &= C_1\chi_2 + C_2(\chi_{H_1} - \chi_{H_2} + \chi_{H_3} - \chi_{H_4}) \\ \phi_3 &= C_1\chi_3 + C_2(\chi_{H_1} - \chi_{H_2} - \chi_{H_3} + \chi_{H_4}) \\ \phi_4 &= C_1\chi_4 + C_2(\chi_{H_1} + \chi_{H_2} - \chi_{H_3} - \chi_{H_4}) \end{aligned} \qquad \{VII\text{-}38\}$$

Instead of dealing with all the LMO specified in equation {VII-37} let us examine Ψ_1, the first LMO, only.

$$\Psi_1 = (\phi_1\phi_2\phi_3\phi_4) \begin{pmatrix} 1/2 \\ 1/2 \\ 1/2 \\ 1/2 \end{pmatrix} \qquad \{VII-39\}$$

As specified earlier in equation {VII-38}, we substitute the expressions for the set of ϕ into equation {VII-39}.

$$\Psi_1 = 1/2\phi_1 + 1/2\phi_2 + 1/2\phi_3 + 1/2\phi_4 = 1/2\{\phi_1 + \phi_2 + \phi_3 + \phi_4\}$$

$$= 1/2\{C_1\chi_1 + C_2(\chi_{H_1} + \chi_{H_2} + \chi_{H_3} + \chi_{H_4}) + \qquad \{VII-40\}$$

$$+ C_1\chi_2 + C_2(\chi_{H_1} - \chi_{H_2} + \chi_{H_3} - \chi_{H_4}) +$$

$$+ C_1\chi_3 + C_2(\chi_{H_1} - \chi_{H_2} - \chi_{H_3} + \chi_{H_4}) +$$

$$+ C_1\chi_4 + C_2(\chi_{H_1} + \chi_{H_2} - \chi_{H_3} - \chi_{H_4})\}$$

When the additions are performed, terms involving χ_{H_2}, χ_{H_3} and χ_{H_4} cancel and we are left with the expression

$$\Psi_1 = C_1\{1/2(\chi_1 + \chi_2 + \chi_3 + \chi_4)\} + 2C_2\chi_{H_1} \qquad \{VII-41\}$$

The hybrid structure is clearly recognizable (cf. Figure VII-1) in the first term or the above expression, and we may rewrite it conveniently in the form

$$\Psi_1 = C_1\lambda_1 + 2C_2\chi_{H_1} \qquad \{VII-42\}$$

This may now be compared to the equation given at the beginning of this section for the first C-H LMO, viz.

$$\Psi_1 = a_1\lambda_1 + b_1\chi_{H_1} \qquad \{VII-43\} \text{ and } \{VII-33a\}$$

It must be admitted that the successful equivalence achieved in this case is a result of the high symmetry of CH_4.

For a less symmetrical molecule, e.g. CH_3OH, this would not be possible. Furthermore the concept of hybridization can only be defined for a minimal basis set (i.e. 2s, $2p_x$, $2p_y$, $2p_z$) and is not applicable to more extended basis sets, even in relatively simple cases where only two functions are used to simulate the traditional orbitals (e.g. 2s, 2s', $2p_x$, $2p_x'$, $2p_y$, $2p_y'$, $2p_z$, $2p_z'$).

For this reason the external localization, in which we specify the bonds, lone pairs, etc. and begin with hybrids that are already localized in a specific region of space, is of limited value.

The most objective localization procedure (intrinsic localization) that could be employed would involve the construction of orbitals which are separated from each other as much as possible without having to stipulate in advance the location of these orbitals in space. Such a localization would require only that the definition of "separation" be decided upon.

We may measure the "degree of separation" in a number of ways such as stereochemically and energetically and the two most commonly used methods adopt these two different measures. In the Boys' method of localization the sum of distances between all the centroids of electron pairs (as defined by the orbitals) is maximized while in the Edmiston-Ruedenberg method of localization separation of the electron pairs (as defined by the orbitals) involves maximization of the "total self-repulsion", i.e. the diagonal elements of the electron-electron repulsion. We will discuss this latter method in some detail below.

The electron-electron repulsion term, i.e. the two-electron contribution (E_2) to the total energy ($E = E_o + E_1 + E_2$) results from the two electron operator ($\hat{H}_2 = \sum'_{i,j} r_{ij}^{-1}$) of the total Hamiltonian ($\hat{H} = \hat{H}_o + \hat{H}_1 + \hat{H}_2$).

Considering the case of closed electron shells only we may write:

$$E_2 = 2 \sum_i \sum_j J_{ij} - \sum_i \sum_j K_{ij} \qquad \{VII-44\}$$

Because the diagonal elements of both the coulomb and exchange integrals are equal ($J_{ii} = K_{ii}$) they can be combined as indicated

$$E_2 = \sum_{i} J_{ii} + 2\sum_{\substack{ij \\ (j\neq i)}}\sum J_{ij} - \sum_{\substack{ij \\ (j\neq i)}}\sum K_{ij} \qquad \{VII-45\}$$

It is important to note that in the previous equation both terms are invariant under any unitary transformation of the MO basis (i.e. their numerical values are identical for both delocalized and any localized MO). However, in the latter equation, neither of the terms is invariant under a unitary transformation of the basis set. In fact, it is just this characteristic property which has been used as the localization criterion in the method of Edmiston and Ruedenberg. The object of this localization method is to maximize the "self-repulsion", i.e. the first term on the right hand side of the latter equation. This term is referred to as the localization sum.

$$J_o = \sum_{i} J_{ii} = \sum_{i} K_{ii} \qquad \{VII-46\}$$

The larger the J_o, the more localized the orbitals. The unitary matrix that maximizes the diagonal elements while simultaneously minimizing the off-diagonal elements of the $\underline{\underline{K}}$ matrix may be obtained by a Jacobi type diagonalization:

$$\underline{\underline{K}}^{\Psi} = \underline{\underline{U}}^{+} \underline{\underline{K}}^{\phi} \underline{\underline{U}} \qquad \{VII-47\}$$

This matrix $\underline{\underline{U}}$ is to be used to transform the CMO to the LMO in a unitary (orthogonal) transformation

$$\Psi = \phi\underline{\underline{\Psi}} \qquad \{VII-48\}$$

Some results ($\underline{\underline{K}}^{\phi}$, $\underline{\underline{K}}^{\Psi}$ and $\underline{\underline{U}}$) are summarized for the ammonia molecule in the following three tables (VII-1 to VII-3).

TABLE VII-1 EXCHANGE INTEGRALS ($\underline{\underline{K}}^{\phi}$) OVER CMO FOR NH_3

	$1a_1$ CORE	$2a_1$ BOND	e_1 BOND	e_2 BOND	$3a_1$ LONE PAIR
$1a_1$	4.1222				
$2a_1$	0.0483	0.6348			
e_1	0.0147	0.1296	0.5878		
e_2	0.0147	0.1296	0.0411	0.5878	
$3a_1$	0.0251	0.0945	0.0273	0.0273	0.4650

LOCALIZATION SUM = 6.3967

TABLE VII-2 EXCHANGE INTEGRALS ($\underline{\underline{K}}^{\Psi}$) OVER LMO FOR NH_3

	CORE	BOND	BOND	BOND	LONE PAIR
CORE	4.2057				
BOND	0.0113	0.7567			
BOND	0.0113	0.0256	0.7567		
BOND	0.0113	0.0256	0.0256	0.7567	
LONE PAIR	0.0159	0.0351	0.0351	0.0351	0.7415

LOCALIZATION SUM = 7.2173

TABLE VII-3 TRANSFORMATION MATRIX ($\underline{\underline{U}}$)
FROM CMO TO LMO FOR NH_3

	$1a_1$ CORE	$2a_1$ BOND	e_1 BOND	e_1 BOND	$3a_1$ LONE PAIR
CORE	0.9926	-0.1146	0.0000	0.0000	-0.0402
BOND	0.0540	0.5387	0.8165	0.0000	-0.2006
LONE PAIR	0.0773	0.3412	0.0000	0.0000	0.9368

4. Orbital Density Contours

The Chemist's phraseology that "an electron pair occupies a molecular orbital" is not rigorous although it comes as close to reality* as is possible if the orbitals are energy localized according to the Edmiston-Ruedenberg procedure as described in the previous section.

However this traditional identification of a molecular orbital with an electron pair made a demand for plotting orbital densities either in the form of contours or as pseudo three dimensional objects.

The electron density D_p associated with the pth MO is defined as $|\phi|^2$ or, more precisely, as $\phi\phi\dagger$:

$$D_p(x,y,z) = \phi_p(x,y,z)\ \phi_p^\dagger(x,y,z) \qquad \{VII\text{-}49\}$$

where $\phi_p^\dagger$ is the transpose of ϕ_p. The density may be calculated at any point (x,y,z) of the three dimensional physical space in terms of the AO basis $\{\eta\}$ and the MO coefficient matrix $\underline{\underline{C}}$:

$$D_p \equiv \phi_p\phi_p^\dagger = f_p(\eta_1\eta_2 \cdots \eta_N) \begin{pmatrix} C_{1p} \\ C_{2p} \\ \vdots \\ C_{Np} \end{pmatrix} (C_{1p}C_{2p} \cdots C_{Np}) \begin{pmatrix} \eta_1 \\ \eta_2 \\ \vdots \\ \eta_N \end{pmatrix} =$$

$$= f_p(\eta_1\eta_2 \cdots \eta_N) \begin{pmatrix} \rho^p_{11} & \rho^p_{12} & \cdots & \rho^p_{1N} \\ \rho^p_{21} & \rho^p_{22} & \cdots & \rho^p_{2N} \\ \vdots & \vdots & & \vdots \\ \rho^p_{N1} & \rho^p_{N2} & \cdots & \rho^p_{NN} \end{pmatrix} \begin{pmatrix} \eta_1 \\ \eta_2 \\ \vdots \\ \eta_N \end{pmatrix}$$

$$\{VII\text{-}50\}$$

---

*In other words when the total molecular wavefunction is constructed from Edmiston-Ruedenberg type (energy optimized) LMO the Hartree-type ordinary orbital product approximates the Hartree-Fock type antisymmetrized orbital product as far as possible.

where $\underline{\underline{\rho}}^p$ is the density matrix of the pth MO. When summed up over all occupied MO, it yields the total density matrix $\underline{\underline{\rho}}$ with elements defined as:

$$\rho_{ij} = \sum_{p=1}^{M} \rho^p{}_{ij} \qquad \{VII\text{-}51\}$$

Thus D_p may be written in the following abbreviated notation:

$$D_p(x,y,z) = f_p \, \underline{\eta}(x,y,z) \, \underline{\underline{\rho}}^p \, \underline{\eta}(x,y,z) =$$

$$= f_p \sum_{i=1}^{N} \sum_{j=1}^{N} \eta_i(x,y,z) \, \rho^p{}_{ij} \, \eta_j(x,y,z) \qquad \{VII\text{-}52\}$$

Note that the factor f_p is the integrated spin part of the orbital, which is 2 if the orbital is doubly occupied, 1 if it is singly occupied and 0 if it is empty. The sum of the M doubly occupied MO densities is then

$$D(x,y,z) = \sum_{p=1}^{M} D_p(x,y,z) = 2 \sum_{i=1}^{M} \sum_{j=1}^{N} \eta_i(x,y,z) \, \rho_{ij} \eta_j(x,y,z)$$

$$\{VII\text{-}53\}$$

The problem is simply to calculate the individual orbital electron density values D_p at every intersection of a given mesh around the molecule. In practice, 1600 points, that is, a 40 x 40 mesh provide a fine enough grid for a 10 x 10 $bohr^2$ (nearly 5 x 5 $Å^2$) area. Special purpose programs may be used to interpolate electron densities so that the density contours may be recorded by a two dimensional (X,Y) plotter connected to the computer.

If the coefficient matrix $\underline{\underline{C}}$ transforms the AO basis to the CMO basis then the densities of the CMO are generated. If the coefficient matrix however connects the AO to LMO then the LMO densities are obtained.

CMO densities associated with FCH_2-OH are shown in Figure VII-6 and the LMO densities for FCH_2-OH are illustrated in Figure VII-7.

5. The "Sizes" and "Shapes" of Localized MO

As mentioned previously an electron pair "a" may be identified with a localized molecular orbital ψ_a. The expectation values of the first (r) and the second (r^2) moment operators, in terms of these LMO, may be used to define the centroids of charge and the sizes and shapes of the electron pairs respectively.

The centroid of charge, with respect to an arbitrary coordinate system, is given in terms of the components (x_a, y_a, z_a) of the first moment $\langle r \rangle_o$:

$$x_a \equiv \langle x \rangle_o = \langle \psi_a | x | \psi_a \rangle$$

$$y_a \equiv \langle y \rangle_o = \langle \psi_a | y | \psi_a \rangle \qquad \{VII\text{-}54\}$$

$$z_a \equiv \langle z \rangle_o = \langle \psi_a | z | \psi_a \rangle$$

where the subscript o indicates that the centroid of charge is expressed with respect to the origin of the arbitrary coordinate system. The following geometrical relationship holds for the distance of the centroid of charge (R_a) measured from the arbitrary origin:

$$R_a = |\langle r \rangle_o| = \langle x \rangle_o^2 + \langle y \rangle_o^2 + \langle z \rangle_o^2 = x_a^2 + y_a^2 + z_a^2 \qquad \{VII\text{-}55\}$$

The size of an electron pair "a" as represented by LMO ψ_a may then be defined as the expectation value of a spherical quadratic (second moment) operator (r^2) evaluated at the centroid of charge (R_a) defined by the coordinates x_a, y_a, z_a:

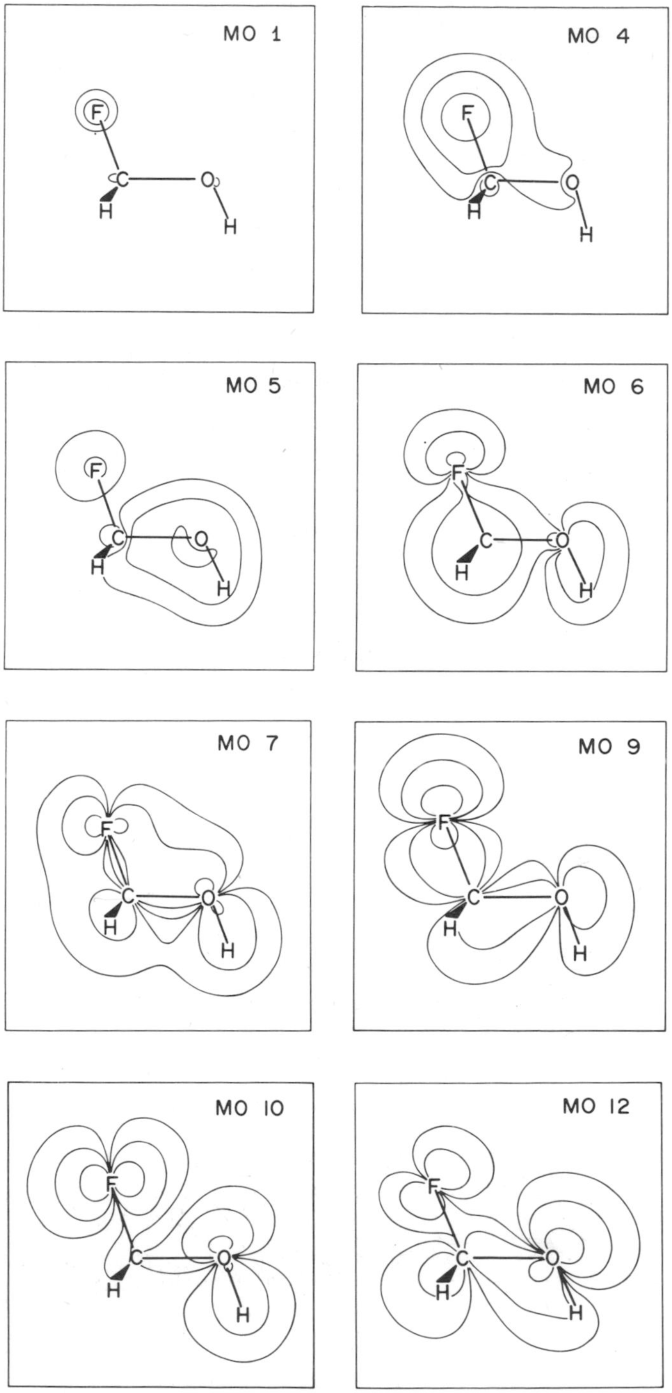

Figure VII-6. CMO electron density contours of fluoromethanol (staggered conformation). [C and O cores are not shown, all other CMO with electron density in the FCOH plane are included. Outmost contour is 0.002 c/bohr^3, next 0.02 e/bohr^3 and innermost 0.2 e/bohr^3.]

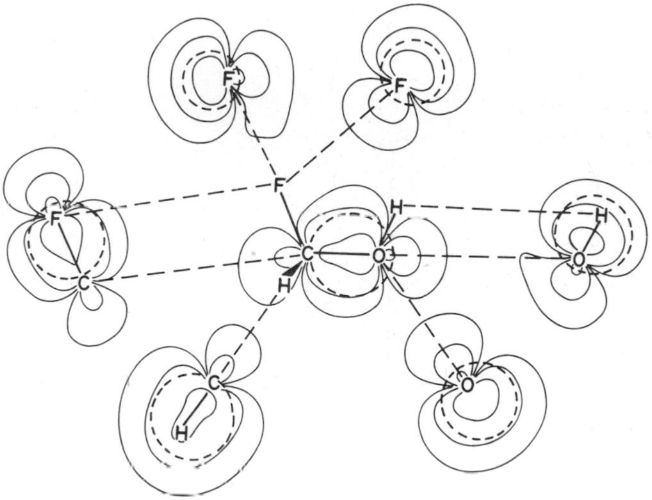

Figure VII-7. Valence LMO electron densities
of fluoromethanols (staggered conformation).

$$\langle r^2\rangle = \langle (x - x_a)^2\rangle + \langle (y - y_a)^2\rangle + \langle (z - z_a)^2\rangle = \qquad \{VII\text{-}56\}$$

$$= \langle x^2\rangle_o - 2x_a\langle x\rangle_o + x_a^2 + \langle y^2\rangle_o - 2y_a\langle y\rangle_o + y_a^2 + \langle z^2\rangle_o - 2z_a\langle z\rangle + z_a^2 =$$

$$= \langle x^2\rangle - 2x_a^2 + x_a^2 + \langle y^2\rangle_o - 2y_a^2 + y_a^2 + \langle z^2\rangle_o - 2z_a^2 + z_a^2$$

or:

$$\langle r^2\rangle = \langle x^2\rangle_o + \langle y^2\rangle_o + \langle z^2\rangle_o - (x_a^2 + y_a^2 + z_a^2) = \langle r^2\rangle_o - \langle r\rangle_o^2$$

{VII-57}

where $\langle r^2\rangle_o$ is the second moment of a given LMO, ψ_a, with respect to the arbitrary origin:

$$\langle r^2\rangle_o = \langle \psi_a | r^2 | \psi_a\rangle \qquad \{VII\text{-}58\}$$

It is more practical however to collect the x, y and z components as shown below, in order to have an explicit expression for the components labelled $\langle x^2\rangle$, $\langle y^2\rangle$ and $\langle z^2\rangle$:

$$\langle r^2\rangle = \langle r^2\rangle_o - \langle r_o\rangle^2$$

$$= \{\langle x^2\rangle_o - \langle x\rangle_o^2\} + \{\langle y^2\rangle_o - \langle y\rangle_o^2\} + \{\langle z^2\rangle_o - \langle z\rangle_o^2\}$$

$$= \langle x^2\rangle + \langle y^2\rangle + \langle z^2\rangle. \qquad \{VII\text{-}59\}$$

The shape of an electron pair "a" may be identified with the three components of the size defined in equation {VII-59}, and these are characteristic of an ellipsoid. However, the arbitrary coordinate system (x,y,z) may not be in alignment with the major and two minor axes of the ellipsoid. Consequently it may be desirable to rotate the arbitrary coordinate system to a new one (x', y', z') which is now parallel to the major and minor axes of the ellipsoid. In this new coordinate system the size may be written in terms of its new components,

$$\langle r^2\rangle = \langle x'^2\rangle + \langle y'^2\rangle + \langle z'^2\rangle \qquad \{VII\text{-}60\}$$

which will uniquely define the shape of the electron pair but not alter the numerical value of the size.

Both the spherical average, $\langle r^2 \rangle$, which is a measure of size, and its components $\langle x'^2 \rangle$, $\langle y'^2 \rangle$ and $\langle z'^2 \rangle$ which are related to the shape (i.e. to the length of the major and minor axes of the ellipsoid), are graphically illustrated together with the centroid of charge $\langle r \rangle_o$, in Figure VII-8.

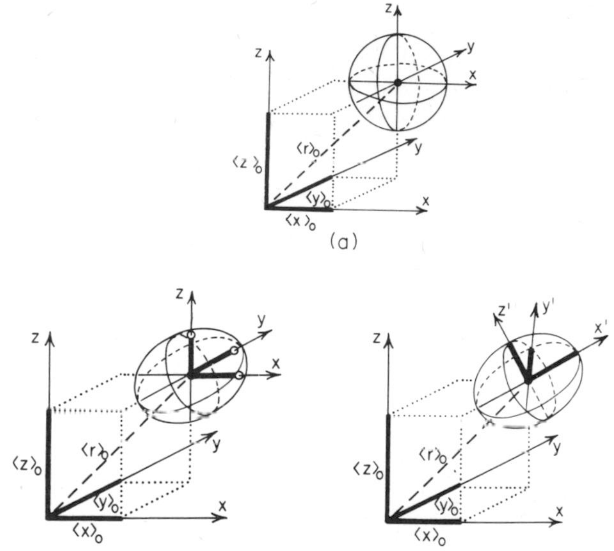

Figure VII-8. Spherical and ellipsoidal representations of the second moment components of LMO.

CHAPTER VIII

LIMITATIONS OF MOLECULAR ORBITAL THEORIES

1. The Concept of the Hartree-Fock Limit (HFL)

The experimental total energy of an atom is the negative of the sum of all of its ionization potentials. For example, the total energy of Li is

$$E = -\sum_{i=1}^{3} I_i = -7.4788 \text{ hartree} \qquad \{VIII-1\}$$

Calculation of this quantity is the ultimate objective of quantum mechanics but this is not yet possible for molecules of chemical interest because the Hamiltonian usually employed does not contain terms to describe relativistic effects. One therefore obtains a total energy that is higher than the experimental energy. The limiting energy value that can be computed with the non-relativistic Hamiltonian is referred to as the non-relativistic limit (NRL). The difference between the experimental energy and that associated with the NRL is the relativistic energy. This is small for light atoms (Li → Ne) and is presumed to remain nearly constant when the atoms in question form a molecule.

When the total single determinant wavefunction is constructed from the most accurate set of MO a limiting total energy value is reached which is higher than the energy associated with the NRL. This is referred to as the Hartree-Fock limit (HFL) and the energy difference between the two limits ($\Delta E = E_{NRL} - E_{HFL}$) is called the correlation energy.

The Hartree-Fock (HF) energies, the correlation energies, the relativistic corrections, and the total energies (the sum of these three components) of the first 18 atoms are summarized in Table VIII-1. Note that the values given in this Table for the total energies are somewhat different from those in the earlier equation {VII-56}. This is not a discrepancy between theory and experiment but is due to different sources of experimental data, used for the determination of the total energy.

The relativistic correction is the energetic consequence of the relativistic effects suffered by the electrons as they

TABLE VIII-1 THEORETICAL COMPONENTS OF ATOMIC ENERGY VALUES

Z	Atom	HF[a]	Corr[b]	Rel[c,d]	Total
1	H	-0.500000	-	-	-0.5000
2	He	-2.861679	-0.0421	-0.0001[e]	-2.9039
3	Li	-7.432726	-0.0455	-0.0006	-7.4788
4	Be	-14.573014	-0.0943	-0.0022[e]	-14.6695
5	B	-24.529049	-0.1248	-0.0061	-24.6599
6	C	-37.688584	-0.1581	-0.0138	-37.8605
7	N	-54.400904	-0.1883	-0.0274	-54.6166
8	O	-74.809347	-0.2575	-0.0494	-75.1162
9	F	-99.409290	-0.3236	-0.0829	-99.8158
10	Ne	-128.546980	-0.3927	-0.1313[e]	-129.0710
11	Na	-161.858570	-0.4030	-0.2008	-126.4624
12	Mg	-199.614320	-0.4490	-0.2954	-200.3587
13	Al	-241.876300	-0.4820	-0.4219	-242.7802
14	Si	-288.854200	-0.5220	-0.5856	-289.9618
15	P	-340.718570	-0.5610	-0.7927	-342.0723
16	S	-397.504690	-0.6400	-1.0530	-399.1977
17	Cl	-459.481810	-0.7120	-1.3750	-461.5788
18	Ar	-526.817050	-0.7910	-1.7660	-529.3741

[a]E. Clementi, J. Chem. Phys. 38, 1001 (1963).

[b]E. Clementi, J. Chem. Phys. 39, 175 (1963).

[c]E. Clementi and A. D. McLean, Phys. Rev. 133 (No. 2A), A419 (1964).

[d]E. Clementi, J. Mol. Spect. 12, 18 (1964).

[e]E. Clementi, J. Chem. Phys. 38, 2248 (1963).

pass the nucleus. Consequently, this term is a function of Z. It is comparatively small for the light atoms (e.g. up to Ne) but the heavier elements (e.g. S, P and Cl) have an appreciable relativistic energy (cf. Table VIII-1).

The correlation energy can be associated with the consequences of electron pairing (see Ch. III-5). Since the number of electron pairs increases with Z, the heavier atoms have a larger correlation energy than the lighter ones. The actual value depends, in addition to Z, on the pairing scheme that exists within the atom in question. For example, the increase in correlation energy from He to Li or from Ne to Na is small since the incoming electron in unpaired. However, the increase in correlation energy from Li to Be and Na to Mg is larger because for these atoms a new electron pair is formed. Similarly the increase in correlation energy is steeper for O, F and Ne than for B, C and N because the electron pairing in O, F and Ne is more extensive than in B, C and N. These differences are seen in the following figure (Figure VIII-1).

To compute an energy which is lower than the HFL requires a wavefunction that transcends the HF wavefunction by taking into account the consequences of electron pairing. These more sophisticated wavefunctions (such as configuration interaction :CI) will be treated later in the present chapter.

2. Estimation of the Molecular Hartree-Fock Limit

The total energy (E_{tot}) for any chemical system may be written in terms of the previously discussed components viz. the Hartree-Fock energy (E_{HF}), the correlation energy (E_{cor}) and the relativistic correction (E_{rel}):

$$E_{Tot} = E_{HF} + E_{Cor} + E_{Rel} \qquad \{VIII-2\}$$

The SCF energy (E_{SCF}) that is computed by non-empirical MO theory, constitutes an upper bound on E_{HF}. Only at the HFL will E_{SCF} equal E_{HF}. However to achieve this result it is necessary to employ a complete basis set of infinite dimension (Hilbert space). Since this is not possible for large molecules one is always forced to use a truncated basis. The degree of truncation is important because of the

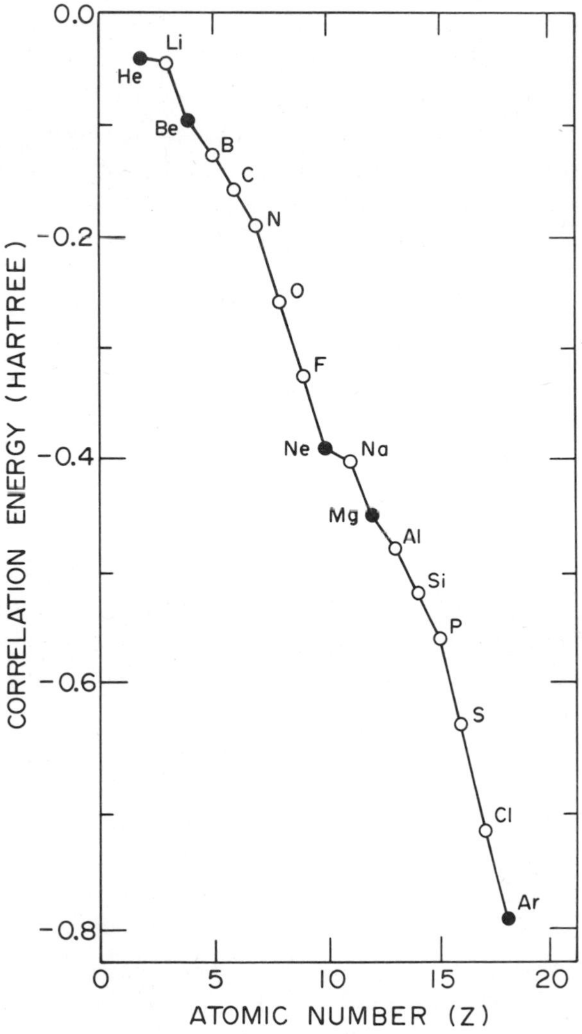

Figure VIII-1 Atomic Correlation Energy as a Function of Atomic Number

expectation that a fairly extensive and well balanced basis may produce an E_{SCF} that approximates E_{HF} to a pre-defined number of significant figures.

For atoms (see Table VIII-1) and for diatomic molecules Hartree-Fock accuracy can be obtained, as will now be illustrated for the case of CO. The minimum number of atomic orbitals that may be chosen for the basis (i.e. the minimal basis set) is 1s, 2s, $2p_x$, $2p_y$ and $2p_z$, on carbon and the same number of orbitals on oxygen. This amounts to a total of 10 atomic orbitals in the basis set. Although the SCF energy computed with this minimal basis set is 1.4 hartree (ca. 250 kcal/mole) above the HFL, this energy nevertheless represents 98.7% of the total energy (-112 hartree) of the system. It is this relatively low percentage error in the total energy that is used to justify the use of minimal basis set calculations to study real chemical problems (see also Ch. IX). If one wishes to have E_{SCF} closer to E_{HF} a larger AO basis must be employed. The most frequently used set of orbitals is that normally referred to as the "double zeta" AO basis. (The Greek letter zeta: ζ denotes the orbital exponent which is treated as a variable parameter.) In a double zeta basis two AO are used to describe each orbital that was included in the minimal basis so that one now has 1s, 1s', 2s, 2s', $2p_x$, $2p_x'$, $2p_y$, $2p_y'$, $2p_z$, $2p_z'$. The unprimed and primed orbitals have different exponents, ζ and ζ', the larger corresponding to a "tight" orbital and the smaller to a "loose" orbital. Although the double zeta basis for CO now includes 20 AO the E_{HF} cannot be reached until higher angular momentum functions (d, f, ... etc.) are added to the basis. To reproduce the energy of the HFL to 3 decimal places requires 32 AO. The following table (Table VIII-2) summarizes the orbital and total energy values of CO that have been computed with different basis sets.

It is customary to assume that the relativistic correction (E_{rel}) of the atoms is not changed when these form a molecule, i.e. that

TABLE VIII-2

ORBITAL AND TOTAL ENERGY VALUES* AS COMPUTED BY NON-EMPIRICAL MO CALCULATIONS ON CO WITH DIFFERENT AO BASIS SETS

	10 AO (min basis)	32 AO
5σ	-0.4842	-0.5530
$1\pi_x$, $1\pi_y$	-0.5582	-0.0377
4σ	-0.7279	-0.8024
3σ	-1.4812	-1.5192
2σ	-11.2856	-11.3593
1σ	-20.6679	-20.6612
TOTAL	-111.3910	-112.7860†

*The energy values are quoted in Hartree a.u.
†Assumed to approach the HF limit

$$E_{rel}^{mol} \simeq \sum_{i}^{\text{all atoms}} (E_{rel}^{atom})_i \qquad \{VIII-3\}$$

For carbon monoxide this amounts to

$$-0.063 = (-0.014) + (-0.049) \qquad \{VIII-3a\}$$

where the first and second terms on the right refer to E_{rel} for carbon and oxygen respectively. Thus, in the energetic relationship

$$E_{Tot} = E_{HF} + E_{Cor} + E_{Rel} \qquad \{VIII-4\}$$

we now have a value of -112.786 for E_{HF} and -0.063 for E_{Rel}. If E_{Cor} could now be determined we would have E_{Tot} as the sum of the three theoretical components. However it is easier to determine E_{Tot} from experimental data and then, by difference, to compute E_{Cor}.

$$E_{Cor} = E_{Tot} - (E_{HF} + E_{Rel}) \qquad \{VIII\text{-}5\}$$

The equation for determination of E_{Tot} is

$$E_{Tot} = E_{Atom} + E_{Diss} + E_{zpv}$$

where E_{Atom} is the sum of the experimental energy values of each of the constituent atoms, E_{Diss} is the dissociation energy, or more precisely the energy of atomization as estimated from the heats of formation and E_{zpv} is the molecular zero point vibrational energy as calculated from the fundamental vibrational frequencies,

$$E_{zpv} = \frac{1}{2} hc \sum_{i=1}^{3N-6} \nu_i \; (cm^{-1}) \qquad \{VIII\text{-}6\}$$

For CO, Table VIII-3 displays the experimental and theoretical partitioning of the total energy and Figure VIII-2 illustrates the interrelationships of the various terms.

TABLE VIII-3 A BREAKDOWN OF THE TOTAL ENERGY OF CO INTO EXPERIMENTAL AND THEORETICAL COMPONENTS

Experimental	E(Hartree)	Theoretical	E(Hartree)
Sum of Atomic Energies	-112.977	Hartree-Fock Energy	-112.786
Dissociation Energy	-0.413	Correlation Energy	-0.551
Zero Point Energy	-0.010	Relativistic Energy	-0.063
TOTAL ENERGY	-113.400		-113.400

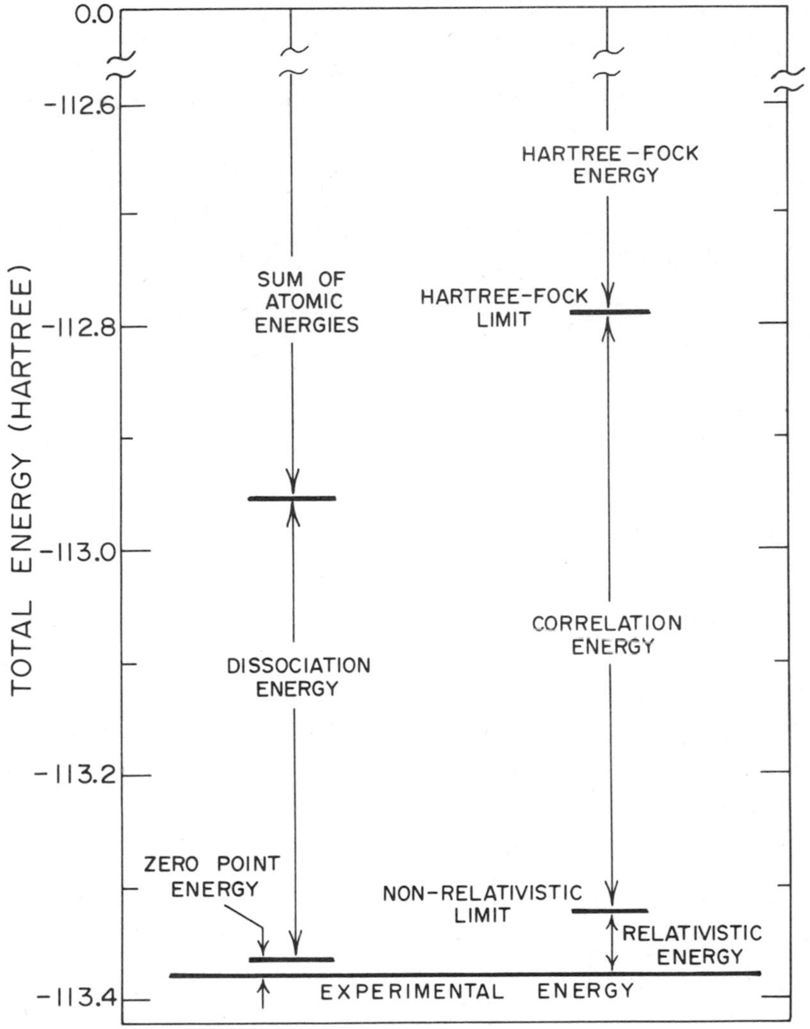

Figure VIII-2 Breakdown of total energy for C≡O to experimentally observable and quantum chemically calculable fractions

For molecules that contain more than two atoms computation of the HF energy is much more difficult and it is then customary to find an approximate value of the correlation energy (E_{Cor}) and with this and E_{Tot} to estimate the HFL.

An appropriate approximation to E_{Cor} treats this quantity as the sum of the atomic correlation (of Table VIII-1) plus an average value for each new electron pair (chemical bond)

$$E_{Cor}{}^{mol} = \sum_i (E_{Cor}{}^{Atom})_i + n \times \Delta E_{Cor}{}^{bond} \qquad \{VIII\text{-}7\}$$

the term $\Delta E_{Cor}{}^{bond}$ can be estimated by comparison with appropriate systems. This leads to different values for F-H, O-H, N-H, etc., and one then employs an average value. The value

$$E_{Cor}{}^{bond} = 0.065 \text{ hartree} \qquad \{VIII\text{-}8\}$$

may be employed with reasonable confidence. An alternative method derives the molecular correlation energy as the sum of correlation energies of appropriate fragments of the molecule. For example episulfide may be formed by addition of a sulphur atom to ethylene.

$$CH_2 = CH_2 + S \rightarrow CH_2 \overset{S}{-} CH_2 \qquad \{VIII\text{-}9\}$$

The E_{Cor} for episulfide would then be

$$E_{Cor}{}^{C_2H_4S} = E_{Cor}{}^{C_2H_4} + E_{Cor}{}^{S} + (2\, E_{Cor}{}^{\sigma\text{-bond}} - E_{Cor}{}^{\pi\text{-bond}})$$

$$= -0.534 - 0.640 - 0.065 = -1.239 \text{ hartree} \qquad \{VIII\text{-}10\}$$

With the aid of this approximated molecular correlation energy and the approximate molecular relativistic energy (cf.

equation {VIII-3}) the HF energy may be calculated from the experimental total energy (E_{Tot}):

$$E_{HF}^{Mol} \simeq E_{Tot}^{Mol} - (E_{Cor}^{Mol} + E_{Rel}^{Mol}) \quad \{VIII\text{-}11\}$$

$$E_{HF}^{C_2H_4S} \simeq -477.976 - (-1.239 - 1.077) \quad \{VIII\text{-}11a\}$$

These relationships together with computed molecular SCF energies are summarized in Figure VIII-3.

Once an approximate E_{HF} is found this value may be used as a "primary standard" to judge the accuracy of the various computed E_{SCF} values. The larger the AO basis the closer the E_{SCF} tends to E_{HF}

$$\lim_{N\to\infty} E_{SCF} = E_{HF} \quad \{VIII\text{-}12\}$$

The following figure (Figure VIII-4) illustrates this principle for LiH.

3. <u>Beyond the Hartree-Fock Limit</u>

Although the correlation energy (E_{corr}) usually amounts to less than 1% of the total experimental energy of a molecule, for many systems this amount nevertheless is larger than or comparable to the energy associated with physical or chemical phenomena. Thus for ammonia the correlation energy is estimated to be 206.5 Kcal/mole (0.329 hartree) but the barrier to pyramidal inversion is only 5.8 Kcal/mole and the proton affinity is 206.4 Kcal/mole. It is generally believed that these quantities are reproduced fairly accurately at the HFL, i.e. that there is a cancellation of errors. [For the two quantities mentioned this means that E_{Cor}(planar) $\approx$ E_{Cor}(pyramidal) and $E_{Cor}(NH_4^+) \approx E_{Cor}(NH_3)$.] For phenomena in which electrons are unpaired (e.g. excitation, dissociation) the correlation energy cannot be ignored and it is necessary to go beyond the HFL. A wavefunction that does this is

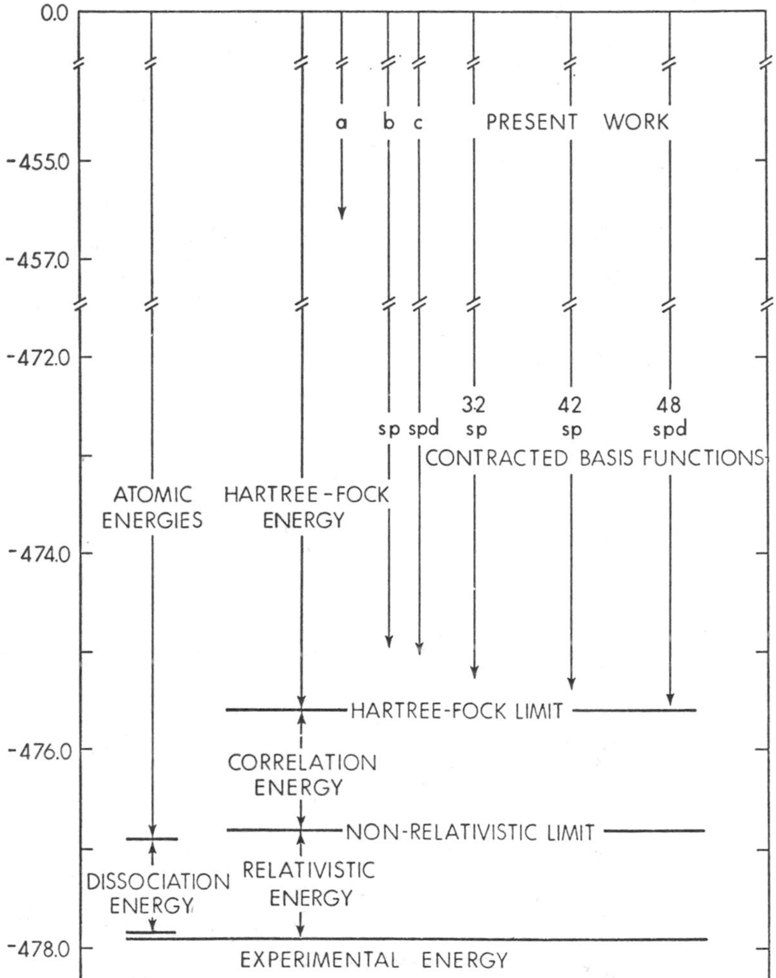

Figure VIII-3 A breakdown of total energy for episulfide (C_2H_4S) to experimentally observable and quantum chemically calculable fraction. The results obtained by the different atomic basis sets are shown on the upper right hand side of the figure.

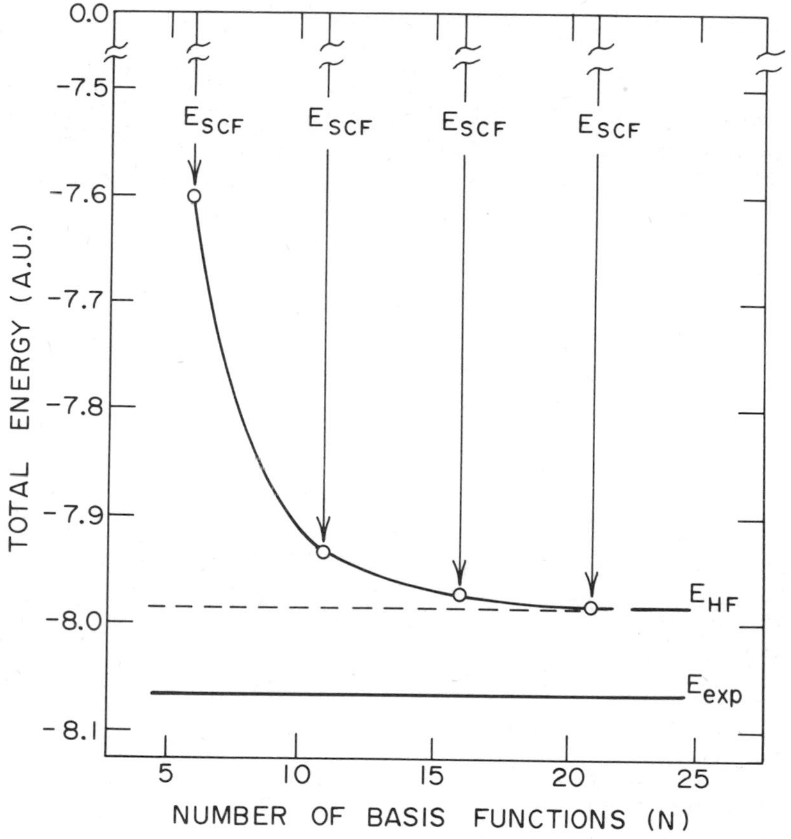

Figure VIII-4 The convergence of E_{SCF} to E_{HF} with increasing basis set size for the ground state of LiH. [cf. J. Chem. Phys. 44, 1849 (1966)]

referred to as a correlated wavefunction. Its aim is to recover at least a portion of this systematic error.

From a Mathematical standpoint the correlation energy is the result of a systematic error in the wavefunction computed by the HF method.

This error is the result of our construction of a 2M x 2M determinantal wavefunction.

$$\Phi_o(1,2 \ldots 2M) = \det \left| \phi_1(1)\alpha(1)\phi_1(2)\beta(2) \ldots \phi_M(2N-1)\alpha(2M-1)\phi_M(2M)\beta(2M) \right| \quad \{VIII\text{-}13\}$$

using only M m.o. from the very extensive set of N m.o. (N > M) obtained from the HF-SCF. This is illustrated in equation {VIII-14} where $\{\eta\}$ is the set of AO and represents the set of MO in column vector notation

$$\begin{pmatrix} \phi_1 \\ \phi_2 \\ \vdots \\ \phi_M \\ \hline \vdots \\ \phi_N \end{pmatrix} = \begin{pmatrix} a_{11} & a_{12} & \cdots & a_{1N} \\ a_{21} & a_{22} & \cdots & a_{2N} \\ \vdots & \vdots & & \vdots \\ a_{M1} & a_{M2} & \cdots & a_{MN} \\ \hline \vdots & \vdots & & \vdots \\ a_{N1} & a_{N2} & \cdots & a_{NN} \end{pmatrix} \begin{pmatrix} \eta_1 \\ \eta_2 \\ \vdots \\ \eta_N \end{pmatrix} \quad \{VIII\text{-}14\}$$

Even if $\{\eta\}$ had represented a complete set [i.e. $\{\phi\}$ also represents a complete set] the Slater determinant is computed from much less than the complete set of $\{\phi\}$. For example we might have employed 50 basis functions (N = 50) for a calculation on NH_3 even though there are only 5 occupied MO (N = 5). The remaining 45 empty MO would be discarded (cf. the partitioning in equation {VIII-14} in constructing the Slater determinant.

From a physical standpoint the correlation problem is related to the position of one electron with respect to any other.

The non-relativistic Hamiltonian includes the reciprocal values of all r_{ij},

$$H(1,2, \ldots) = \sum_{i=1}^{\text{all electrons}} h(i) + \sum_{i,j}^{\text{all electron pairs}}{}' \frac{1}{r_{ij}} \qquad \{VIII-15\}$$

the terms in r_{ij}^{-1} constituting the mutual repulsion between electrons i and j.

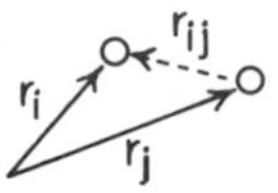

Figure VIII-5 Vector model of electron-electron repulsion

Since $r_{ij}^{-1} \to \infty$ as $r_{ij} \to 0$ approach of two electrons to each other is not favorable and each is surrounded by a "Coulomb hole". As r_i approaches r_j (i.e. $r_{ij} \to 0$ as shown in Figure VIII-5) the motion of the two electrons becomes "correlated".

Since the Pauli principle has been incorporated in the wavefunction in the form of an antisymmetrized orbital product (Slater determinant):

$$\Phi_o(1,2 \ldots) = \mathrm{Det}|\overbrace{\phi_1(1)\alpha(1)}^{u_1(1)} \quad \overbrace{\phi_1(2)\beta(2)}^{u_2(2)} \ldots \quad = A[u_1(1)u_2(2) \ldots] \qquad \{VIII-16\}$$

(where $u_i(i)$ represents for the ith spin-orbital), the probability of finding two electrons at the same position is zero or in other words the 2-electron probability density vanishes for $r_i = r_j$ in the case of two electrons having the same spin ("Fermi hole"). For this reason the correlation problem can be thought of as the search for a proper treatment of the "Coulomb" correlation between electrons having antiparallel spin. The correlation between electrons of antiparallel spin may involve numerous interactions. This is illustrated for 2 and 3 pairs of electrons in Figure VIII-6.

One way to deal with the correlation problem is to include a function of r_{12} explicitly in the total wavefunction which will not permit r_{12} to become zero. Thus instead of equation {VIII-16} we may write the following

$$\Psi(1,2) = \hat{A}[u_1(1)u_2(2)]\ f(1,2) \qquad \{VIII\text{-}17\}$$

For systems that contain more than 2 electrons this explicit correlated wavefunction is not practical and the correlation problem is usually tackled by some implicit method.

The most general "implicit" method (where r_{ij} is not included explicitly in the wavefunction) is known as the method of configuration interaction (CI).

The previously unused (virtual) molecular orbitals are substituted into the expression for Φ_o (this is possible because of the nodal properties of these virtual MO which is a consequence of their orthogonality to the occupied MO)

$$\Phi_o = [u_1(1)\ ..\ u_i(i)\ ..\ u_j(j)\ ..\ u_k(k)\ ..\ u_N(N)] \qquad \{VIII\text{-}18\}$$

Substitution may take place at 1, 2 or more sites. Thus, for a triple substitution, the wavefunction associated with the new configuration would be

$$\Phi_u(1,2,\ ...,\ N) \equiv \Phi^{...a...b...c}_{...i...j...k} = \hat{A}[u_1(1)\ ..\ u_a(i)\ ..\ u_b(j)\ ..\ u_c(k)\ ...\ u_N(N)] \qquad \{VIII\text{-}19\}$$

where spin-orbitals a, b, c have replaced spin-orbitals i, j, k.

The exact wavefunction is then written as the linear combination of these different substituted configurations.

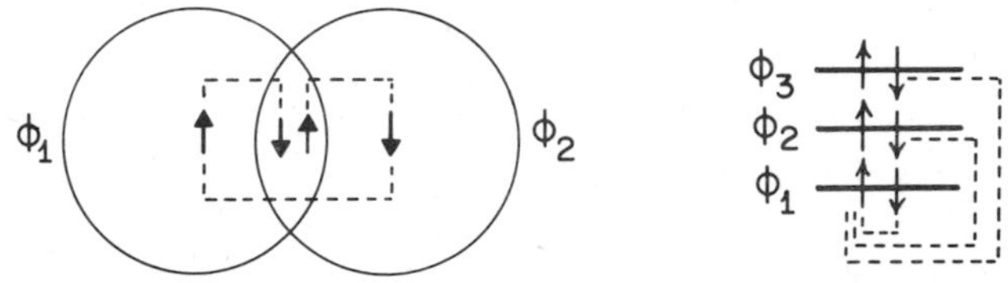

Figure VIII-6 Schematic representations of the electron correlation problem

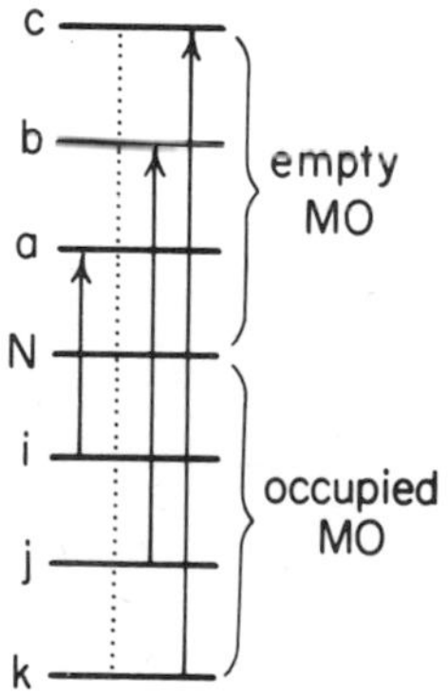

Figure VIII-7 Schematic representation of substitution (excitation) for the generation of configurational wavefunctions

$$\Psi^{CI} = C_o \, \Phi_o^{SCF} + \sum_{i,a} C_i^a \Phi_i^a + \sum_{ij,ab} C_{ij}^{ab} \Phi_{ij}^{ab} + \ldots \qquad \{VIII\text{-}20\}$$

where the set of many electron functions $\{\Phi\}$ represent configurations. The general approach will be illustrated for the case of H_2 (a 2 orbital-2 electron problem).

The ground electronic configuration is represented by the single Slater determinant $\Phi(1,2)$.

$$\Phi_o(1,2) = \frac{1}{\sqrt{2}} \begin{vmatrix} \phi_1(1)\alpha(1) & \phi_1(1)\beta(1) \\ \phi_1(2)\alpha(2) & \phi_1(2)\beta(2) \end{vmatrix} \quad \text{i.e.} \qquad \{VIII\text{-}21\}$$

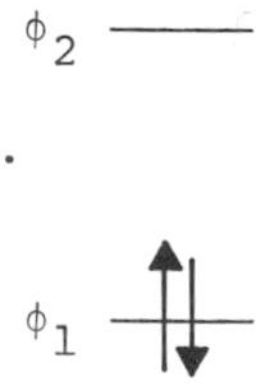

which incorporates only the lowest MO(ϕ_1).

Single substitution involves the replacement of ϕ_1 in one of the 2 columns by ϕ_2 but, for a closed electronic shell each MO occurs twice so that there are two possibilities for replacement. These involve substitution in the column containing α spin and in the column having the β spin associated with the same spatial function

{VIII-22}

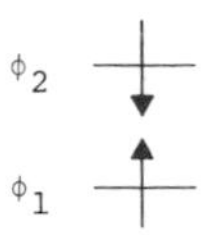

i.e.

$$\begin{vmatrix} \phi_1(1)\alpha(1) & \phi_2(1)\beta(1) \\ \phi_1(2)\alpha(2) & \phi_2(2)\beta(2) \end{vmatrix} \quad \text{and} \quad \begin{vmatrix} \phi_2(1)\alpha(1) & \phi_1(1)\beta(1) \\ \phi_2(2)\alpha(2) & \phi_1(2)\beta(2) \end{vmatrix}$$

i.e.

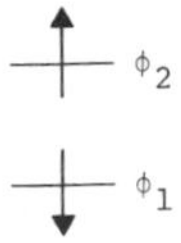

Neither of these determinants has physical significance but with a linear combination of the two both types of spin arrangement may be equally important. The linear combinations having positive and negative signs yield wavefunctions of triplet and singlet multiplicity respectively.

$$\Phi_1(1,2) = \frac{1}{\sqrt{2}}\left\{\frac{1}{\sqrt{2}}\begin{vmatrix} \phi_1(1)\alpha(1) & \phi_2(1)\beta(1) \\ \phi_1(2)\alpha(2) & \phi_2(2)\beta(2) \end{vmatrix} - \frac{1}{\sqrt{2}}\begin{vmatrix} \phi_2(1)\alpha(1) & \phi_1(1)\beta(1) \\ \phi_2(2)\alpha(2) & \phi_1(2)\beta(2) \end{vmatrix}\right\}$$

{VIII-23}

The three configurations Φ_0, Φ_1, Φ_2 are now treated as the many electron basis for the calculation of a state wave-function (just as atomic orbitals constituted the one-electron basis functions for the calculation of MO). For example, for the ground electronic state Ψ_o we may write

$$\Psi_o(1,2) = C_o\ \Phi_o(1,2) + C_{10}\Phi_1(1,2) + C_{20}\Phi_2(1,2) \qquad \{VIII-24\}$$

However, since one may obtain as many as 3 state functions from three basis functions, we write

$$(\Psi_o(1,2)\Psi_1(1,2)\Psi_2(1,2)) = (\phi_o(1,2)\phi_1(1,2)\phi_2(1,2)) \begin{pmatrix} c_{00} & c_{01} & c_{02} \\ c_{10} & c_{11} & c_{12} \\ c_{20} & c_{21} & c_{22} \end{pmatrix}$$

{VIII-25}

This linear transformation can be viewed as the rotation of a 3 dimensional vector space into another vector space, as illustrated graphically in the next figure.

Computation of the coefficient matrix $\underline{\underline{C}}$ constitutes an eigen-value problem

$$\underline{\underline{C}}^{\dagger}\ \underline{\underline{H}}\ \underline{\underline{C}} = \underline{\underline{C}}^{\dagger}\ \underline{\underline{S}}\ \underline{\underline{C}}\ \underline{\underline{E}} \qquad \{VIII-26\}$$

in which $\underline{\underline{H}}$ is the matrix representative of the many-electron Hamiltonian H(1,2, ...) over the configurational basis chosen (including single, double, ... etc. substitution).

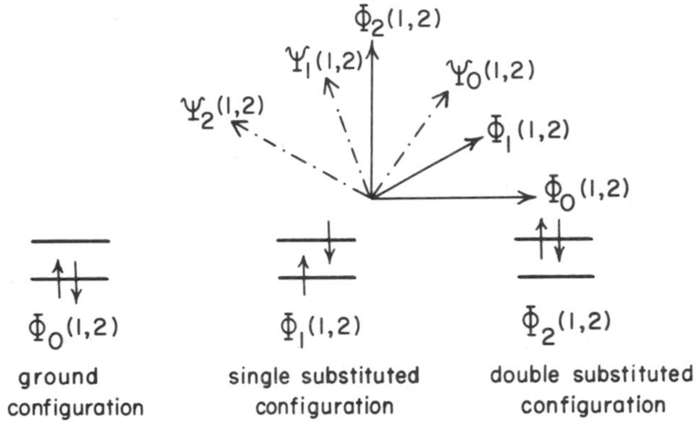

Figure VIII-8 A vector model depicting the linear transformation of a 3D configurational vector space (configurations → states) for the case of H_2

$$\underline{\underline{H}} = \left| \begin{array}{ccc} \overbrace{\langle \Phi_o|\hat{H}|\Phi_o\rangle}^{\text{H.F.}} & \overbrace{\langle \Phi_o|\hat{H}|\Phi_o\rangle \;\ldots\ldots}^{\text{single substitution}} & \overbrace{\langle \Phi_o|\hat{H}|\Phi_{ij}^{ab}\rangle \;\ldots\ldots}^{\text{double substitution}} \\ & \langle \Phi_i^{a}|H|\Phi_i^{a}\rangle \;\ldots & \langle \Phi_i^{a}|\hat{H}|\Phi_{ij}^{ab}\rangle \;\ldots \\ & & \langle \Phi_{ij}^{ab}|\hat{H}|\Phi_{ij}^{ab}\rangle .. \end{array} \right| \quad \{VIII\text{-}27\}$$

It is readily appreciated that the size of the basis { (1,2,3, ...)}, i.e. the number of configurations that can be constructed, increases greatly as the number of available orbitals increases. For example a full CI (including all possible substitutions for 10 electrons and 10 MO involves the order of 10^5 configurations. This means that the $\underline{\underline{H}}$ matrix has a dimension 100,000 x 100,000 and it is not possible to diagonalize such a matrix. It is evident that CI calculations cannot be performed routinely and are usually performed only in a limited way.

An example of the use of CI in a conformational problem is the calculation of barrier heights discussed below. The orbitals used for CH_3^- and NH_3 are shown schematically in Figure VIII-9. These MO yielded a total of 5260 configurations. From this set 911 were selected by perturbation theory. A CI calculation performed over these 911 configurations, lowered the total energy appreciably, and recovered about half of the correlation energy (cf. Figure VIII-9). However, it was important to select the best 911 configurations (i.e. those that contribute most) at every geometry (solid curve) because if the same 911 configurations were used at each geometry the barrier became anomalously high (broken curve) as indicated in Figure VIII-10. The overall energy change is shown in Figure VIII-11, together with some computed results which are also summarized in Table VIII-4.

In closing it might be appropriate to take this opportunity in order to clarify that most of the chemist's language is based on a HF scheme (i.e. double occupancy, distribution of electrons in molecular orbitals, excitation of electrons by "jumping" from one orbital to a different one, atomic configurations like $1s^2$, $2s^2$, $2p^6$, $3s^2$, $3p^5$ for chlorine, etc.).

The language "works" because the HF wavefunction is a very good approximation (99.5%) but it is not everything. All that chemical language vanishes and means nothing when a more sophisticated wavefunction is used. It is like a "Law of Complementarity": The simpler the wavefunction, the easier to interpret and explain qualitatively the conclusions; on the other hand the better the wavefunction, the more it becomes pure mathematical entity and it is more difficult to talk about in a lecture room.

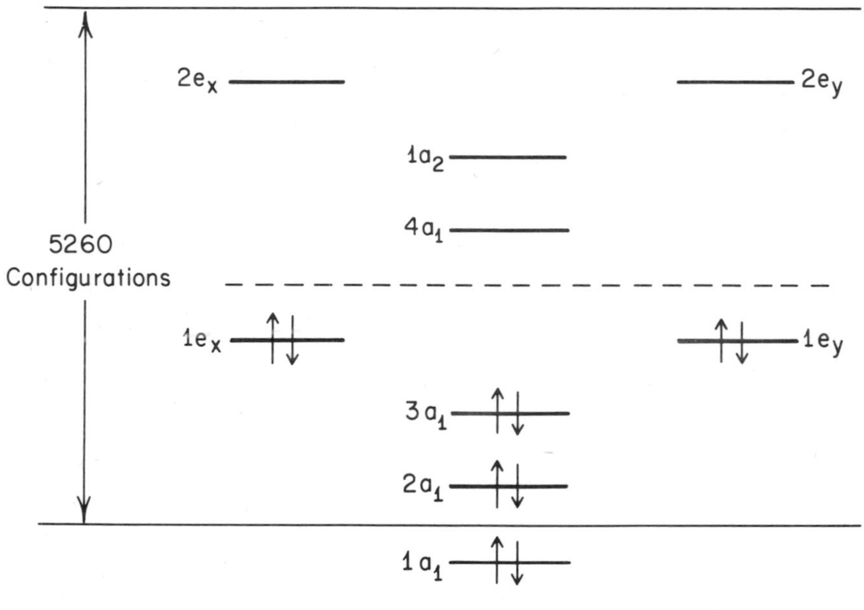

Figure VIII-9 Molecular orbitals involved in the CI calculations of CH_3^- and NH_3

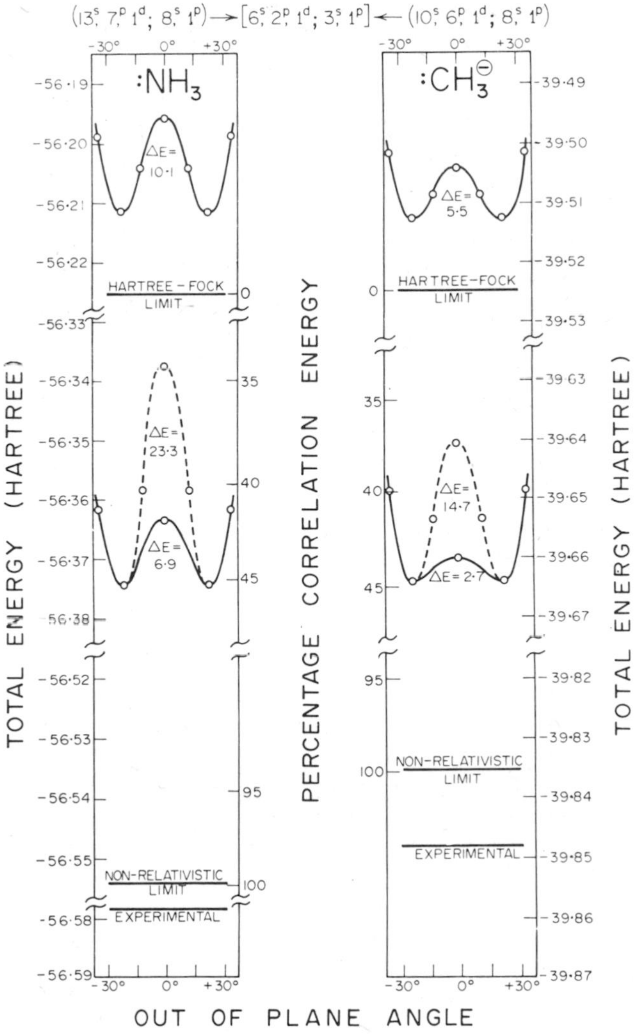

Figure VIII-10 The variation of CH_3^- and NH_3 total energies along the inversion coordinate as computed from SCI and CI wavefunctions

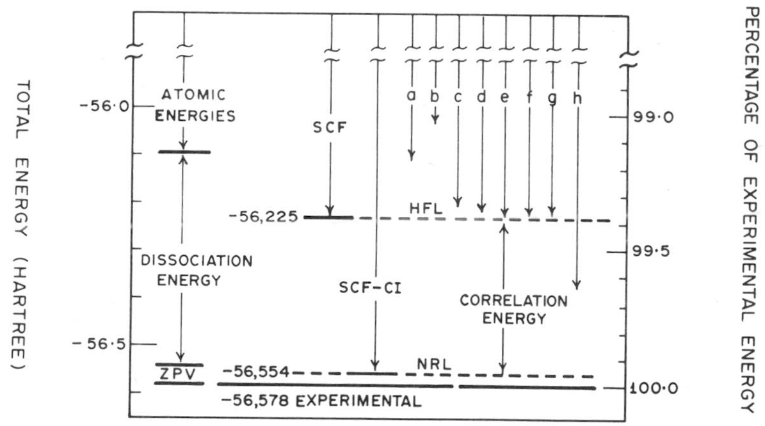

Figure VIII-11 A breakdown of the total energy for NH_3 to experimentally observable and quantum chemically calculable fractions. The computed energy values (a-h) are summarized in Table VIII-4.

TABLE VIII-4

A COMPARISON OF SCF AND CI ENERGIES OF NH_3

Author	Code for Figure 1	Reference in J. Chem. Phys.	Basis[a]	Method[b]	E(Hartree)
Kaldor *et al.*	a	45, 888 (1966)	13 STO	SCF	-56.0992
Clementi	b	46, 3851 (1967)	53 GTF	SCF	-56.0108
Ritchie *et al.*	c	47, 564 (1967)	60 GTF	SCF	-56.2015
Clementi *et al.*	d	49, 4916 (1968)	67 GTF	SCF	-56.2109
Rauk *et al.*	e	52, 4133 (1970)	91 GTF	SCF	-56.2219
Stevens	f	55, 1729 (1971)	37 STO	SCF	-56.2211
Kari *et al.*	g	56, 4337 (1972)	73 GTF	SCF	-56.2117
				CI	-56.3747

[a]STO Slater type orbitals; GTF Gaussian type functions

[b]SCF Self-Consistent Field; CI Configuration interaction (911 configurations)

CHAPTER IX

APPLICATIONS OF MO THEORY TO CLOSED SHELL PROBLEMS

1. Primary Molecular Properties as Observables

All of those observables that are computed as the expectation values of an operator are termed primary properties. The application of MO theory or in general Quantum Mechanics requires the computation of observables from the many electron wavefunction Φ_ν for a given state ν

$$\Phi_\nu = \Phi_\nu(1,2,3,...) \qquad \{IX-1\}$$

where 1,2,3,... symbolize the coordinates of electrons 1,2,3,... The observable (Ω) is written as the expectation value of the appropriate operator ($\hat{\Omega}$) that describes the physical quantity in question over the many electron wavefunction

$$\Omega_\nu \equiv \langle\Phi_\nu(1,2,3,...)|\hat{\Omega}_{1,2,3,...}|\Phi_\nu(1,2,3,...)\rangle \qquad \{IX-2\}$$

In MO theory, the wavefunction is a single Slater determinant if we are dealing with a closed electronic shell (singlet) ground state, radicals or molecular ions (doublet states) and even for a triplet (ground or excited) state. For singlet excited states, however, one needs at least a pair of Slater determinants as discussed in the previous Chapter. Thus, for all our purposes the wavefunction is an antisymmetrized orbital product.

As far as the operator $\hat{\Omega}_{1,2,3,...}$ is concerned it may be written as a sum of operators: The first term $\hat{\Omega}_o$ (a no electron operator) operates on the nuclear wavefunction. The other two terms

$$\hat{\Omega} = \hat{\Omega}_o + \hat{\Omega}_i + \hat{\Omega}_{ij} \qquad \{IX-3\}$$

act on the electronic wavefunction. These electronic operators ($\hat{\Omega}_i$ and $\hat{\Omega}_{ij}$) may be classified as one electron operators, i.e. operators that are dependent on the coordinates of a single electron ($\hat{\Omega}_i$) and two electron operators that are dependent on the coordinates of two electrons simultaneously ($\hat{\Omega}_{ij}$).

All quantum chemical operators may be subdivided into these two major classes. One of the two classes includes all three terms of the above equation and the corresponding quantity is not trivial to compute because the two electron terms involve calculation of a great many two electron integrals of the type

$$\langle \phi_p(1)\phi_q(2)|\hat{\Omega}_{12}|\phi_r(1)\phi_s(2)\rangle \qquad \{IX-4\}$$

where the $\{\phi_n\}$ are the individual molecular orbitals. However, these operators are very important because a number of physical properties (observables) are described by such operators. Most important is the Hamiltonian operator (which includes a two electron operator ($\hat{H}_{ij} = 1/r_{ij}$)) responsible for the computation of molecular energies

$$E = \langle \Phi|\hat{H}_o + \sum_i \hat{H}_i + \sum_{ij}{}' \hat{H}_{ij}|\Phi\rangle =$$

$$= \langle \Phi|\hat{H}_o|\Phi\rangle + \sum_i \langle \Phi|\hat{H}_i|\Phi\rangle + \sum_{ij}{}' \langle \Phi|\hat{H}_{ij}|\Phi\rangle = \qquad \{IX-5\}$$

$$= E_o + \sum_i E_i + \sum_{ij}{}' E_{ij}$$

The first term (E_o) is the nuclear-nuclear repulsion (a no electron property), the second term corresponds to the one electron energy (kinetic + nuclear-electron attraction) and the last term represents the two electron energy (electron-electron repulsion). Other molecular properties that involve two electron operators are also important (e.g. molecular polarizability).

The other class of quantum chemical operators include only the first two terms in equation {IX-3} and we refer to the properties (P) that we may compute with the aid of such operators as one electron properties:

$$P = \langle \Phi | \hat{P}_o + \sum_i \hat{P}_i | \Phi \rangle$$

$$= \langle \Phi | \hat{P}_o | \Phi \rangle + \sum_i \langle \Phi | \hat{P}_i | \Phi \rangle \qquad \{IX-6\}$$

$$= P_o + \sum_i P_i$$

The individual P_i in the above sum have the general form

$$\langle \phi_q(1) | \hat{P}_1 | \phi_r(1) \rangle$$

where $\{\phi_n\}$ are again the individual molecular orbitals and $\hat{P}_1$ is a special case of $\hat{\Omega}_i$.

Many properties fall in this category but among the most noteworthy is the dipole moment μ. The dipole moment operator is the most popular among computational quantum chemists because the accuracy of the computed μ is very sensitive to the quality of the wavefunction. It also has great utility in molecular spectroscopy in connection with the selection rules for the electric dipole allowed electronic transitions.

The computation of the dipole moment value (μ^{oo}) for the electronic ground state (Φ_o) may be written as:

$$\mu^{oo} = \langle \Phi_o(1,2,3,\ldots) | \hat{\mu}_o + \sum_i \hat{\mu}_i | \Phi_o(1,2,3,\ldots) \rangle \qquad \{IX-7\}$$

$$= \langle \Phi_o(1,2,3,\ldots) | \hat{\mu}_o | \Phi_o(1,2,3,\ldots) \rangle + \sum_i \langle \Phi_o(1,2,3,\ldots) | \hat{\mu}_i | \Phi_o(1,2,3,\ldots) \rangle$$

$$= \mu_o^{oo} + \sum_i \mu_i^{oo}$$

The dipole moment for the ν-th excited state ($\mu^{\nu\nu}$) may be similarly defined

$$\mu^{\nu\nu} = \langle\Phi_\nu(1,2,3,\ldots)|\hat{\mu}_o + \sum_i \hat{\mu}_i|\Phi_\nu(1,2,3,\ldots)\rangle$$

$$= \mu_o^{\nu\nu} + \sum_i \mu_i^{\nu\nu} \qquad \{IX\text{-}8\}$$

The transition dipole moment, $\mu^{o\nu}$, of the molecule between the ground and excited state requires the computation of the following integral:

$$\mu^{o\nu} = \langle\Phi_o(1,2,3,\ldots)|\hat{\mu}_o + \sum_i \hat{\mu}_i|\Phi_\nu(1,2,3,\ldots)\rangle$$

$$= \mu_o^{o\nu} + \sum_i \mu_i^{o\nu} \qquad \{IX\text{-}9\}$$

Other one electron operators were used in the definition of "sizes" and "shapes" earlier (cf. Chapter VII) and fall in this category together with higher moments (quadrupole, octupole, etc.) as well as others (e.g. diamagnetic susceptibility, etc.) to be discussed later.

2. Definition of Closed Shell Problems

Before making an attempt to discuss the application of the closed shell Hartree-Fock (i.e. SCF) method to organic chemistry we have to make a clear definition of just what constitutes a closed shell problem and what types of chemical systems will be classified as open shell problems. A necessary but not sufficient condition for a molecule to be classified as a closed shell is that it has an even number of electrons. In other words compounds with an odd number of electrons are, by definition, open shell systems such as CH_3 and NO_2 illustrated in Figure IX-1. However, compounds with even numbers of electrons may be classified as either a closed or an open shell system as illustrated for some selected homonuclear

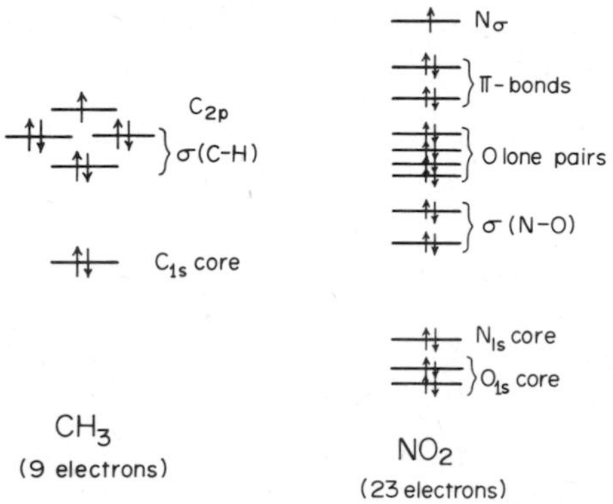

Figure IX-1. Open shell systems with an odd number of electrons

diatomics in Figure IX-2. Nevertheless, organic molecules very often have even numbers of electrons and represent closed shell systems. Some organic molecules with an even number of electrons are also suspected to be of triplet multiplicity in their lowest (i.e. ground) electronic state but this can

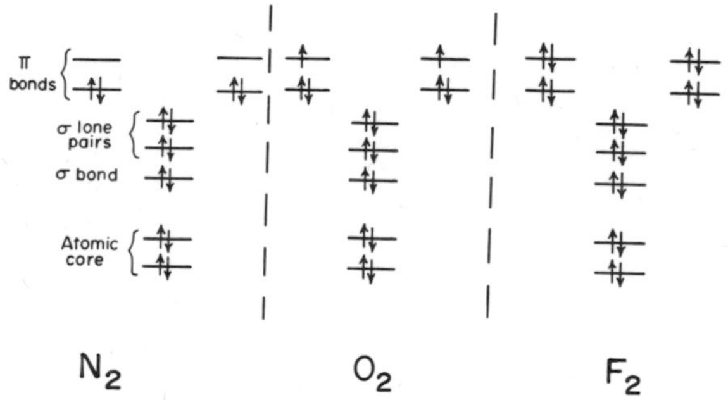

Figure IX-2. Selected homonuclear diatomic molecules. Open and closed shells.

occur only if in the most favoured conformation a complete electron pairing is impossible as may be the case for $H_2C = SiH_2$ {IX-10}. This particular problem will be discussed in

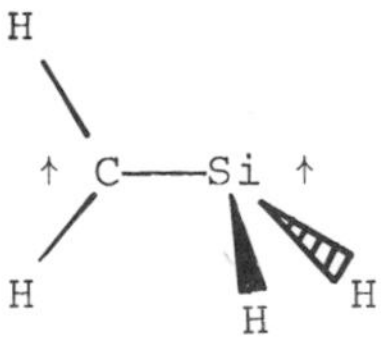

{IX-10}

some detail in Chapter XI.

However, apart from a few exceptions most organic molecules fall into the category of closed shell systems. This means that during a conformational change or unimolecular reaction, a closed shell chemical system will retain its identity as a closed electronic shell insofar as we do not attempt to unpair electrons in the course of stereochemical or reactive change. Of course, only heterolytic bond cleavages qualify for this classification. A typical example of homolytic bond breaking (i.e. electron unpairing) is the cis-trans isomerization of olefins that proceeds, through a triplet state:

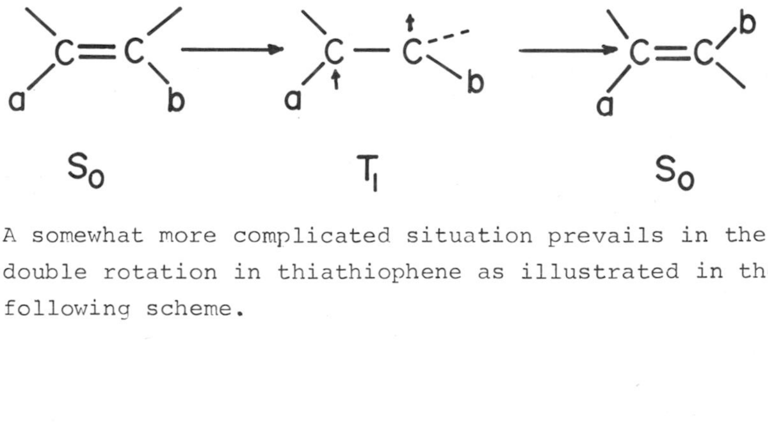

{IX-11}

A somewhat more complicated situation prevails in the case of double rotation in thiathiophene as illustrated in the following scheme.

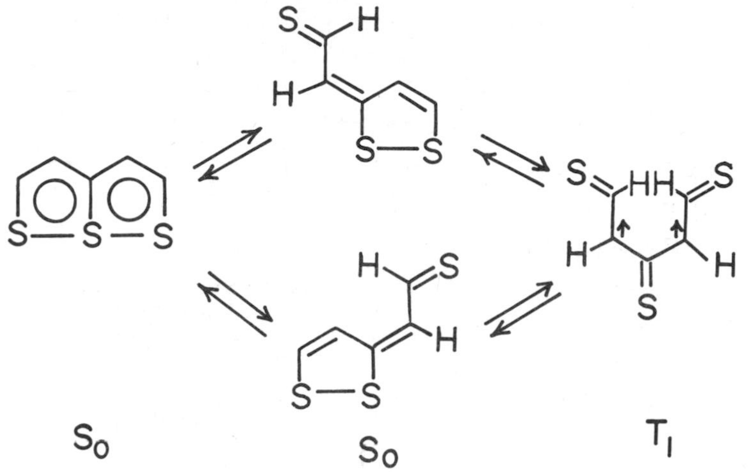

{IX-12}

However, in contrast to the limited number of situations exemplified above most conformational changes and unimolecular reactions proceed with the conservation of the number of electron pairs and hence represent closed shell problems.

On the other hand electronic (VIS and UV) excitation always involve the unpairing of at least one electron pair consequently electronic excitation and photochemistry as a whole represent open shell problems. In addition, photoelectron spectroscopy (ESCA) that involves the generation of molecular ions (radical cations) is also an open shell problem since it involves a transition from the ground singlet state (S_o) to the ground or one of the excited states of the molecular ions of doublet multiplicity (D_i^{+}) (cf. Figure IX-3).

For analysing bimolecular reactions we must consider both reagent and reactant. If both of them are in ground states of singlet multiplicity then again we have a closed shell problem.

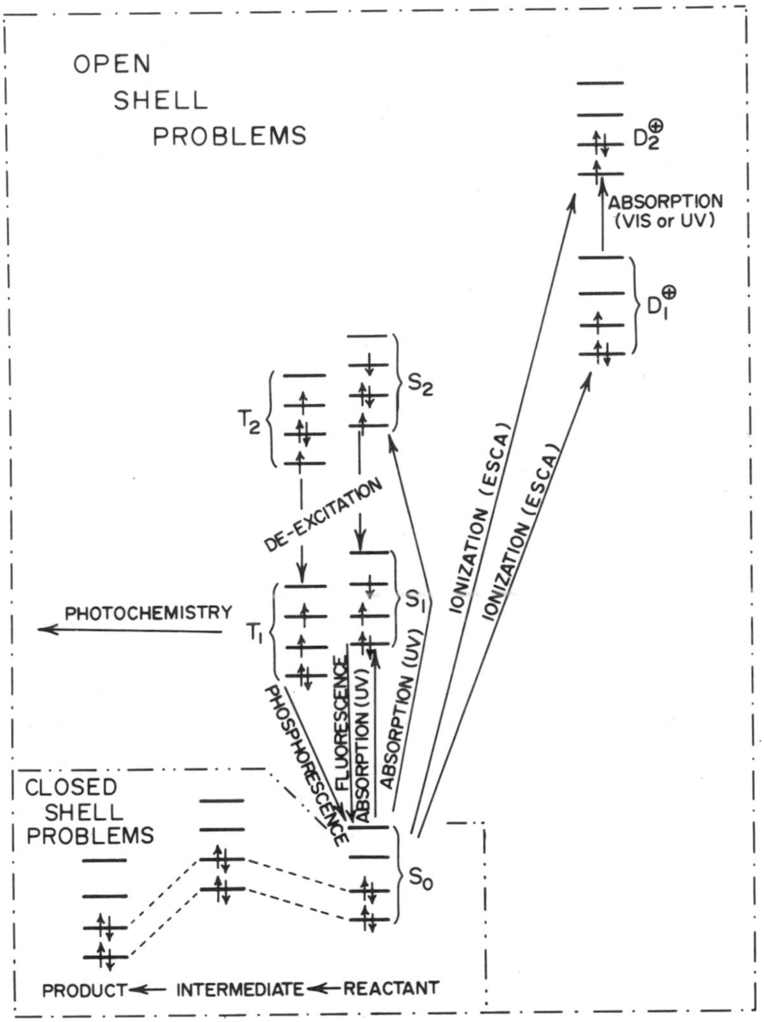

Figure IX-3. Open and closed shell physical and chemical processes.

3. Geometry Optimization

Although the geometries of most stable molecules are well known from experimental (spectroscopic) studies to obtain consistent theoretical results it is customary to carry out geometry optimizations. The results obtained are basis set dependent and only in the limit of a very large basis set is the experimental geometry expected to match the calculated geometry. The calculation of molecular energies with small or limited basis sets at experimental geometries is certain to yield energies higher than the minimum attainable with the given basis set. Thus, the procedure of full geometry optimization should, in principle, be carried out for all molecular calculations. Let us consider briefly what such a procedure involves.

To every geometry, defined by a set of internal coordinates $\{x_i\}$, there is an associated energy value E.

$$E = E(x_1, x_2, \ldots, x_{3N-6}) \qquad \{IX\text{-}13\}$$

For the optimum geometry $\{x_i^{opt}\}$ there is an optimum (i.e. minimum) energy value.

$$E^{opt} = E(x_1^{opt}, x_2^{opt}, \ldots, x_{3N-6}^{opt}) \qquad \{IX\text{-}14\}$$

The search for the optimum geometry, i.e. the choice of $\{x_i^{opt}\}$ from an infinite number of sets of $\{x_i\}$, is termed geometry optimization.

The search for an optimum geometry implies an <u>a priori</u> knowledge about the functional dependence of energy E on the independent internal coordinates $\{x_i\}$. Considering a one-dimensional problem, i.e. a diatomic molecule, the energy depends on one internal parameter only, the bond length of the diatomic molecule: $x_1 = r$. The stretching potential

$$E = E(r) \qquad \{IX\text{-}15\}$$

can be approximated by a quadratic (i.e. parabolic) function, close enough to the minimum:

$$E = E^{opt} + a(r - r^{opt})^2 \qquad \{IX-16\}$$

The key expression in the above statement is "close enough" because the real function {IX-15} and the approximate quadratic function {IX-16} osculate only at the minimum and only nearly osculate "close enough" to the minimum. The deviation between the two functions becomes apparent when we look at the regions not "close enough" to the minimum as illustrated by Figure IX-4. The constant "a" in equation {IX-16} is related

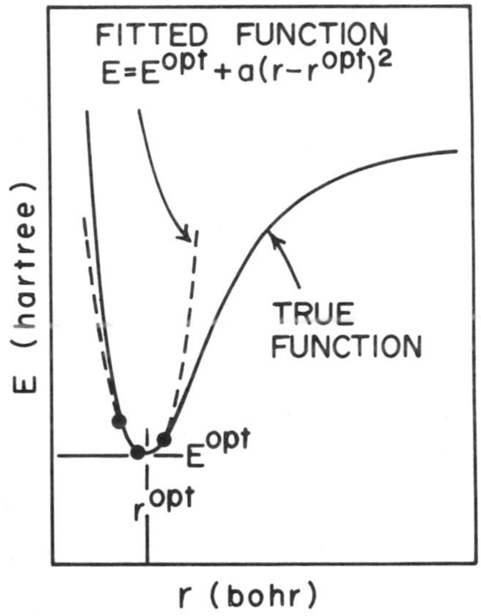

Figure IX-4. The osculation of a fitted quadratic function and the true function at the minimum. (The three solid dots close to the minimum are the computed points.)

to the harmonic force constant "k" according to the following equation:

$$\frac{d^2E}{dr^2} = 2a = k \qquad \{IX-17\}$$

For problems involving more than one dimension we have to generalize the above procedure which means that we have to express E as a quadratic function of a series of 3N-6 variables measured from the minimum:

$$(x_1-x_1^{opt}),(x_2-x_2^{opt}),\ldots,(x_{3N-6}-x_{3N-6}^{opt}). \qquad \{IX-18\}$$

If we denote the vector shown in {IX-18} as $(x-x^{opt})$ we may construct equation {IX-19} which is the generalized form of equation {IX-16}

$$E = E^{opt} + (x-x^{opt})\,\underline{\underline{A}}\,(x-x^{opt})^{\dagger} \qquad \{IX-19\}$$

The matrix elements of $\underline{\underline{A}}$ are also related to

$$\underline{\underline{A}} = \begin{pmatrix} a_{11} & a_{12} & a_{13} & \cdots\cdots \\ & a_{22} & a_{23} & \cdots\cdots \\ & & a_{33} & \\ & & & \ddots \end{pmatrix} \qquad \{IX-20\}$$

force constants. Each of the diagonal elements is associated with the force constant of one mode only while the off-diagonal

$$\frac{\partial^2 E}{\partial(x_i-x_i^{opt})^2} = 2a_{ii} = k_{ii} \qquad \{IX-21\}$$

elements are related in an analogous fashion to interaction force constants:

$$\frac{\partial^2 E}{\partial(x_i-x_i^{opt})\,\partial(x_j-x_j^{opt})} = 2a_{ij} = k_{ij}. \qquad \{IX-22\}$$

To fit equation {IX-19} to the true hypersurface {IX-13} one needs to determine all variable parameters in {IX-19}. This means E^{opt} and all the different elements of matrix $\underline{\underline{A}}$ have to be determined in the fitting procedure, i.e. during the geometry optimization.

Since $\underline{\underline{A}}$ is a real symmetric matrix $a_{ij} = a_{ji}$, therefore, only elements of a hemi-matrix, say the upper triangular of $\underline{\underline{A}}$, need to be determined. This requires $(1/2)(3N-6)[(3N-6)+1] = (1/2)(3N-6)^2 + (1/2)(3N-6)$ elements of $\underline{\underline{A}}$, considerably less than $(3N-6)^2$. For a fit, therefore, one needs to compute a minimum of $1 + (1/2)(3N-6)^2 + (1/2)(3N-6)$ points on the energy hypersurface "close enough" to the minimum. The accuracy of the fit will in fact depend on the closeness of the computed points to the unknown minimum. If one has more than the minimum number of points then, of course, a least squares fit is possible. However, the minimum number of points required for an ordinary fit is normally already too large.

The process of geometry optimization may be illustrated for CH_2. For CH_2, or any other triatomic molecule of AX_2 formula, there are only two geometrical parameters to be optimized, the bond length r (both AX bonds assumed equal) and the bond angle α

$$E = E(r,\alpha) \qquad \{IX-23\}$$

The surface that is associated with the above equation is shown schematically in Figure IX-5. The quadratic equation to be fitted to the above surface may be constructed according to equation {IX-19}

$$E = E^{opt} + \left((r-r^{opt}),\ (\alpha-\alpha^{opt})\right) \begin{pmatrix} a_{11} & a_{12} \\ a_{21} & a_{22} \end{pmatrix} \begin{pmatrix} (r-r^{opt}) \\ (\alpha-\alpha^{opt}) \end{pmatrix} \qquad \{IX-24\}$$

where $a_{21} = a_{12}$ and a_{11}, a_{22} as well as a_{12} are related to the stretching, bending and interaction force constants as discussed earlier:

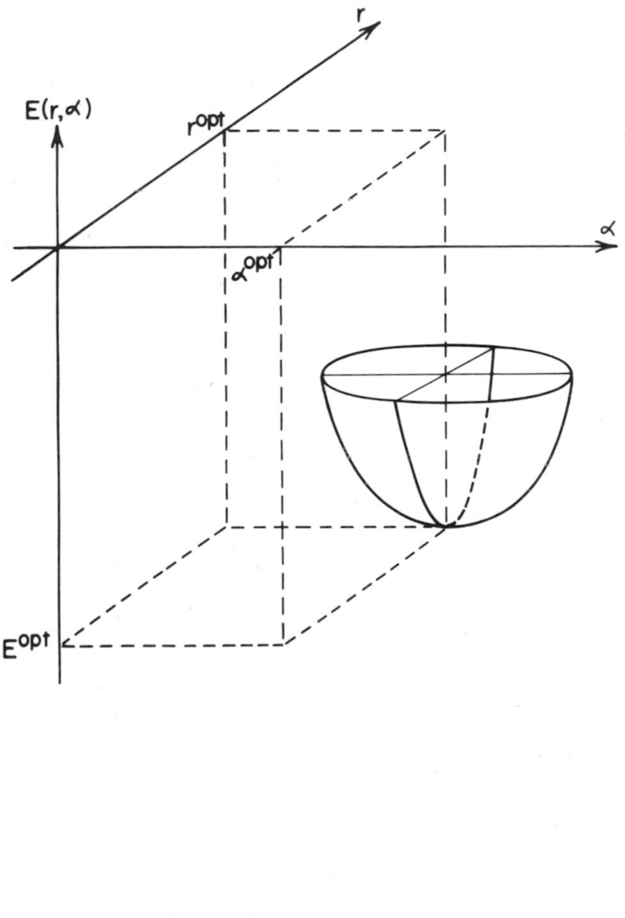

Figure IX-5. Geometry optimization for a triatomic molecule, AX_2.

$$a_{11} = \frac{1}{2} k_{rr}$$

$$a_{22} = \frac{1}{2} k_{\alpha\alpha} \qquad \{IX-25\}$$

$$a_{12} = \frac{1}{2} k_{r\alpha}.$$

In order to determine E^{opt}, r^{opt} and α^{opt} we have to fit the above equation to the computed points. Since there are six parameters (E^{opt}, r^{opt}, α^{opt}, a_{11}, a_{22} and a_{12}) we have to have six energy values (E_1, E_2, E_3, E_4, E_5 and E_6) computed at six different geometries [(r_1,α_1), (r_2,α_2), (r_3,α_3), (r_4,α_4), (r_5,α_5) and (r_6,α_6)].

Because of the great many points needed to fit a hypersurface of the type {IX-19} it is customary to approximate the off-diagonal matrix elements of $\underline{\underline{A}}$. The simplest such approximation would be setting $k_{ij} = 0$ (i.e. $a_{ij} = 0$) for $i \neq j$. The tenuous justification of this approximation is the fact that interaction force constants are usually an order of magnitude smaller than diagonal force constants. Clearly this method is somewhat pedestrian in its approach but it is frequently used. It fits a one-dimensional quadratic equation of one geometrical variable. This means, using the above example of CH_2, that three points are needed for the optimization of the bond length r

$$E = E_r^{opt} + a_r(r-r^{opt})^2 \qquad \{IX-26\}$$

and three points are needed for the optimization of the bond angle α

$$E = E_\alpha^{opt} + a_\alpha(\alpha-\alpha^{opt})^2 \qquad \{IX-27\}$$

The process is illustrated schematically for CH_2 in Figure IX-6. For every parameter one needs 3 points plus a final point to reconfirm the minimum. For a general case involving 3N-6 independent variables we need therefore 1 + 3(3N-6) points which is less than the requirement of the full process.

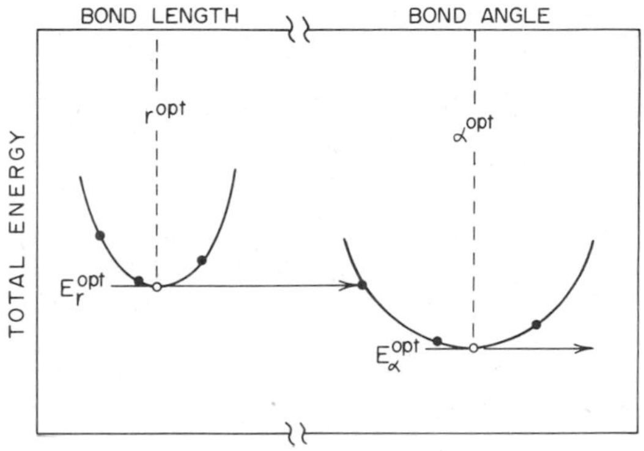

Figure IX-6. Geometry optimization of CH_2 by a "one dimensional" method.

In the full process one can also devise a method for choosing the points so that they may be as close to the minimum as possible. Here in this approximate method we have control over the three points along the one mode that we happen to optimize at a time. Consequently, iterative optimization may have to be imposed. If one starts in the vicinity of the minimum point the set obtained after the first series of optimization: $\{x_i^{opt(1)}\}$, i.e. $r^{opt(1)}$ and $\alpha^{opt(1)}$, is close enough to the true minimum. However, a second cycle will yield a set: $\{x_i^{opt(2)}\}$, i.e. $r^{opt(2)}$ and $\alpha^{opt(2)}$, that is even closer. This process of sequential or iterative optimization

$$\{x_i^{opt(1)}\} \rightarrow \{x_i^{opt(2)}\} \rightarrow \ldots \rightarrow \{x_i^{opt}\} \qquad \{IX-28\}$$

is illustrated by Figure IX-7. Of course, for each cycle of {IX-28} one needs 1 + 3(3N-6) points. With care, reliable results may be obtained as illustrated for $C_2H_2F^+$ in Figure IX-8.

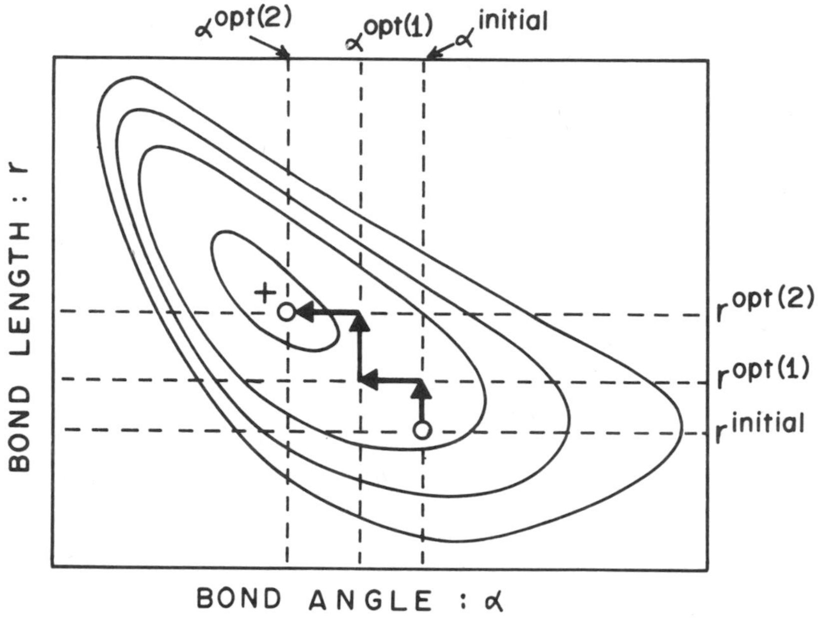

Figure IX-7. The general process of sequential or iterative optimization.

4. Molecular Energy and Stability

So far we did not discuss the utility of E^{opt} which is associated with $\{x_i^{opt}\}$. It becomes important when we compare two minima, say A and B. The two minima (A and B) may be located in two regions of the same surface or they may belong to different surfaces. The two surfaces may be associated with two different electronic states of the same molecule or with two different species (e.g. protonated and unprotonated forms). The energy values (E_A^{opt} and E_B^{opt}) associated with these two states (A and B) may be used to calculate the relative stability (ΔE), i.e. the energy that separates A and B. This is illustrated in Figure IX-9, where the superscript "opt" is dropped

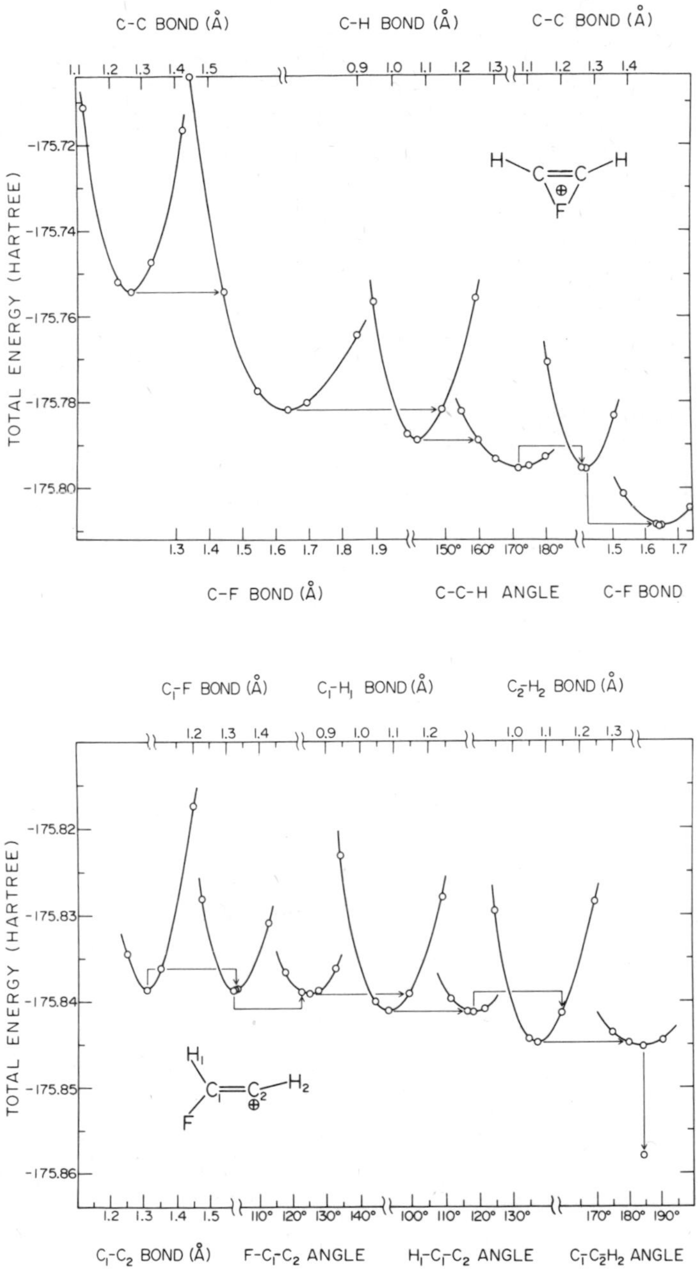

Figure IX-8. Sequential or iterative method of geometry optimization illustrated for $C_2H_2F^+$.

from E_A and E_B. The sign of ΔE is defined in the thermodynamic sense by the direction of the process:

$$\text{For } A \rightarrow B \quad \Delta E = E_B - E_A$$
$$\text{For } B \rightarrow A \quad \Delta E = E_A - E_B \qquad \{IX\text{-}29\}$$

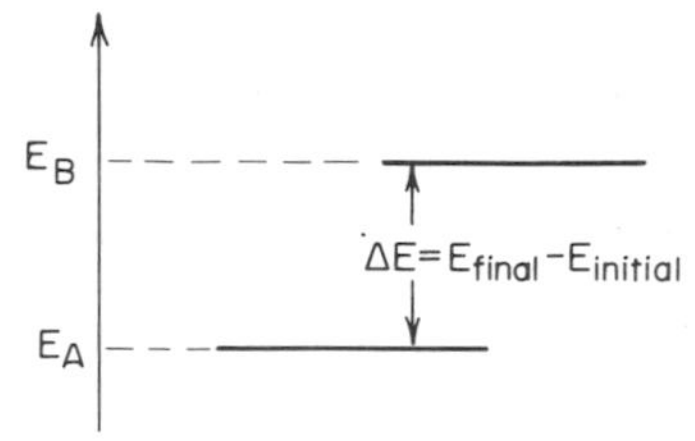

Figure IX-9. The relative stability of two "states" A and B.

(i) Thermodynamic and Kinetic Approaches

All chemical processes may be subdivided into two categories: thermodynamic (i.e. static) and kinetic (i.e. dynamic). If a chemical change is fully analyzed in both its static and dynamic aspects the change is then well understood. This is equivalent to the statement that a chemical change is fully characterized by its energy hypersurface. As discussed earlier, an energy hypersurface is the total molecular energy as a function of 3N-6 independent variables:

$$E = E(x_1, x_2, \ldots, x_{3N-6}) \qquad \{IX\text{-}30\}$$

where N is the number of constituent atoms. A three-dimensional cross-section of the energy hypersurface is an energy surface of two independent variables:

$$E = E(x_1, x_2) \qquad \{IX\text{-}31\}$$

An energy surface for a hypothetical process is illustrated in Figure IX-10. Two dimensional cross-sections are also included in the upper part of Figure IX-10.

$$\begin{aligned} E &= E(x_1) \\ E &= E(x_2) \end{aligned} \qquad \{IX\text{-}32a,b\}$$

The surface shown in Figure IX-10 illustrates a single intermediate, i.e. a three minima surface of a unimolecular (conformational or reactive) process. The three minima correspond to the initial state I (reactant), the intermediate state M, and the final state F (product):

$$I \rightarrow M \rightarrow F \qquad \{IX\text{-}33\}$$

The inter-relationship between any two of the minima (I and M, I and F, or M and F) represents a thermodynamic problem and the energy separation between any of the above pairs measures the thermodynamic stability of one species of the pair against the other species of the same pair:

$$\begin{aligned} \Delta E_{I\rightarrow M} &= E(M) - E(I) \\ \Delta E_{M\rightarrow I} &= E(I) - E(M) \\ \hline \Delta E_{M\rightarrow I} &= -\Delta E_{I\rightarrow M} \end{aligned} \qquad \{IX\text{-}34a,b,c\}$$

$$\begin{aligned} \Delta E_{I\rightarrow F} &= E(F) - E(I) \\ \Delta E_{F\rightarrow I} &= E(I) - E(F) \\ \hline \Delta E_{F\rightarrow I} &= -\Delta E_{I\rightarrow F} \end{aligned} \qquad \{IX\text{-}35a,b,c\}$$

$$\begin{aligned} \Delta E_{M\rightarrow F} &= E(F) - E(M) \\ \Delta E_{F\rightarrow M} &= E(M) - E(F) \\ \hline \Delta E_{F\rightarrow M} &= -\Delta E_{M\rightarrow F} \end{aligned} \qquad \{IX\text{-}36a,b,c\}$$

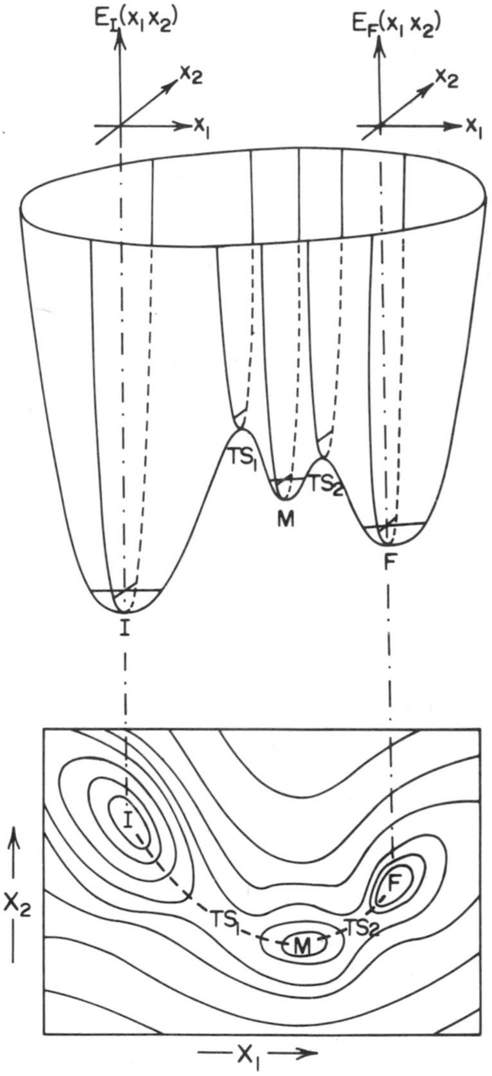

Figure IX-10. Energy surface for a hypothetical process. Three- and two-dimensional representations.

In addition to the thermodynamic analysis of a chemical change we may proceed to the kinetic analysis of the same problem. This requires some knowledge (geometry and energy) about the transition state as implied by the following scheme:

$$I \rightarrow TS_1 \rightarrow M \rightarrow TS_2 \rightarrow F \qquad \{IX\text{-}37\}$$

and also depicted by Figure IX-10. Clearly the generation of a surface or more precisely a hypersurface would give all the information about the chemical system that undergoes a conformational or reactive unimolecular process. The following two sections will discuss both the thermodynamic and kinetic aspects of a few selected processes.

(ii) Molecular Geometry and Stereochemistry

The term "molecular geometry" is used in connection with the thermodynamic approach and "stereochemistry" in connection with the kinetic approach to the same problem. However, in both these approaches the geometry (or geometry change) and the energy (or energy change) are chemically important.

The optimized geometries of $H_2C{=}O$ and $H_2C{=}OH^+$ as well as the change in optimized geometries in the course of protonation (as exemplified by the stretched C-O bond)

$$CH_2 = O \xrightarrow{H^+} CH_2{=}OH^+ \qquad \{IX\text{-}38\}$$

are indicated by Figure IX-11. The optimized energies E^{opt} (CH_2O) and E^{opt} (CH_2OH^+) are also marked on the figure and used in the calculation of the thermodynamic proton affinity (PA) value:

$$PA(CH_2O) \equiv \Delta E = E^{opt}_{CH_2OH^+} - [E^{opt}_{CH_2O} + E_{H^+}] \qquad \{IX\text{-}39\}$$

Note that on the quantum chemical scale $E_{H^+} = 0$ by definition leading to the simplification

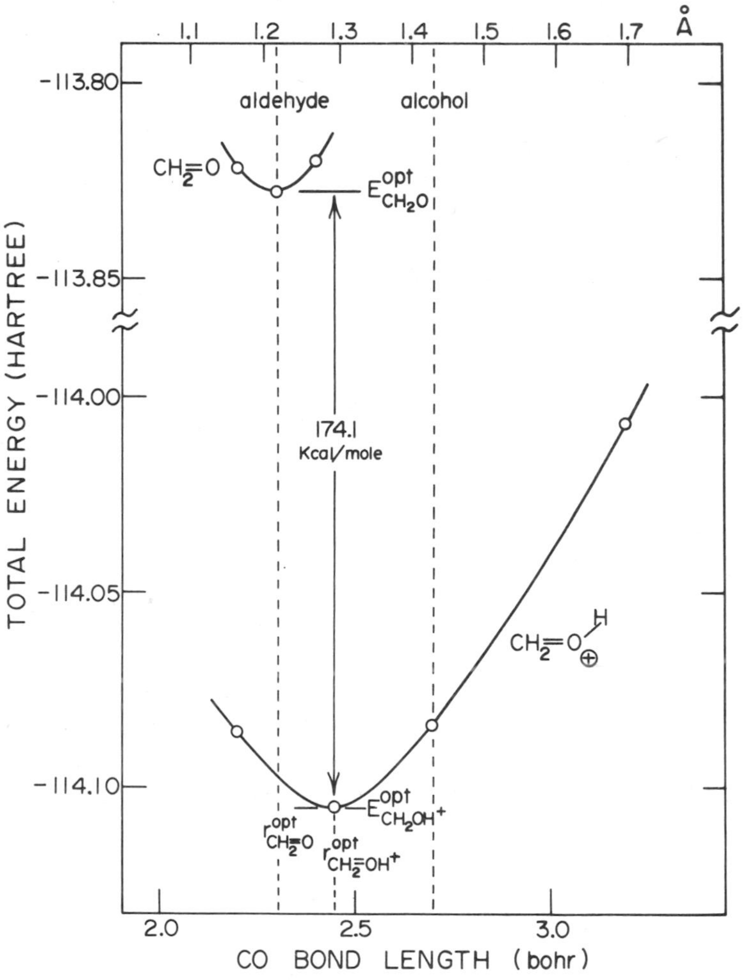

Figure IX-11. Variation of the total energy of $CH_2=O$ and $CH_2=OH^+$ as a function of the C-O bond length.

$$PA(CH_2O) \equiv E^{opt}_{CH_2OH^+} - E^{opt}_{CH_2O} \qquad \{IX-40\}$$

The site of protonation is sometimes in question. The different proton affinities associated with the formation of the different tautomeric forms may be used as a measure of stability. This situation may be illustrated by the protonation of $^-:CH_2NO_2$ leading to two tautomeric neutral compounds and governed by two different proton affinities, one associated with protonation at carbon: PA(C) and the other with protonation at oxygen: PA(O):

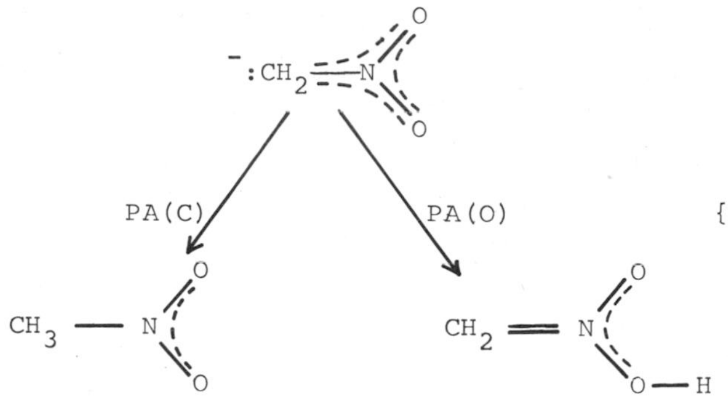

{IX-41}

The energetics of the above process is shown in Figure IX-12. The difference in the proton affinities are also shown in the figure indicating that the nitro-form is considerably more stable than the aci-form of nitromethane.

The sites of protons in a given molecule (proton tautomerism) can be studied by the relative stabilities of the various tautomers. The thermodynamic stabilities of some tautomeric forms of cytosine and thymine are shown in Figure IX-13. Not only heats of tautomerization but heats (i.e. energies) of isomerizations may also be studied by MO-theory. The isomerization scheme (including tautomerization) of CH_3NO is outlined in Figure IX-14 and the energetics of these processes illustrated by Figure IX-15. It is clear from this figure that formamide, the simplest peptide bond containing compound, is the most stable of all CH_3NO isomers.

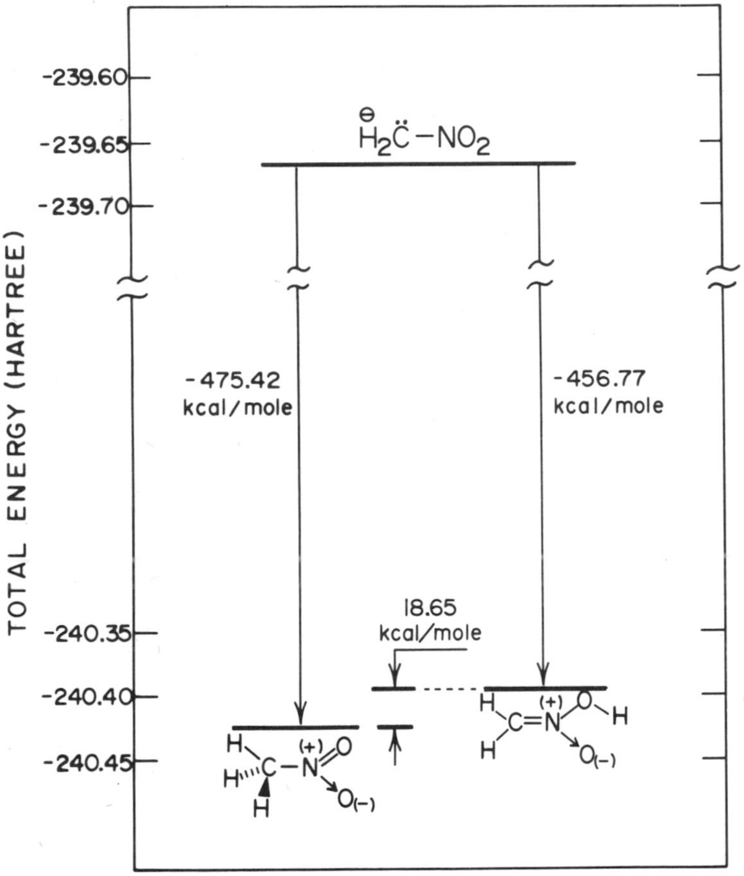

Figure IX-12. Proton affinities for protonation at carbon and nitrogen in nitromethane anion.

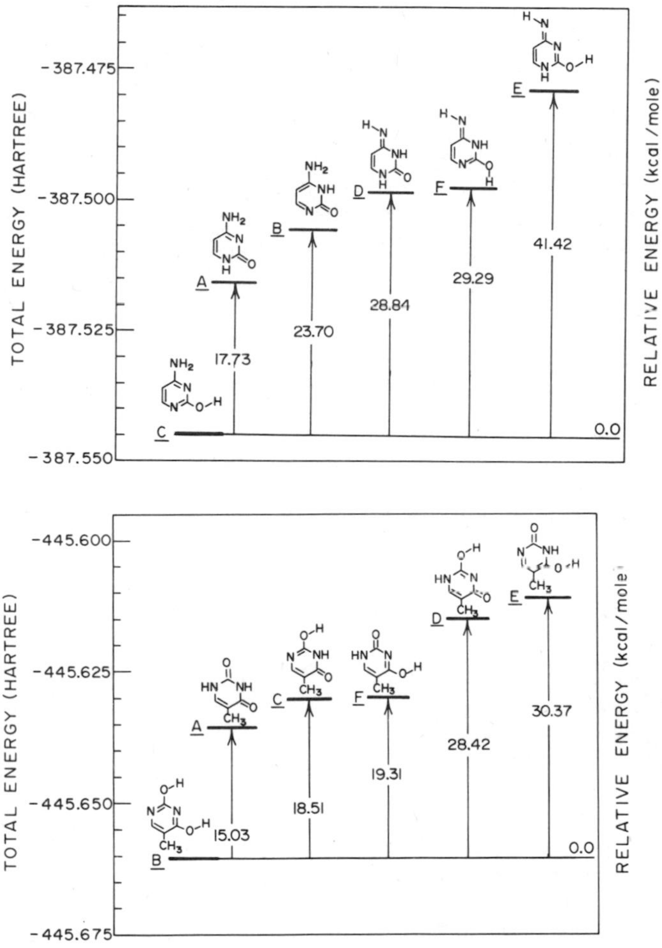

Figure IX-13. Computed thermodynamic stabilities of some tautomeric forms of cytosine and thymine.

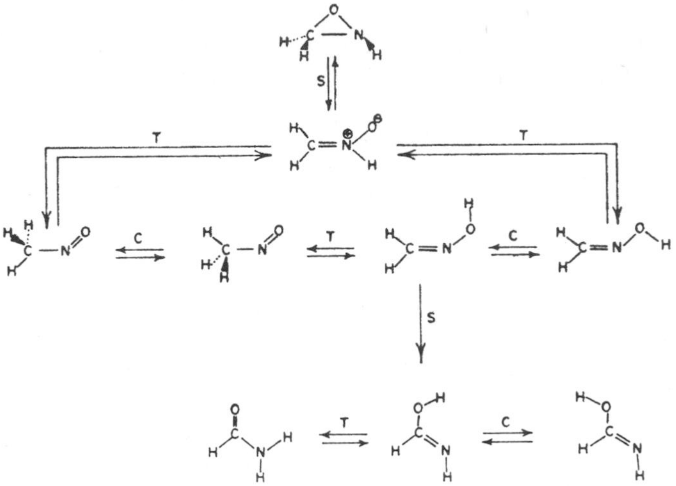

Figure IX-14. Conformational (C), tautomeric (T) and structural (S) isomerism of formamide.

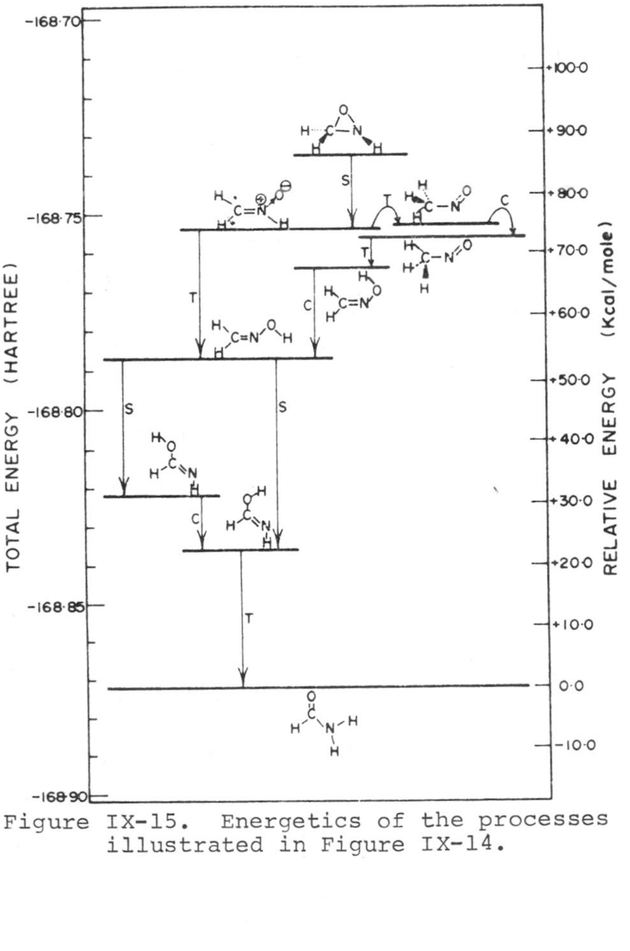

Figure IX-15. Energetics of the processes illustrated in Figure IX-14.

Needless to say very many additional examples might be quoted but the above systems probably serve to illustrate the thermodynamic stabilities and molecular geometries of stable species.

When we compute and plot total energies as a function of molecular geometries we are bound to find points other than the minima on the energy hypersurface. Besides the minima the most important distinct point on a surface or hypersurface is the saddle point associated with the transition state of a conformational change. The exploration of the minima and transition states on a conformational surface or conformational hypersurface leads us to the area of theoretical stereochemistry. In addition to these distinct points the maxima of conformational hypersurfaces are also of interest.

These distinct points (minima, saddle points, maxima) may conveniently be characterized in terms of the force constant or the second derivative of the energy (as defined by equation {IX-21}) with respect to some "coordinate" x_i.

<u>For a Minimum (min):</u>

$$\left(\frac{\partial^2 E}{\partial x_i^2}\right)_{x_i = x_i^{min}} = k_{ii} > 0 \qquad 1 \leq i \leq (3N-6) \qquad \{IX-42\}$$

for all i (i.e. all "coordinates").

<u>For a Saddle Point (sad):</u>

$$\left(\frac{\partial^2 E}{\partial x_i^2}\right)_{x_i = x_i^{sad}} = k_{ii} > 0 \qquad 1 \leq i \leq (3N-7) \qquad \{IX-43\}$$

for all i (i.e. for all coordinates) except one - which may be taken to be the last (i.e. the (3N-6)th) coordinate

$$\left(\frac{\partial^2 E}{\partial x_i^2}\right)_{x_i = x_i^{sad}} = k_{ii} < 0 \qquad i = 3N-6 \qquad \{IX-44\}$$

the force constant is negative corresponding to an imaginary vibrational frequency.

Continuing the enumeration of the number of positive and negative force constants we come to the definition of maxima.

For a Maximum (max):

All force constants are negative, i.e.

$$\left(\frac{\partial E}{\partial x_i^{\,2}}\right)_{x_i = x_i^{max}} = k_{ii} < 0 \qquad 1 \leq i \leq (3N-6)$$

{IX-45}

Thus, a minimum appears to be a minimum no matter which cross-section of the hypersurface we are investigating. A maximum also appears as a maximum regardless of the cross-section chosen for investigation. However, a saddle point may appear as a minimum or a maximum depending on the chosen cross-section. This phenomenon is illustrated (Figure IX-16) by the symmetric conformation surface of CH_3^-. The transition state is a maximum along the angle change: ϕ (i.e. $k_\phi < 0$) but it is a minimum along the C-H stretch: r (i.e. $k_r > 0$).

Rotation-inversion phenomena may be illustrated by $^{-}{:}CH_2{-}OH$. The topomerization of $^{-}{:}CH_2{-}OH$ is shown in Figure IX-17 and the actual conformation (rotation-inversion) surface

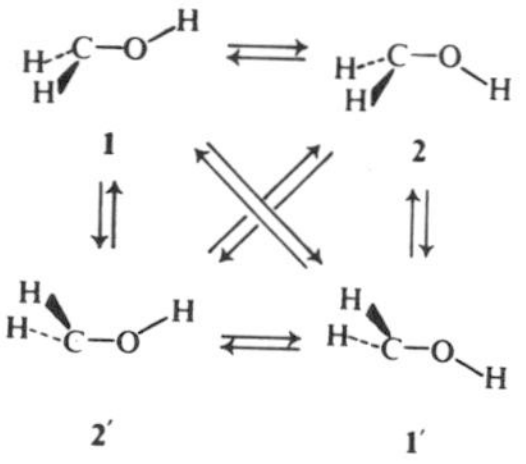

Figure IX-17. Topomerization of $^{-}CH_2OH$. Horizontal arrows: rotation along the C-O bond or in-plane inversion at oxygen. Vertical arrows: pyramidal inversion at carbon. Diagonal arrows: rotation-inversion.

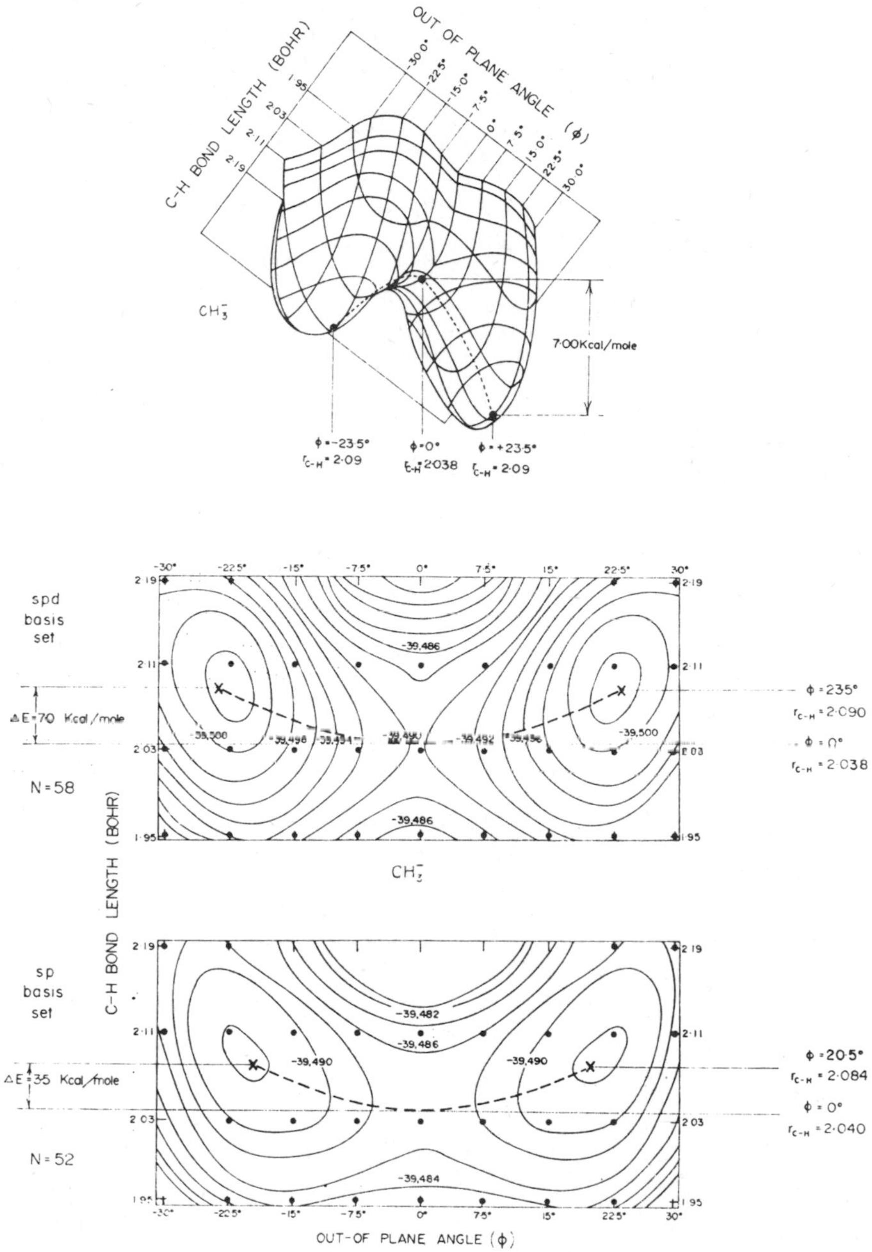

Figure IX-16. Symmetric conformation surface of CH_3^-. Note that the transition state is a maximum along the angle change but a minimum along the CH stretch.

is given in Figure IX-18. As may be seen, the carbanion

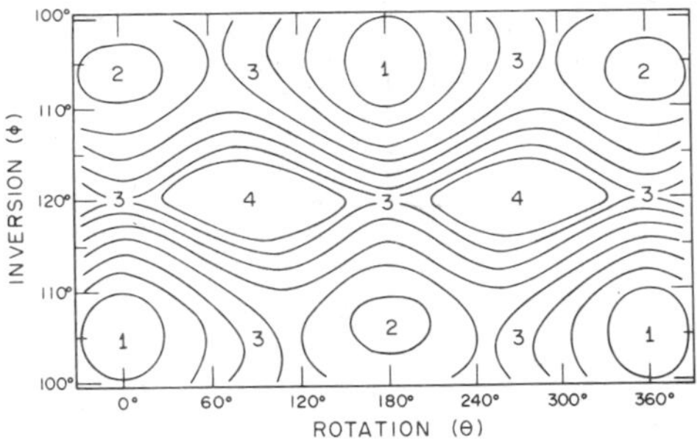

Figure IX-18. Rotation-inversion surface (E(θ,ϕ)) of $^{-}CH_2OH$ at the methanol C-O bond length (1.428 Å). [For 1 and 2 see Fig. IX-17, 3 are transition states, 4 are maxima.]

attached to the lone pair containing heteroatom assumes a pyramidal conformation, thus the absolute minimum of the surface corresponds to structure {IX-46}.

{IX-46}

When the $^{-}:CH_2$ moiety is attached to a NO_2 group the carbanion centre becomes flat and the topology of the corresponding conformational surface quite different as shown in Figure IX-19.

Many conformational properties, both thermodynamic and kinetic of a system may be deduced (e.g. chirality, dynamic nmr, etc.) when a conformational surface or hypersurface is computed at the ab initio level.

(iii) Reactive Intermediates and Reaction Mechanisms

The thermodynamic aspect of this area of chemical reactivity may be labelled "reactive intermediates". Within this framework, the thermodynamic stability of postulated or possible reaction intermediates are computed and compared. Consider the acid catalysed unimolecular hydrolysis of amides as exemplified by the following scheme:

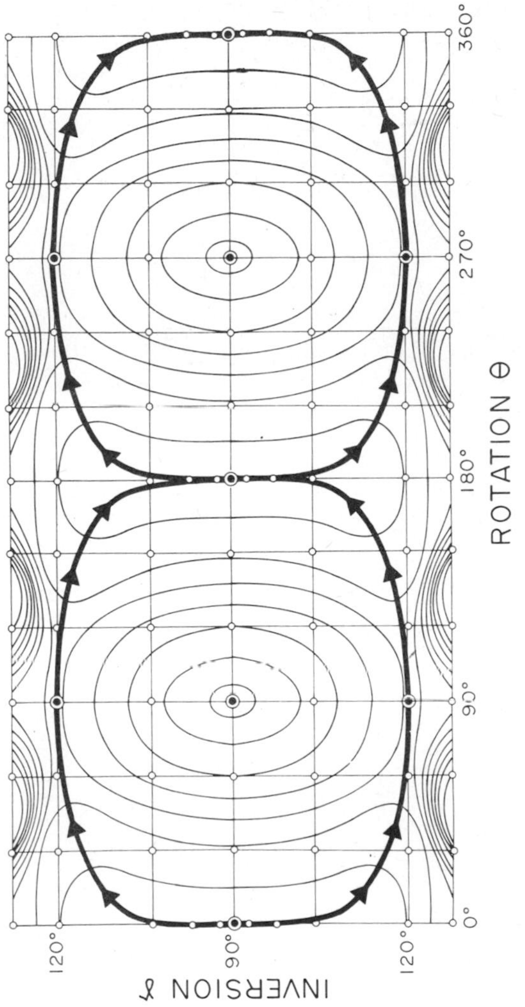

Figure IX-19. Topology of the rotation-inversion conformational energy surface of $^{-}$:CH_2-NO_2.

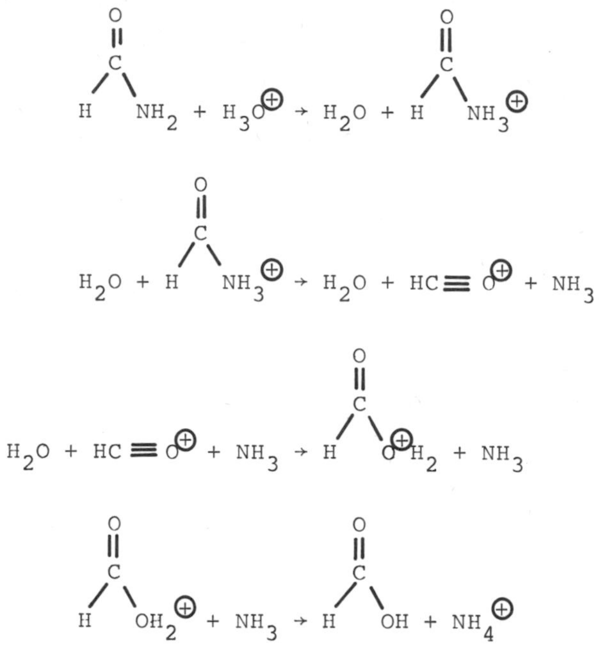

$$H-\underset{NH_2}{\overset{O}{\underset{\|}{C}}}\!\!\!\diagdown \quad + \ H_3O^{\oplus} \rightarrow H_2O + H-\underset{NH_3^{\oplus}}{\overset{O}{\underset{\|}{C}}}$$

$$H_2O + H-\underset{NH_3^{\oplus}}{\overset{O}{\underset{\|}{C}}} \rightarrow H_2O + HC\!\equiv\!O^{\oplus} + NH_3$$

$$H_2O + HC\!\equiv\!O^{\oplus} + NH_3 \rightarrow H-\underset{OH_2^{\oplus}}{\overset{O}{\underset{\|}{C}}} + NH_3 \qquad \{IX\text{-}47\}$$

$$H-\underset{OH_2^{\oplus}}{\overset{O}{\underset{\|}{C}}} + NH_3 \rightarrow H-\underset{OH}{\overset{O}{\underset{\|}{C}}} + NH_4^{\oplus}$$

The thermodynamics of the reaction mechanism (cf. broken arrows in Figure IX-20) implies a barrier between each

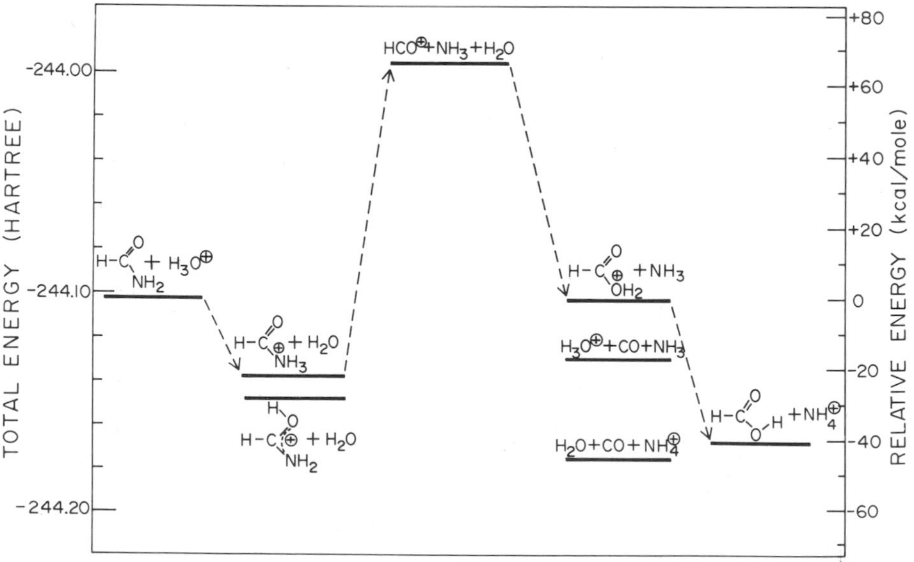

Figure IX-20.   Thermodynamic energy levels of the $A_{Ac}1$ hydrolysis intermediates of formamide.

pair of successive intermediates that must be higher than the thermodynamic separation of the two intermediate states corresponding to two adjacent local minima on the energy hyper-surface.

There may be several proposed structures for a key intermediate and the relative energies may be used to choose between several contenders. Consider for example the Prins' reaction:

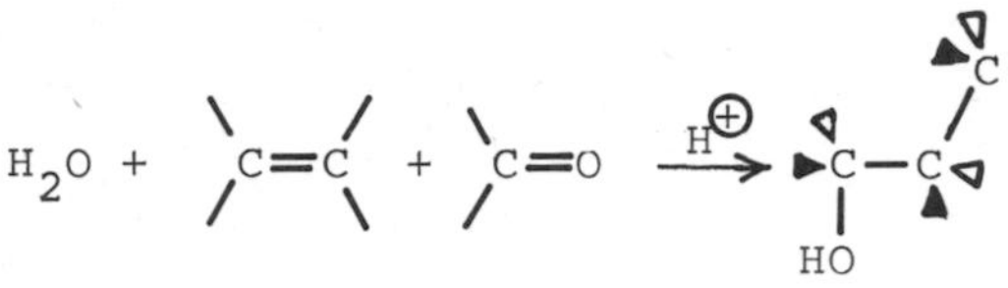

$$H_2O + \quad \overset{\backslash}{\underset{/}{C}}=\overset{/}{\underset{\backslash}{C}} + \quad \overset{\backslash}{\underset{/}{C}}=O \quad \xrightarrow{H\oplus} \quad \overset{C-OH}{\underset{HO}{\overset{|}{C}-C}} \qquad \{IX\text{-}48\}$$

The key intermediate of this reaction may be studied theoretically with prototype reagents:

$$H_2O + C_2H_4 + H_2CO \xrightarrow{H\oplus} H_2O + C_3H_7O\oplus \longrightarrow HO - (CH_2)_3 - OH + H\oplus \qquad \{IX\text{-}49\}$$

key
intermediate

The energies of the possible structures of $C_3H_7O^+$ (cf. Figure IX-21) suggest that $\underline{O}$-protonated oxatane is the most stable of

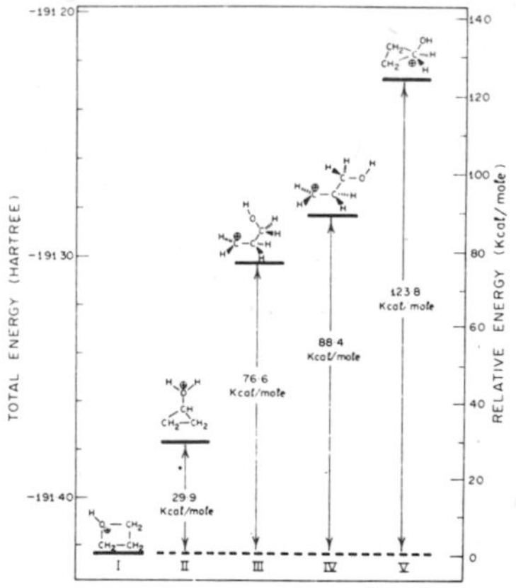

Figure IX-21. Computed energies for possible
structures of $C_3H_7O^+$.

all contenders.

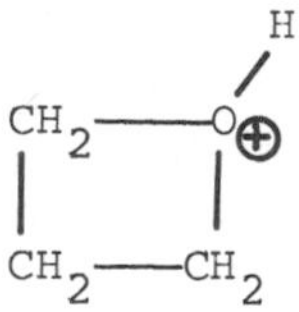

{IX-50}

The question of key intermediate sometimes may be more subtle than that discussed above. Even if there is considerable knowledge regarding the intermediate such as that involved in the electrophilic addition of sulphonyl halides to olefins:

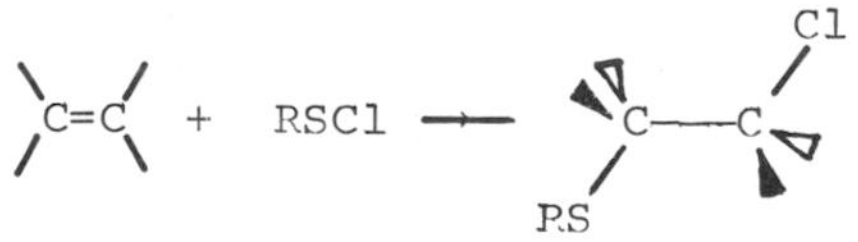

{IX-51}

the cyclic intermediate demanded by the stereospecific anti-addition might be either in the form of an ion-pair or in the form of a covalent intermediate:

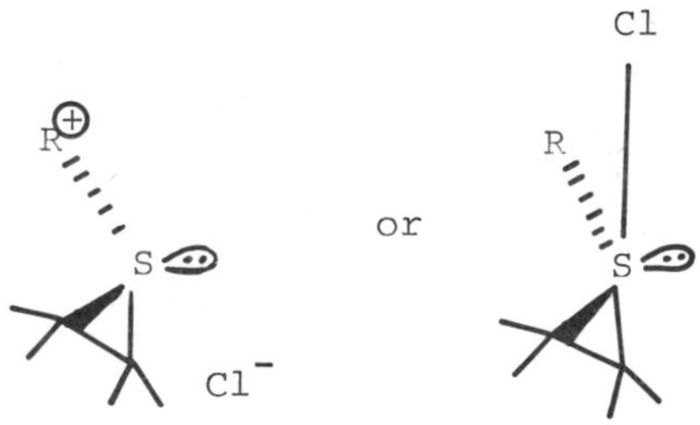

{IX-52}

Figure IX-22 shows the relative stability of these two alternatives. It cannot be emphasized too strongly that all these conclusions are for isolated molecules and thus approximate the gas phase. Solvation may very well change significantly the stabilities of various structures obtained by MO calculations. Molecular orbital treatments of solvent effects are rare and usually limited to the inclusion of a few molecules of solvent.

Figure IX-23 shows a portion of a reaction profile with three points marked I, II and III. They are the computed

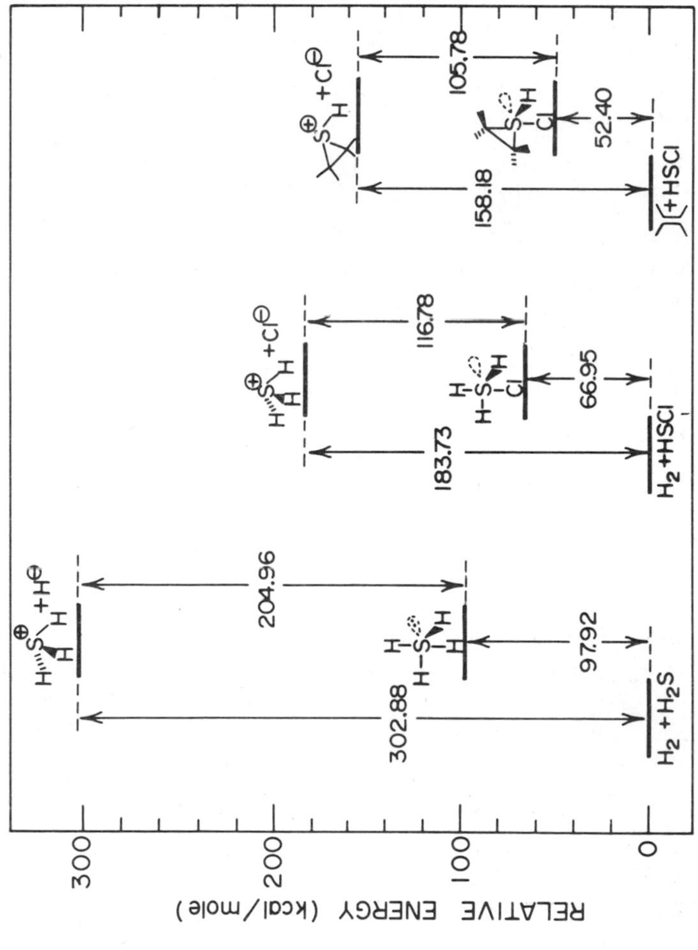

Figure IX-22. Relative energies of C_2H_2SHCl isomers.

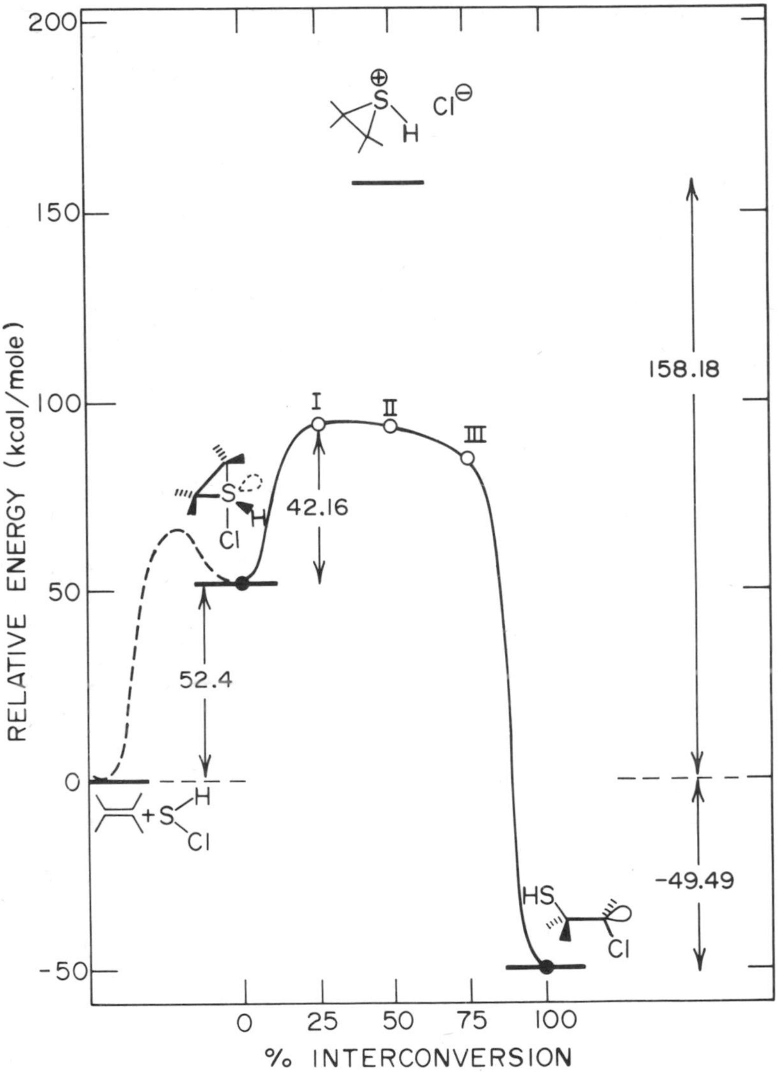

Figure IX-23. Approximate reaction profile of sulfurane → addition product of C_2H_4SHCl.

energy values associated with nominal 25, 50 and 75% of an assumed reaction coordinate. Sometimes these nominal percentages are obtained not on the basis of intuition but as a linear combination of the geometries of the intermediate and product structures. This procedure assumes that the reaction is least motion controlled thus the transition state lies along the line of least motion (i.e. the line of linear combination of geometries).

Of course, the above assumption is not necessarily correct and in fact reactions may be classified as least motion and non-least motion controlled. For the former, the transition state lies along the shortest route (least motion) on the reaction surface which is a straight line (cf. linear combination of geometries). These reaction surfaces are illustrated schematically in Figure IX-24.

(iv) The Balance of Electronic Attraction and Nuclear Repulsion

According to the Born-Oppenheimer approximation the molecular total energy (E_{Tot}) is the sum of the electronic attraction (E_e) and the nuclear repulsion (E_n).

$$E_{Tot} = E_e + E_n \qquad \{IX\text{-}53\}$$

In a conformational change (rotation about a single bond or inversion at a pyramidal centre), the barrier (ΔE_{Tot}) is also expressible in terms of the changes in the electronic (ΔE_e) and nuclear (ΔE_n) components

$$\begin{aligned}\Delta E_{Tot} &= E_{Tot}^{final} - E_{Tot}^{initial} = \\ &= (E_e^{final} + E_n^{final}) - (E_e^{initial} + E_n^{initial}) = \\ &= (E_e^{final} - E_e^{initial}) + (E_n^{final} - E_n^{initial}) = \\ &= \Delta E_e + \Delta E_n \qquad \{IX\text{-}54\}\end{aligned}$$

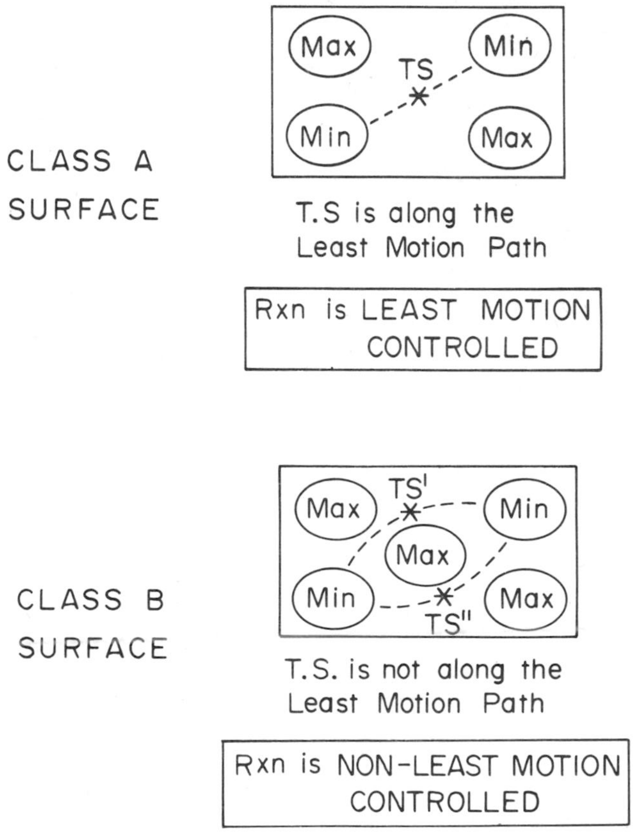

Figure IX-24. Topology of energy surfaces for least motion and non-least motion controlled processes.

Consequently we need to calculate the electronic and nuclear components according to equation {IX-53} for both the initial state and the final state.

Due to the Born-Oppenheimer approximation these two components (E_e and E_n) may be computed separately. The nuclear component (E_n) is calculated according to the Coulomb Law of classical electrostatics:

$$E_n = \sum_{I=1} \sum_{J>I} \frac{Z_I Z_J}{R_{IJ}} \qquad \{IX-55\}$$

while the electronic component (E_e) is computed as the expectation value of the many electron hamiltonian: $\hat{H}(1,2,3,...)$ over a many electron molecular wavefunction $\Phi(1,2,3,...)$ which is usually a single Slater determinant in the MO theory:

$$E_e = \langle\Phi(1,2,3,...)|\hat{H}(1,2,3,...)|\Phi(1,2,3,...)\rangle \qquad \{IX-56\}$$

Clearly the computation of E_e (and of course ΔE_e) according to equation {IX-56} is much more difficult than the calculation of E_n (cf. equation {IX-55}). Consequently, the knowledge of the experimental barrier height (ΔE_{Tot}) would allow us to estimate ΔE_e without computing equation {IX-56} for two geometries provided the geometries of both initial and final states are known:

$$\Delta E_e = \Delta E_{Tot} - \Delta E_n \qquad \{IX-57\}$$

As an example for the rotational mode of motion consider $^-CH_2OH$:

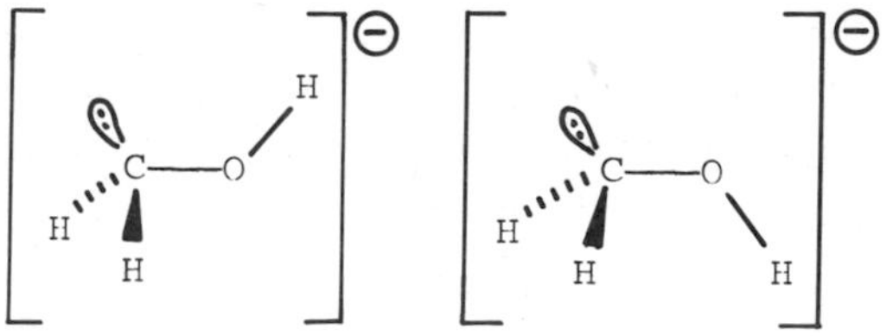

{IX-58}

and for the inversion mode of motion consider NH_3:

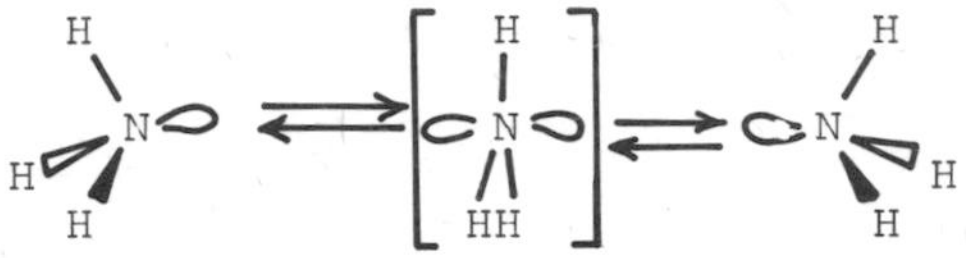

{IX-59}

The potential curves for the above torsional and inversion modes are shown in Figures IX-25 and IX-26 respectively.

Considering these potential curves we may infer three rules by inspection:

Rule 1: "In molecular conformational changes the nuclear and electronic components of the potential curve always have opposite phase."

Rule 2: "In any region of a potential curve, the variation of the total energy of a molecular system during conformational change is in phase either with the nuclear repulsion (nuclear dominant) or with the electronic attraction (electronic dominant)."

Rule 3: "Whenever there is a change from nuclear dominance to electronic dominance in a conformational process, this change will be found at an energy extremum."

These rules are applicable to conformational surfaces such as rotation-inversion surfaces as well as conformational hypersurfaces. However, one has to limit the applicability of these rules to the reaction coordinates since these lines correspond to the paths of shallowest ascent, i.e. the paths of greatest stability. In the language of conformational analysis the reaction coordinate is frequently referred to as the line of relaxed motion while cross-sections parallel to the individual coordinates (e.g. rotation, inversion, etc.) are referred to as rigid motions. The interdependence of these motions is schematically illustrated in Figure IX-27.

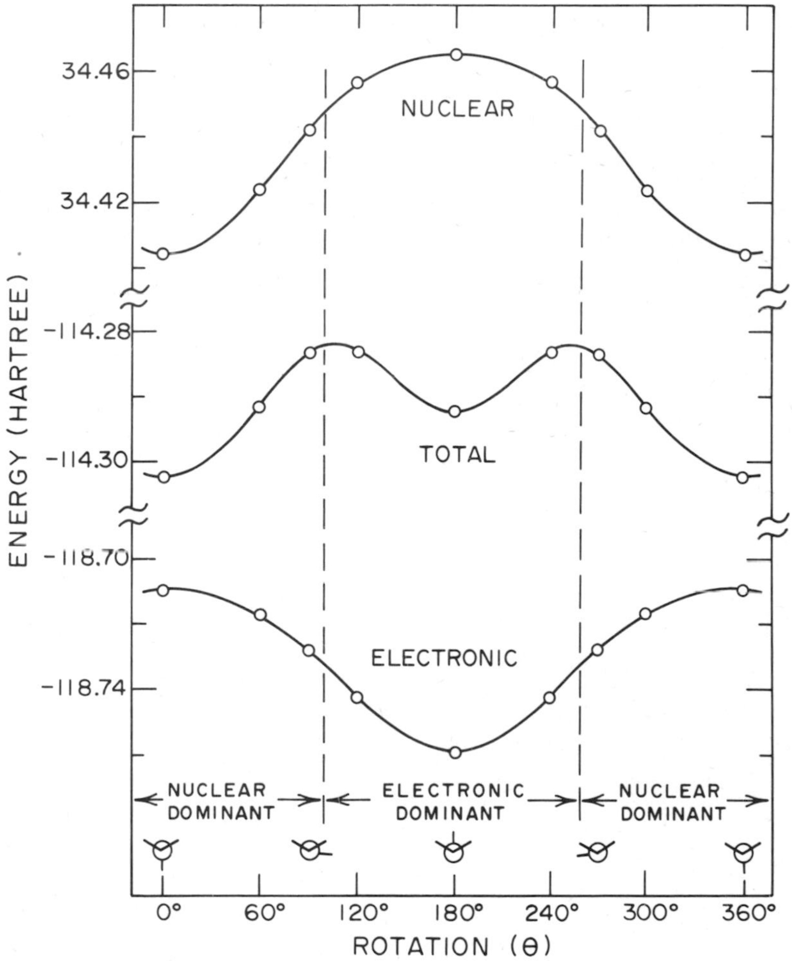

Figure IX-25. Potential curve for the torsional motion in $^{-}CH_2OH$.

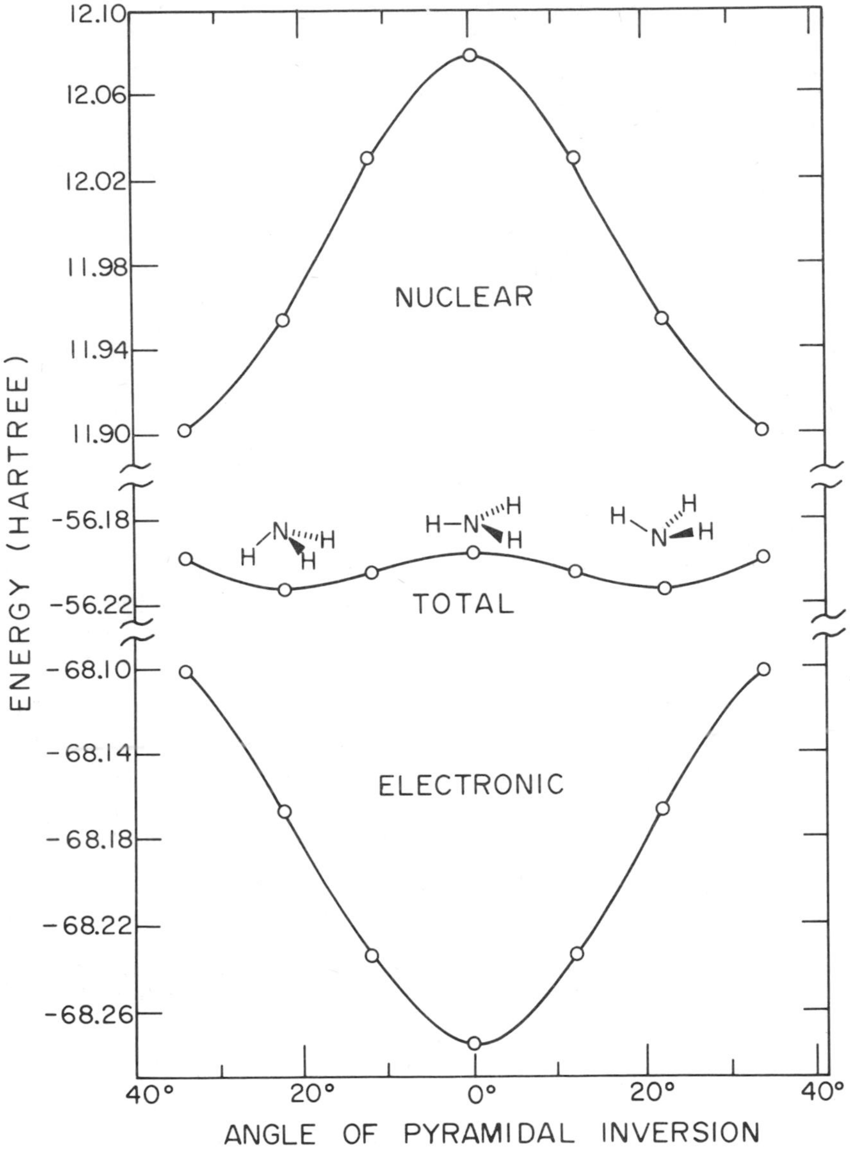

Figure IX-26. Potential curve for the inversion mode in NH_3.

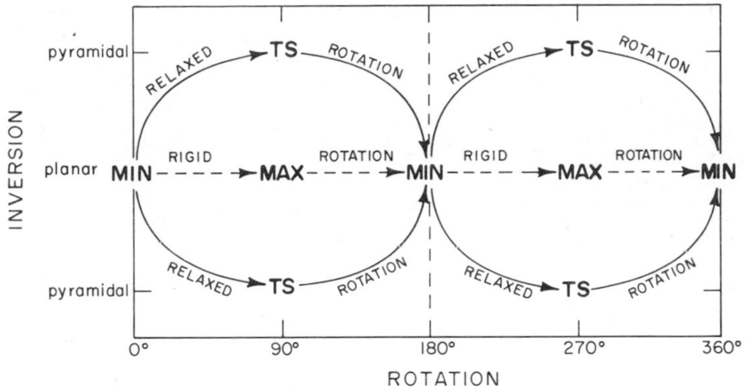

Figure IX-27. The interdependence of rigid and relaxed motions.

5. One Electron Properties

Molecular one electron properties (P) are computed from one electron operators ($\hat{P}$) and molecular, i.e. many electron wavefunctions (Φ):

$$P = \langle\Phi|\hat{P}|\Phi\rangle \qquad \{IX-60\}$$

Within the MO theory the many-electron wavefunction is usually a single Slater determinant:

$$\Phi = \hat{A}[\phi_1(1)\alpha(1)\phi_1(2)\beta(2)\ldots\phi_M(2M-1)\alpha(2M-1)\phi_M(2M)\beta(2M)] \qquad \{IX-61\}$$

while the one electron operator has the following explicit form:

$$\hat{P} = \hat{P}_o + \sum_i \hat{P}_i \qquad \{IX-62\}$$

The summation in {IX-62} is over all the electrons but if we substitute equations {IX-61} and {IX-62} into equation {IX-60} we obtain

$$P = P_o + 2\sum_k^M \langle\phi_k|\hat{P}_k|\phi_k\rangle \qquad \{IX-63\}$$

where the summation is over all occupied MO and the factor 2 is for the double occupancy of each MO.

The various one electron operators are summarized in Table IX-1. The symbols that appear to have found general acceptance are presented in the Table. All the operators are given in atomic units and consequently all properties are calculated in atomic units. Conversion to CGS units is performed on the final results using the conversion factors given in Table IX-2. It should be noted that most operators (cf. Table IX-1) involve both a nuclear and an electronic term in accordance with equation {IX-62}.

TABLE IX-1. ONE ELECTRON OPERATORS

Property	Symbol[a]	Operator[b,c,d] Nuclear	Electronic
Dipole Moment (First Moment)	$\mu_\alpha(N)$	$\sum_p (r_p - N)$	$-2 \sum_i (r_\alpha - N_\alpha)$
Second Moment	$Q_{\alpha\beta}(C)$	$\sum_p Z_p (r_{p\alpha}-C_\alpha)(r_{p\beta}-C_\beta)$	$-2 \sum_i (r_\alpha-C_\alpha)(r_\beta-C_\beta)$
Third Moment	$R_{\alpha\beta\gamma}(C)$	$\sum_p Z_p (r_{p\alpha}-C_\alpha)(r_{p\beta}-C_\beta)(r_{p\gamma}-C_\gamma)$	$-2 \sum_i (r_\alpha-C_\alpha)(r_\beta-C_\beta)(r_\gamma-C_\gamma)$
Quadrupole Moment	$\Theta_{\alpha\beta}(C)$	$\frac{1}{2} \sum_p Z_p [3(r_{p\alpha}-C_\alpha)(r_{p\beta}-C_\beta) - r^2_{pC}\delta_{\alpha\beta}]$	$-\frac{1}{2} \sum_i [3(r_{i\alpha}-C_\alpha)(r_{i\beta}-C_\beta) - r^2_i \delta_{\alpha\beta}]$
Octupole Moment	$\Omega_{\alpha\beta\gamma}(C)$	$\frac{1}{2} \sum_p Z_p [5(r_{p\alpha}-C_\alpha)(r_{p\beta}-C_\beta)(r_{p\gamma}-C_\gamma) - r^2_{pC}\delta_{\alpha\beta}\delta_{\beta\gamma}\delta_{\alpha\gamma}]$	$-\frac{1}{2} \sum_i [5(r_{i\alpha}-C_\alpha)(r_{i\beta}-C_\beta)(r_{i\gamma}-C_\gamma) - r^2_i \delta_{\alpha\beta}\delta_{\beta\gamma}\delta_{\alpha\gamma}]$
Potential	$V(N)$	$+ \sum_p{}' Z_p (1/\lvert r_p - N_p \rvert)$	$-2 \sum_i 1/\lvert r-N \rvert$
Electric Field	$E_\alpha(N)$	$- \sum_p{}' Z_p (r_{p\alpha}-N_\alpha)/\lvert r_{p\alpha}-N_\alpha \rvert^3$	$+2 \sum_i (r_\alpha-N_\alpha)/\lvert (r-N) \rvert^3$
Electric Field Gradient	$q_{\alpha\alpha}(N)$	$- \sum_p{}' Z_p [3(r_{p\alpha}-N_\alpha)(r_{p\beta}-N_\beta) - \delta_{\alpha\beta} r^2_{NN'})/r^5$	$+2 \sum_i [3(r_{i\alpha}-N_\alpha)(r_{i\beta}-N_\beta) - \delta_{\alpha\beta} r^2_{iN})/r^5$
Diamagnetic Susceptibility	$\chi^d_{AV}(C)$	———	$+2 \sum_i \lvert r-C \rvert^2$
Diamagnetic Shielding	$\sigma^d_{AV}(N)$	———	$\frac{e^2}{3mc^2} \sum_i \lvert 1/r_{iN} \rvert$

[a] N, nucleus of interest; C, centre of mass

[b] α, β, γ: Cartesian tensor components

[c] p, runs over nuclei with charge Z_p; i, runs over occupied MO's

[d] Σ', omits $r_p = N$

TABLE IX-2

UNITS AND CONVERSION FACTORS
FOR ONE-ELECTRON PROPERTIES

Property	CGS Equivalent to 1 a.u.
Dipole Moment	2.54154×10^{-18} esu-cm or 2.54154 Debye
Second Moment	1.344911×10^{-26} esu-cm^2 or 1.344911 Buckinghams
Third Moment	0.711688×10^{-34} esu-cm^3
Quadrupole Moment	As for Second Moment
Octupole Moment	As for Third Moment
Potential	9.07618×10^{-12} esu cm^{-1}
Electric Field	0.82377×10^{-12} dynes
Electric Field Gradient	0.324123×10^{16} esu cm^{-3}
Diamagnetic Susceptibility	0.89175 erg G^{-2} mole^{-1}

Note that the dipole, quadrupole, octupole, etc. moments are related to the first, second, third, etc. moments:

$$\begin{array}{lcl} \text{First Moment} & \rightarrow & \text{Dipole Moment} \\ \text{Second Moment} & \rightarrow & \text{Quadrupole Moment} \\ \text{Third Moment} & \rightarrow & \text{Octupole Moment} \\ \vdots & & \vdots \end{array} \qquad \{IX\text{-}64\}$$

The choice of the origin for the various tensor operators is dictated by convention. The potentials, diamagnetic nuclear magnetic shieldings, electric fields, and electric field gradients are computed at the positions of the nuclei in the molecule. It should be noted that, for example, in the calculation of the nuclear contribution to the potential at a particular nucleus, the effect of that nucleus is omitted (as indicated by a prime on the summation of the operator given in Table IX-1).

The first non-zero molecular multipole moment tensor of any neutral species (e.g. the dipole moment for ammonia) is independent of the choice of origin. If the centre of mass is chosen to be the origin then an assembly of point charges in a uniform field would experience no torque in the absence of a dipole moment. Similarly, only those molecules whose second moments relative to the centre of mass vanish experience no torque in a field gradient. Therefore, second and all higher moments are usually computed relative to the centres of mass of molecules (e.g. for the pyramidal geometry of ammonia the centre of mass is located at the point 0.00000, 0.00000, -0.12804). Only one independent scalar quantity is required to determine uniquely the molecular electric multipole tensor of rank p for a molecule with an n-fold axis of symmetry ($p > n$). Thus, for pyramidal ammonia (point group C_{3v}) the quadrupole moment may be specified by θ_{zz} since $\theta_{xx} = \theta_{yy} = -1/2\ \theta_{zz}$, where z is the three-fold symmetry axis. The diamagnetic susceptibility is also computed with respect to the centre of mass to allow comparison with experimental measurements.

One electron properties computed at the Hartree-Fock limit (HFL) are expected to nearly reproduce the experimental values of the one electron properties. In very precise work, vibrational and/or relativistic corrections may be necessary. However, near Hartree-Fock (i.e. large basis set SCF) wavefunctions can reproduce values near the HFL. Unfortunately, convergence of one electron property values is not monotonic which implies that a little improvement in the SCF wavefunction may produce an inferior value for a given property.

For example, the dipole moment μ of LiH improves with basis set size. Clearly the sp basis set gives a limit, the spd produces another limit, the spdf basis set is expected to give yet a further limit thus approximating the Hartree-Fock limit for the dipole moment in a successive way [I. G. Csizmadia, J. Chem. Phys. 44, 1849 (1966)].

The multipole moments of NH_3 and the iso-electronic, iso-protic and iso-symmetric CH_3^- are summarized in Table IX-3, as computed [Theoretica Chimica Acta 22, 1 (1971)] with an SCF (a single configuration) and a CI (nearly a 1000 configuration) wavefunction.

6. Other Applications

(i) Derivative Properties and Vibrational Analysis

In contrast to primary molecular properties where the observable (Ω) is the expectation value of a quantum mechanical operator ($\hat{\Omega}$) over a many electron wavefunction:

$$\Omega = \langle\Phi|\hat{\Omega}|\Phi\rangle \qquad \{IX-65\}$$

there exist for derivative properties, i.e. observables related to the derivatives (first or second partial derivatives) of primary properties. For example the force along a direction z is the first partial derivative of the energy with respect to z

TABLE IX-3

MOMENTS OF THE CHARGE DISTRIBUTION OF NH_3 AS CALCULATED FROM THE SCF AND CI WAVEFUNCTIONS[a]

Moment	NH_3			CH_3^-	
	SCF	CI	Expt	SCF	CI
μ [c]	-1.9463	-1.9187[b]	-1.4820[d]	-1.4710	-1.4570
Q_{xx} [c]	-6.2229	-6.3428	-	-11.8878	-11.9782
Q_{zz}	-8.8036	-8.8329	-	-16.0433	-16.1372
$<r^2>$ [f]	7.3899	7.4459[b]	7.1406[b]	11.7960	-11.8537
Θ_{zz} [e]	-2.5807	-2.4901[b]	-1.0[g]	-4.1554	-4.1577
R_{yyy} [h]	0.9175	0.8514	-	-0.2739	-0.1577
R_{zzz}	-0.5820	-0.5560	-	-2.7776	-2.8300
R_{xxz}	-0.6237	-0.5694	-	-0.5062	-0.4976
R_z [h]	-1.8296	-1.6949	-	-3.7901	-3.8252
Ω_{yyy} [h]	2.2936	2.1326	-	-0.6849	-0.7643
Ω_{zzz}	1.2893	1.1524	-	-1.2588	-1.3371

[a] Second and third moments are relative to the centre of mass.

[b] These values reported by Harrison [J. Chem. Phys. 47, 2990 (1967)] to be -2.11, +7.577 and -3.14 respectively.

[c] Dipole moment in Debye 1 au = 2.54154 Debye.

[d] Herzberg, G.: Infrared and Raman spectra. Toronto: D. Van Nostrand Co. Inc., 1945.

[e] Second moments and quadrupole moment in units of 10^{-26} esu-cm^2, 1 au = 1.344911 x 10^{26} esu/cm^2.

[f] In units of 10^{-16} cm^2, 1 au = 0.280016 x 10^{-16} cm^2, $<r^2>$ CH_3 nuclear = 3.5055.

[g] Stogryn, D. E., Stogryn, A. P., Mol. Phys. 11, 371 (1966).

[h] Third moment and octopole moment in units of 10^{-34} esu-cm^3, 1 au = 0.711688 x 10^{-34} esu/cm^3.

$$F_z = \frac{\partial E}{\partial z} \qquad \{IX\text{-}66\}$$

while the force constant is the second partial derivative of energy

$$k_z = \frac{\partial^2 E}{\partial z^2} \qquad \{IX\text{-}67\}$$

Since the energy expression is

$$E = \langle \Phi | \hat{H} | \Phi \rangle = \int \Phi^* \hat{H} \Phi d\tau \qquad \{IX\text{-}68\}$$

the force {IX-66} or force constant {IX-67} is obtained by the rules of differentiation of expression {IX-68}.

Of course, the direction (i.e. the parameter) of derivation is arbitrary but it is customary to choose bonds and bond angles so that traditional force constants may be obtained. These force constants may be transformed by Wilson's FG matrix formalism to obtain fundamental vibrational frequencies.

The derivatives may be obtained by computing a number of points about the equilibrium conformation and fitting a quadratic equation

$$E = E^{opt} = \frac{1}{2} k(r-r^{opt})^2 \qquad \{IX\text{-}69\}$$

to obtain the harmonic force constant k or a more complicated analytic equation that yields force constants which includes anharmonic correction. This process is illustrated for FNO_2 in Acta Phys. Hung. 27, 377 (1969).

Alternatively, one can carry out the differentiation of equation {IX-68} and that process is expected to yield more accurate force constants than the method outlined above that used several computed points. This approach has been used by Pulay for several organic molecules (e.g. NH_3, P. Pulay and W. Meyer, J. Chem. Phys. 57, 3337 (1972)).

(ii) Change of Charge Distribution in Chemical Processes

For any stereochemical or chemical process the molecular charge distribution is expected to change from the
initial state through the transition state to the final state
of the process. The change of charge distribution does not
necessarily parallel the energy profile. Thus one could
envisage a process in which the charge distribution hardly
changes as the transition state is reached (reactant-like
transition state) while in the opposite case most of the
required charge redistribution is achieved by the system
before reaching the geometry of the activated complex
(product-like transition state).

One way the change of charge distribution may be
monitored is the changing values obtained from a Mulliken
population analysis. Figure IX-28 illustrates such results
for the unimolecular rearrangement of {IX-70}.

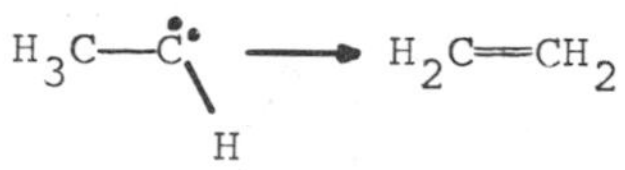

$$H_3C-\overset{\bullet\bullet}{\underset{H}{C}} \longrightarrow H_2C = CH_2 \qquad\qquad \{IX-70\}$$

An alternative way to monitor the charge distribution
is to plot the electron density (ρ) in the form of density
contours. Since the change in ρ is relatively small to the
naked eye the contour plot of the electron density difference
($\Delta\rho$) with respect to the initial state (0% reaction)

$$\Delta\rho(x\%) = \rho(x\%) - \rho(0\%) \qquad\qquad \{IX-71\}$$

may be more illustrative. In addition to the possibility of
monitoring the total molecular ρ or total molecular $\Delta\rho$ one
might plot the density of individual molecular orbitals.
Here the change in density may be enormous or negligible
depending on the nature of the individual MO. Considering the
above process {IX-70} one would expect the pair of carbon
cores and the C-C bonds to remain relatively constant. However, the migrating C-H bond and lone electron pair in the
reactant are expected to undergo a major reorganization in the

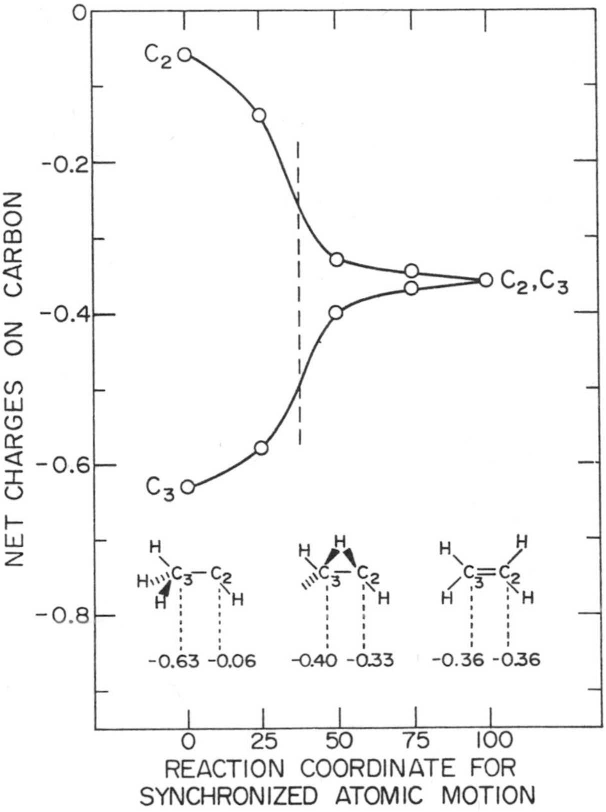

Figure IX-28. Changes in the charge distribution as measured by the Mulliken population analysis for the rearrangement of methylcarbene.

course of the molecular rearrangement to form the π-bond and a new C-H bond. Figure IX-29 shows these MO density plots for the unimolecular rearrangement of {IX-70} along the least motion reaction coordinate. From these plots it is clear that in accordance with Zimmerman's "orbital following" concept the nodal character of the MO is conserved in the course of the reaction. One may view the conservation of the orbital nodal character as related to the conservation of orbital symmetry as pointed out by Woodward and Hoffmann.

Alternatively, one may monitor the charge distribution in the course of a process by computing the "size" of functional groups as a function of the reaction coordinate. This is illustrated for rotation along the C-O bond in FCH_2-OH in Figure IX-30.

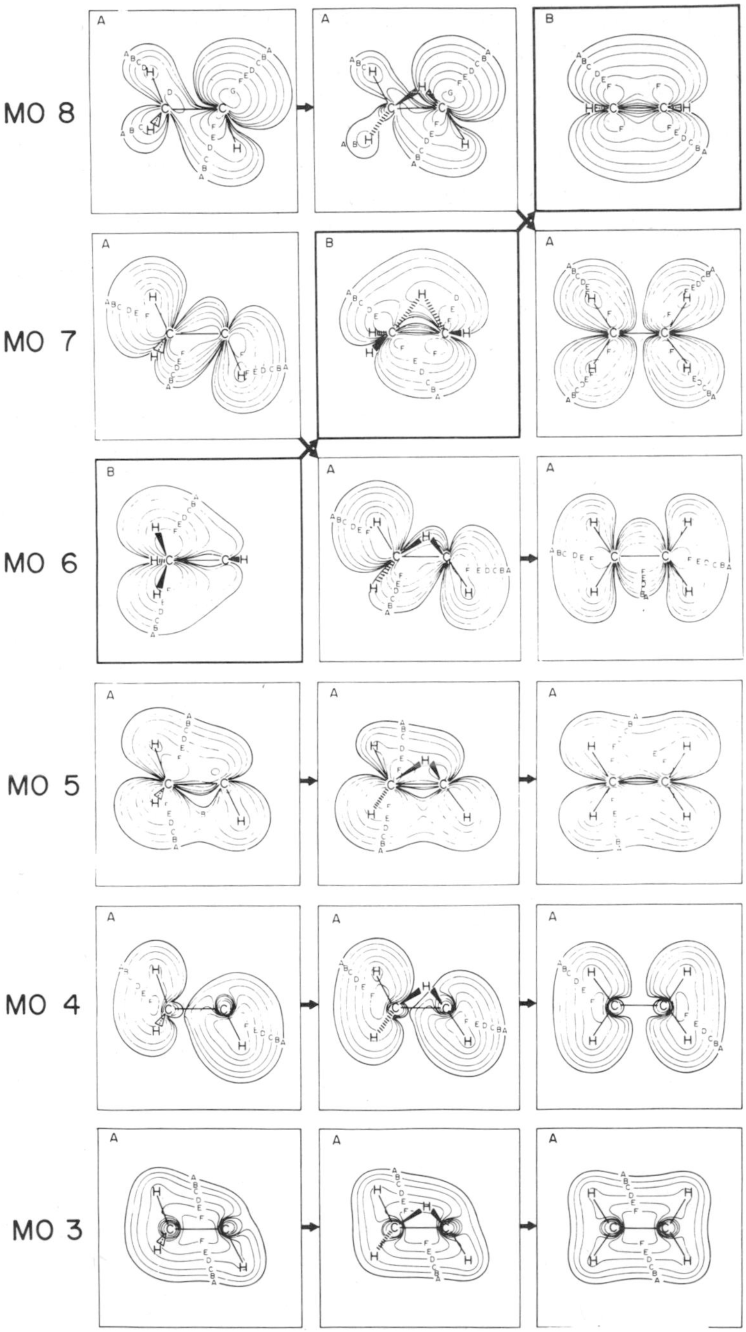

Figure IX-29. Valence MO density plots for the unimolecular rearrangement of methylcarbene along the least motion reaction coordinate.

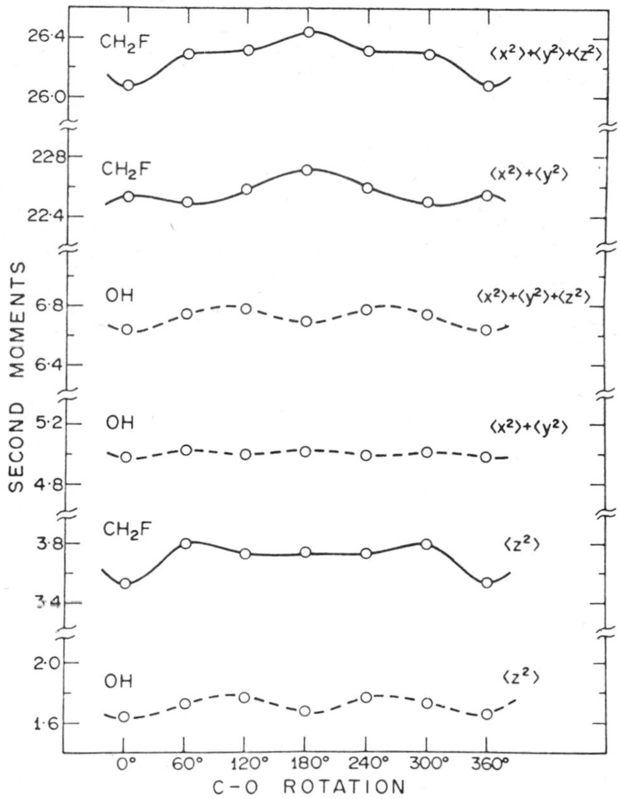

Figure IX-30. The periodic variation of groups sizes ($<x^2> + <y^2> + <z^2>$) for the FCH_2 and OH groups. The components along the rotational axis ($<z^2>$) as well as in the plane perpendicular to the rotational axis ($<x^2> + <y^2>$) are also shown.

SECTION C

THEORY OF OPEN ELECTRONIC SHELLS

CHAPTER X

OPEN SHELL SCF THEORIES

As was pointed out in Chapter IX all systems with an odd number of electrons are classified as open shell systems but some molecules with an even number of electrons may also form open shell systems if the electron pairing within the molecule is not complete (e.g. O_2), usually due to the presence of degenerate orbital levels.

1. Classification of Open Shell Systems

Within the SCF framework, the term "shell " must be precisely defined. What is generally termed a subshell in atomic theory (s, p, d, etc.) is considered to be a shell here. Individual (non-degenerate) orbitals as well as sets of degenerate orbitals are called shells as illustrated in Figure X-1. As this figure indicates, an open shell may be "occupied"

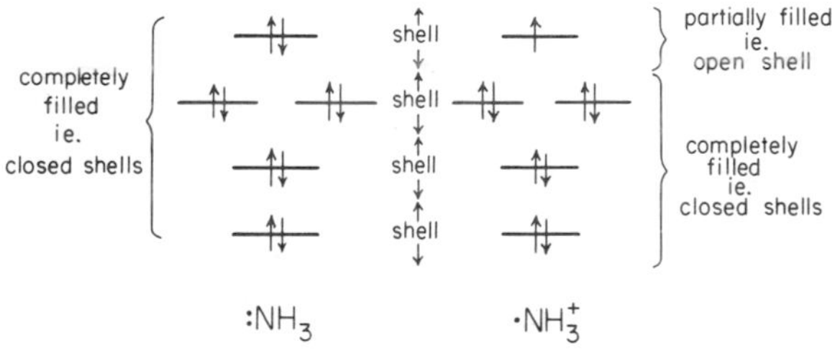

Figure X-1. Definition of shells and an illustration for a closed shell (:NH_3) and an open shell (·$NH_3^{\oplus}$) system.

by only one electron (for an odd number of electrons) if the occupied MO is non-degenerate within the open shell. Once the shell consists of a set of degenerate orbitals the shell may contain more than a single electron and still be classified as an open shell. The number of electrons is smaller than the maximum number of electrons the system may accommodate in the shell. As soon as the maximum number is reached the shell will be completely filled and thus "closed". The oxygen molecule or the acetylene molecule and cations may serve as examples (cf. Figure IX-2 and X-2).

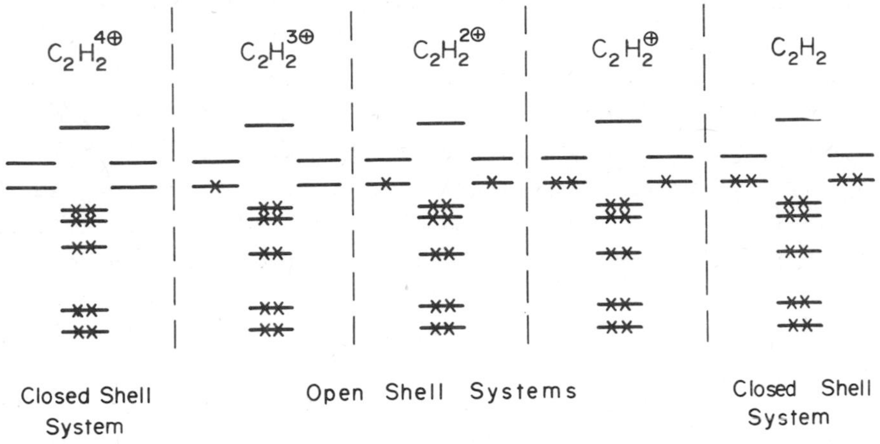

Figure X-2. Closed and open shells of linear acetylene cations.

The spin multiplicity of the system is essential to the classification of open shell systems

$$\text{spin multiplicity} = 2|S_z| + 1 \qquad \{X-1\}$$

where S_z is the expectation value of the total spin angular momentum operator with respect to one arbitrary axis(z) computed by summing the spin eigenvalues (+1/2 and -1/2) of the individual electrons of the system as given by

$$S_z = \sum_{i=1}^{\text{all electrons}} s_z(i) \qquad \{X-2\}$$

Considering the two examples in Figure X-1, NH_3 has five electrons with spin eigenvalue +1/2 and five electrons with spin eigenvalue -1/2. Consequently there is a cancellation in the summation of equation {X-2} and $S_z = 0$. The resultant multiplicity is 1 and we refer to ammonia as a chemical system of singlet multiplicity, or simply as a "singlet". In contrast the $|S_z|$ value we compute for NH_3^+ is 1/2 since only partial cancellation occurs in equation {X-2}. The multiplicity, calculated according to equation {X-1} is 2 and the molecular ion NH_3^+ is of doublet multiplicity, or a "doublet". Clearly, the multiplicity of the system depends on the number of odd (spin unpaired) electrons the molecule may have (cf. Figure X-2) and this dependence is summarized in Table X-1 for the cases of up to 5 unpaired electrons. Note that for the examples in Figure X-3 along the horizontal and vertical lines we go from a singlet to a doublet or from doublet to singlet multiplicity. Higher multiplicities may also be generated if electronic excitation is involved as illustrated by Figure X-4.

In constructing a proper determinantal wavefunction for an open shell system we normally wish to ensure that the wavefunction must be a simultaneous eigenfunction of $\hat{S}_z$ and $\hat{S}_z^2$ spin angular momentum operators. For a doublet state (e.g. a free radical, molecular radical cation or radical anion) we may construct two equivalent wavefunctions (this in fact accounts for the terminology "doublet") as illustrated for a 3-electron system (e.g. Li) by equations {X-3} and {X-4}

ϕ_2

ϕ_1

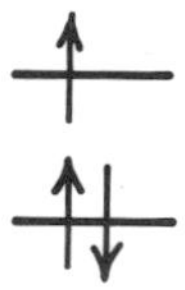

$$\Phi_\alpha = \hat{A}[\phi_1(1)\alpha(1)\phi_1(2)\beta(2)\phi_2(3)\alpha(3)]$$

$$\hat{S}_z\Phi_\alpha = +\,1/2\ \Phi_\alpha \qquad \{X\text{-}3\}$$

TABLE X-1

THE DEPENDENCE OF MULTIPLICITY ON THE ODD NUMBER OF ELECTRONS

Number of odd electrons	Resultant Spin $\lvert S_z \rvert$	Multiplicity
0	0	1 (Singlet)
1	1/2	2 (Doublet)
2	0	1 (Singlet)
	1	3 (Triplet)
3	1/2	2 (Doublet)
	3/2	4 (Quadruplet)
4	0	1 (Singlet)
	1	3 (Triplet)
	2	5 (Quintuplet)
5	1/2	2 (Doublet)
	3/2	4 (Quadruplet)
	5/2	6 (Sextuplet)

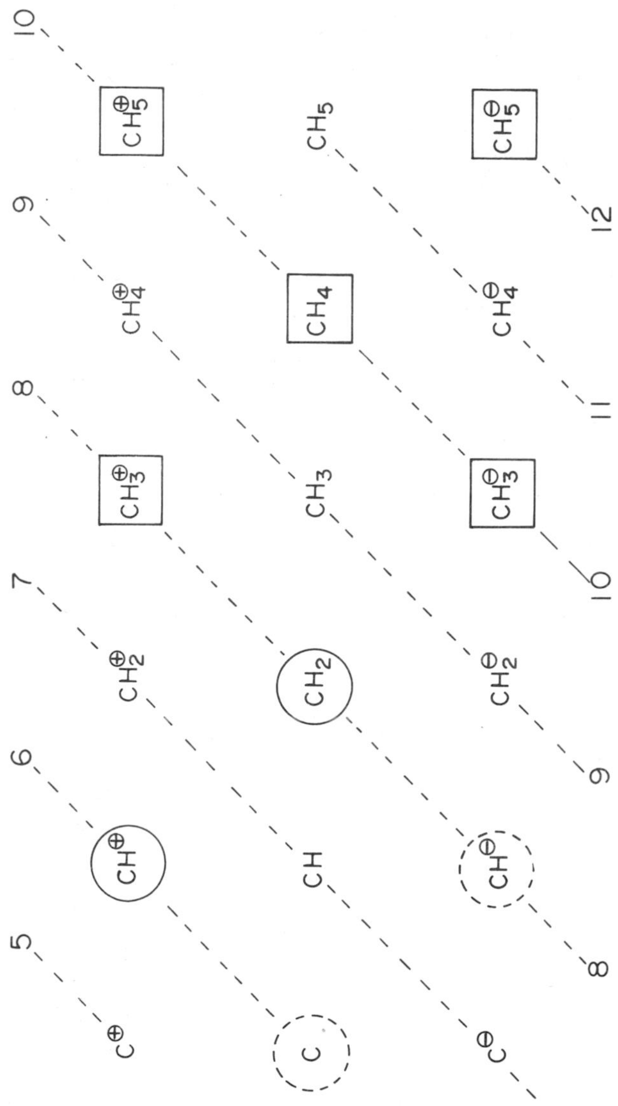

Figure X-3. The hydrides of carbon. Species framed by a square are definitely of singlet multiplicity and do form closed electronic shells. Species framed by a solid circle may be of singlet multiplicity and only then represent closed electronic shells. Species framed by broken circles may be of singlet multiplicity but under no circumstances may be represented by a closed shell formalism. The isoelectronic families are interrelated by broken lines.

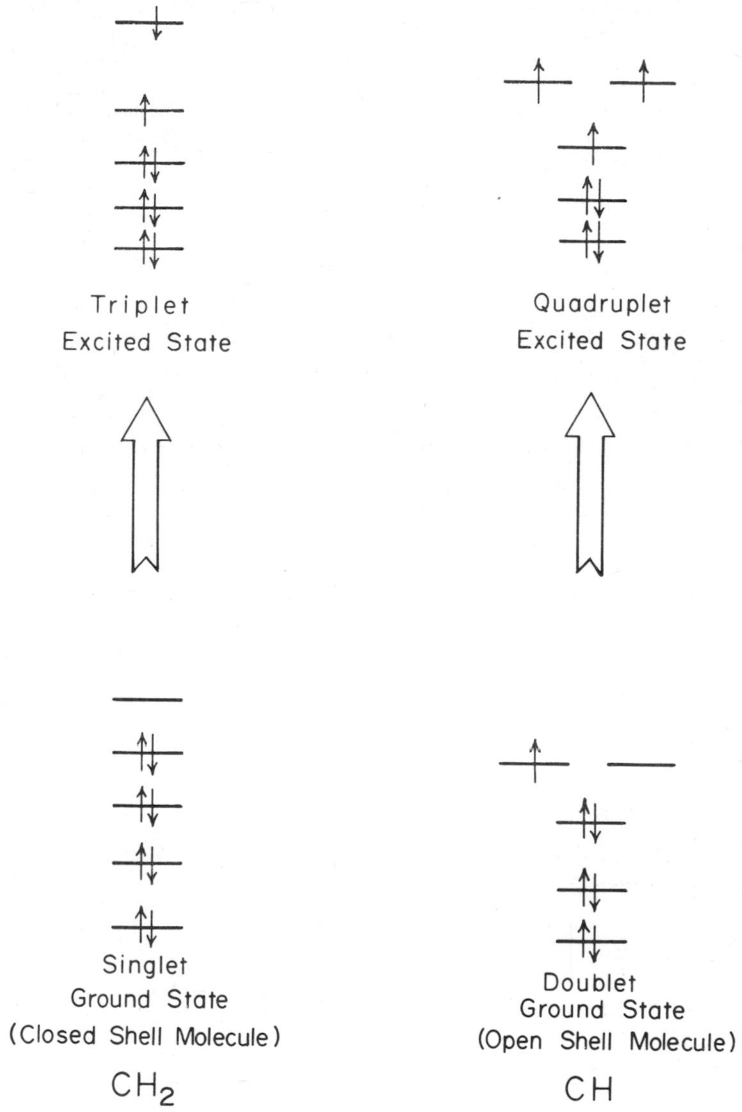

Figure X-4. A schematic illustration of the generation of higher multiplicity during excitation. (Note that while CH_2 pedagogically seems to be an ideal example the triplet state may be more stable than the singlet state at certain geometries.)

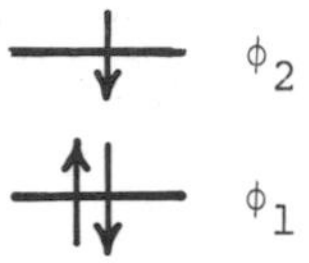

$$\Phi_{\beta} = \hat{A}[\phi_1(1)\alpha(1)\phi_1(2)\beta(2)\phi_2(3)\beta(3)] \qquad \{X\text{-}4\}$$

$$\hat{S}_z\Phi_{\beta} = -\ 1/2\ \Phi_{\beta}$$

These two spin states are equivalent in the absence of a magnetic field (cf. ESR spectroscopy), therefore it is sufficient to treat only one of them and by convention we choose the one that yields the largest spin eigenvalue which in the present case corresponds to the antisymmetrized spin-orbital product (i.e. Slater determinant) expressed in equation {X-3}.

An analogous situation holds for the description of the triplet state. Since the state is of triplet multiplicity we should be able to construct three wavefunctions, again degenerate in the absence of the magnetic field, corresponding to the spin eigenvalues of +1, 0, -1 or of $\alpha\alpha$, $\alpha\beta$, $\beta\beta$ spin configurations respectively (cf. equations {X-5}, {X-6}, {X-7}).

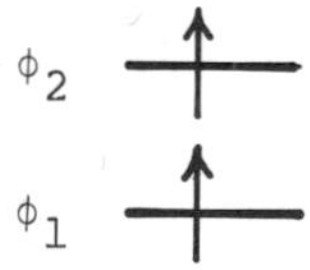

$$\Phi_{\alpha\alpha} = \frac{1}{2}\begin{vmatrix} \phi_1(1)\alpha(1) & \phi_2(1)\alpha(1) \\ \phi_1(2)\alpha(2) & \phi_2(2)\alpha(2) \end{vmatrix} \qquad \{X\text{-}5\}$$

$$\hat{S}_z\Phi_{\alpha\alpha} = \Phi_{\alpha\alpha}$$

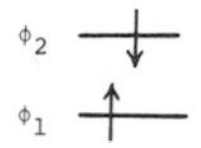

$$\Phi_{\alpha\beta} = \frac{1}{\sqrt{2}}\left\{ \frac{1}{\sqrt{2}} \begin{vmatrix} \phi_1(1)\alpha(1) & \phi_2(1)\beta(1) \\ \phi_1(2)\alpha(2) & \phi_2(2)\beta(2) \end{vmatrix} + \frac{1}{\sqrt{2}} \begin{vmatrix} \phi_2(1)\alpha(1) & \phi_1(1)\beta(1) \\ \phi_2(2)\alpha(2) & \phi_1(2)\beta(2) \end{vmatrix} \right\}$$

$$\hat{S}_z \Phi_{\alpha\beta} = 0 \qquad \{X\text{-}6\}$$

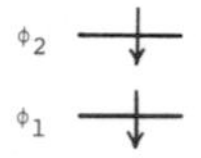

$$\Phi_{\beta\beta} = \frac{1}{\sqrt{2}} \begin{vmatrix} \phi_1(1)\beta(1) & \phi_2(1)\beta(1) \\ \phi_1(2)\beta(2) & \phi_2(2)\beta(2) \end{vmatrix}$$

$$\hat{S}_z \Phi_{\beta\beta} = -\Phi_{\beta\beta} \qquad \{X\text{-}7\}$$

Note that the wavefunction in equation {X-6} is analogous to that of equation {VIII-23} except that both determinants enter with a positive sign (triplet state) while in the singlet state {VIII-23} one of the two determinants entered with a negative sign. Again, by convention, only the function of highest spin eigenvalue (i.e. $\Phi_{\alpha\alpha}$) is used in the SCF calculation.

As far as the $\hat{S}^2$ operator is concerned we may exemplify the situation for a doublet and a triplet state as follows:

$$\hat{S}^2 \phi_\alpha = \frac{1}{2}\left(\frac{1}{2} + 1\right) \Phi_\alpha = 0.75\ \Phi_\alpha$$

$$\hat{S}^2 \Phi_{\alpha\alpha} = 1\ (1 + 1)\ \Phi_{\alpha\alpha} = 2.00\ \Phi_{\alpha\alpha} \qquad \{X\text{-}8\}$$

The 0.750000... and 2.0000... are measures of the spin purity of the state for doublet and triplet multiplicities respectively. The purity of the spin state is guaranteed in the Restricted Hartree-Fock Method (RHF) (see Figure X-5A). However, for the Unrestricted Hartree-Fock Method (UHF) (cf. Figure X-5B) the spin purity of the wavefunction is no longer guaranteed.

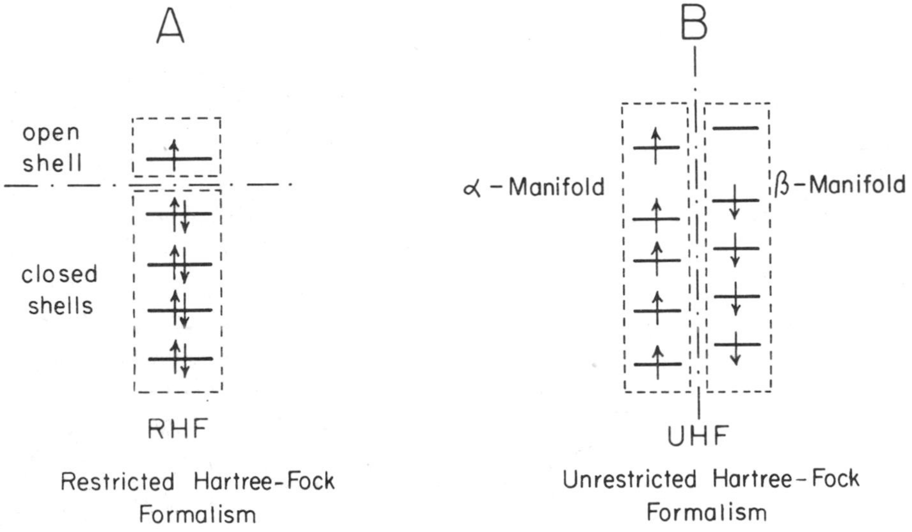

Figure X-5. (A) Restricted Hartree-Fock Formalism
(B) Unrestricted Hartree-Fock Formalism

2. Restricted Hartree-Fock (RHF) Formalism

The MO of the system are divided into closed shell MO (ϕ_c) and open shell MO (ϕ_o) which are restricted to be orthonormal

$$\phi = (\phi_c, \phi_o) \qquad \{X-9\}$$

If the wavefunction is constructed from these MO as a single Slater determinant the molecular electronic energy for an arbitrary open shell may be written as follows:

$$\begin{aligned} E = {} & \underbrace{2\sum_{k=1}^{n_c} H_k + \sum_{k=1}^{n_c}\sum_{\ell=1}^{n_c}(2J_{k\ell} - K_{k\ell})}_{\text{closed shell}} \\ & + \underbrace{2\sum_{m=n_c+1}^{n_o} H_m + \sum_{m=n_c+1}^{n_o}\sum_{m=n_c+1}^{n_o}(2J_{mn} - K_{mn})}_{\text{open shell}} \\ & + \underbrace{\sum_{k=1}^{n_c}\sum_{m=n_c+1}^{n_o}(2J_{km} - K_{km})}_{\text{closed shell-open shell interaction}} \end{aligned} \qquad \{X-10\}$$

where

k,ℓ are the running indices for closed shell orbitals
m,n are the running indices for open shell orbitals
n_c is the number of MO in the closed shell
n_o is the number of MO in the closed and open shell together (i.e. $n_o - n_c$ is the number of MO in the open shell only)

For a relatively large class of systems the energy may be written as follows:

$$
\left.
\begin{aligned}
E = {} & 2 \sum_{k=1}^{n_c} H_k + \sum_{k=1}^{n_c} \sum_{\ell=1}^{n_c} (2J_{k\ell} - K_{k\ell}) \\
& + f\Big[2 \sum_{m=1}^{n_s} H_m + \sum_{m=1}^{n_s} \sum_{n=1}^{n_s} (2aJ_{mn} - bK_{mn})\Big] \\
& + 2f \sum_{k=1}^{n_c} \sum_{m=1}^{n_s} (2J_{km} - K_{km})
\end{aligned}
\right\} \quad \{X\text{-}11\}
$$

where a, b depend on the specific spin state of a given configuration [cf. tabulated values in C.C.J. Roothaan: Rev. Mod. Phys. 32, 179 (1960)]. The fractional occupancy (f) varies between zero and unity ($0 \leq f \leq 1$) where zero is the value for an empty shell and unity is the value of f for a closed shell. For example, doubly ionized methane (CH_4^{2+})(forced to be in a tetrahedral geometry) represents an open shell even though it contains an even number of electrons. Its fractional occupancy is 2/3 (i.e. , that is f = 4/6 = 2/3). Note that n_s is the number of degenerate spatial orbitals within the open shell:

$$
\begin{aligned}
&\text{for triple degeneracy } (t_{2g}) \;\ldots\ldots\; && n_s = 3 \\
&\text{for double degeneracy } (e_g) \;\ldots\ldots\; && n_s = 2 \qquad \{X\text{-}12\} \\
&\text{for no degeneracy } \ldots\ldots\ldots\ldots\ldots && n_s = 1
\end{aligned}
$$

A number of important cases fall within the range of this method including the half closed shell (open shell consisting of a single half occupied orbital) and partially occupied degenerate sets of orbitals such as all the states arising from the p^N configurations for an atom ($1 \leq N \leq 5$) and also the lowest triplet state of a closed shell molecule. A slightly more general expression for the energy allows for the case of two open shells of different symmetry to be handled (e.g. linear molecules of $\sigma\pi^N$ configuration where $1 \leq N \leq 3$)

[cf. S. Huzinaga, Phys. Rev. 120, 866 (1960)].

The two sets of orbitals (ϕ_c and ϕ_o) are eigenfunctions of two molecular Fock operators ($\hat{F}_c$, $\hat{F}_o$)

$$\hat{F}_c \phi_c = \phi_c \varepsilon_c \quad \text{(for closed shells)}$$
$$\hat{F}_o \phi_o = \phi_o \varepsilon_o \quad \text{(for open shells)} \qquad \{X\text{-}13\}$$

where the corresponding Fock matrices are:

$$\underline{\underline{F}}_c = \underline{\underline{H}} + 2\underline{\underline{J}}_c - \underline{\underline{K}}_c + 2\underline{\underline{J}}_o - \underline{\underline{K}}_o + 2\alpha\underline{\underline{L}}_o - \beta\underline{\underline{M}}_o$$
$$\underline{\underline{F}}_o = \underline{\underline{H}} + 2\underline{\underline{J}}_c - \underline{\underline{K}}_c + 2a\underline{\underline{J}}_o - b\underline{\underline{K}}_o + 2\alpha\underline{\underline{L}}_c - \beta\underline{\underline{M}}_c \qquad \{X\text{-}14\}$$

The L and M operators are defined as follows:

$$\hat{L}_c = \sum_k^{n_c} |\phi_k\rangle\langle\phi_k| \, \hat{J}_o$$

$$\hat{M}_c = \sum_k^{n_c} |\phi_k\rangle\langle\phi_k| \, \hat{K}_o$$

$$\hat{L}_o = f \sum_{m=1}^{n_s} |\phi_m\rangle\langle\phi_m| \, \hat{J}_o \qquad \{X\text{-}15\}$$

$$\hat{M}_o = f \sum_{m=1}^{n_s} |\phi_m\rangle\langle\phi_m| \, \hat{K}_o$$

where

$$\alpha = \frac{1 - a}{1 - f}$$

and

$$\beta = \frac{1 - b}{1 - f} \qquad \{X\text{-}16\}$$

The basis of Roothaan's approach to the RHF open-shell problem is to redefine the Fock operator to include open shell-closed shell coupling terms. Other approaches have also been used.

3. Unrestricted Hartree-Fock (UHF) Theory

In the UHF method the α and β spins are separated into two manifolds as illustrated by the vertical line in Figure X-5B. This method in which double occupancy in the "closed shell" part is no longer enforced is also referred to as Different Orbitals for Different Spins (DODS) or Spin Polarized Hartree-Fock.

Again, we have two Hartree-Fock problems, where now one is for the α spin manifold and one for the β spin manifold.

$$\hat{F}^{\alpha}\phi^{\alpha} = \phi^{\alpha}\varepsilon^{\alpha}$$
$$\hat{F}^{\beta}\phi^{\beta} = \phi^{\beta}\varepsilon^{\beta} \qquad \{X\text{-}17\}$$

where

$$\phi^{\alpha} = \eta\underline{\underline{C}}_{\alpha}$$
$$\phi^{\beta} = \eta\underline{\underline{C}}_{\beta} \qquad \{X\text{-}18\}$$

The two Fock matrices may be written:

$$\underline{\underline{F}}^{\alpha} = \underline{\underline{H}} + \underline{\underline{J}} - \underline{\underline{K}}^{\alpha}$$
$$\underline{\underline{F}}^{\beta} = \underline{\underline{H}} + \underline{\underline{J}} - \underline{\underline{K}}^{\beta} \qquad \{X\text{-}19\}$$

where $\underline{\underline{K}}^{\alpha}$ and $\underline{\underline{K}}^{\beta}$ depend on $\underline{\underline{\rho}}^{\alpha}$ and $\underline{\underline{\rho}}^{\beta}$ the respective density matrices:

$$\underline{\underline{\rho}}^{\alpha} = \underline{\underline{C}}_{\alpha}\underline{\underline{C}}_{\alpha}^{+},$$
$$\underline{\underline{\rho}}^{\beta} = \underline{\underline{C}}_{\beta}\underline{\underline{C}}_{\beta}^{+} \qquad \{X\text{-}20\}$$

Elements of the $\underline{\underline{J}}$, $\underline{\underline{K}}^{\alpha}$ and $\underline{\underline{K}}^{\beta}$ matrices are defined as follows:

$$\left.\begin{aligned}
J_{rs} &= \sum_{t}\sum_{n} (\rho^{\alpha}_{tu} + \rho^{\beta}_{tu}) \langle \eta_r(1)\eta_t(2) | \frac{1}{r_{12}} | \eta_s(1)\eta_r(2) \rangle \\
K^{\alpha}_{rs} &= \sum_{t}\sum_{n} \rho^{\alpha}_{tu} \langle \eta_r(1)\eta_t(2) | \frac{1}{r_{12}} | \eta_u(1)\eta_s(2) \rangle \\
K^{\beta}_{rs} &= \sum_{t}\sum_{n} \rho^{\beta} \langle \eta_r(1)\eta_t(2) | \frac{1}{r_{12}} | \eta_u(1)\eta_s(2) \rangle
\end{aligned}\right\} \quad \{X-21\}$$

The total electronic energy of the system may be written as:

$$E = \sum_{i}^{\alpha+\beta} H_i + \frac{1}{2} \sum_{i}^{\alpha+\beta} \sum_{j}^{\alpha+\beta} J_{ij} - \frac{1}{2} \left(\sum_{i}^{\alpha} \sum_{j}^{\alpha} K^{\alpha}_{ij} + \sum_{i}^{\beta} \sum_{j}^{\beta} K^{\beta}_{ij} \right) \quad \{X-22\}$$

As mentioned previously the wavefunction obtained by the UHF method is not a pure spin state and thus is, in a sense, unrealistic since it is a wavefunction "contaminated" by spin states of higher multiplicity. Although methods of spin projection of the desired spin component or of annililation of the next highest spin component (the principal contaminant) are available they are fairly complicated computationally and more importantly, the projected wavefunctions are not variationally optimal.

An important consequence of this contamination is that the energy computed in the UHF approach is not a quantum mechanically rigorous upper bound for the energy of a molecular state. In other words, the energy already incorporates a small (and undesired) fraction of the correlation energy.

CHAPTER XI

LIMITATIONS AND APPLICATIONS OF OPEN SHELL SCF THEORIES

1. Limitations of Open Shell SCF Theories

(a) The Correlation Problem

Whatever was said concerning the electron correlation problem in connection with closed shell systems (cf. Chapter VIII) is also applicable to open shell molecules since in nearly every open shell problem some electrons are paired forming a partial closed electronic shell. Consequently, in a process (stereochemical or reactive) involving open shell (OS) species

$$A_{OS} \rightarrow X^{\dagger}_{OS} \rightarrow B_{OS} \qquad \{XI\text{-}1\}$$

the correlation energy is expected to remain constant or nearly so provided that the scheme of electron pairing does not change greatly during the process.

However, in many instances comparison between closed shell (CS) and open shell (OS) systems is necessary particularly when a process may occur (at least in principle) either through a closed shell or an open shell mechanism.

$$\begin{array}{ccccc} A_{OS} & \rightarrow & X^{\dagger}_{OS} & \rightarrow & B_{OS} \\ \uparrow & & & & \downarrow \\ A_{CS} & \rightarrow & X^{\dagger}_{CS} & \rightarrow & B_{CS} \end{array} \qquad \{XI\text{-}2\}$$

In this case the correlation energy is expected to be nearly constant at one value for the CS process and again approximately constant at a different value for the OS process. But, in moving from one horizontal process to another horizontal process in equation {XI-2} there is expected to be a significant change in correlation energy.

Limiting the discussion to cases where a single electron pair is unpaired on going from the CS to OS or reformed on going from the OS to CS we find that the CS systems have one more electron pair that the OS systems. Consequently, the absolute value of $E_{cor}^{A(CS)}$ and $E_{cor}^{B(CS)}$ is larger than the absolute value of $E_{cor}^{A(OS)}$ and $E_{cor}^{B(OS)}$ respectively.

$$E_{cor}^{A(CS)} = E_{cor}^{A(OS)} + \Delta E_{cor}^{A}$$

$$E_{cor}^{B(CS)} = E_{cor}^{B(OS)} + \Delta E_{cor}^{B}$$

$$\{XI-3a,b\}$$

While the change in the correlation energy of the isoelectronic species is similar in magnitude, the actual value can only be estimated to be in the range zero to $\sim$-30 Kcal/mole.

$$0 \geq \Delta E_{cor}^{A} \approx \Delta E_{cor}^{B} \geq -30 \text{ Kcal/mole} \qquad \{XI-4\}$$

The result of the problem is that the absolute values of $\Delta E_{CS \to OS}^{A}$ and $\Delta E_{CS \to OS}^{B}$ are computed to be smaller (upper part of Figure XI-1) than the corresponding experimental values (lower part of Figure XI-1) and the discrepancy may be anywhere between 0 and 30 Kcal/mole. This is a limitation that must be kept in mind whenever open shell SCF theories are applied and the results compared to other calculations involving closed shell species.

(b)   The Utility of RHF and UHF Theories

The energy of the open shell RHF theory is a quantum mechanically rigorous upper bound and is thus favoured a priori. The disadvantage of this method is that the eigenvalues obtained cannot be interpreted as orbital energies analogous   to the case of closed shell SCF; i.e. Koopmann's theorem at least as originally formulated is not applicable to the RHF results.   Normally, reliable spin densities may not be computed from a RHF wavefunction. For example, the methyl radical, $\cdot CH_3$, would have a full spin at the carbon nucleus and zero spin at each of the three hydrogen nuclei:

$$
\begin{array}{lll}
0.00 \quad \text{H} & & \\
 & \diagdown & \\
1.00 \quad & \text{C}\!\!-\!\!\text{H} & 0.00 \qquad \{XI-5\} \\
 & \diagup & \\
0.00 \quad \text{H} & &
\end{array}
$$

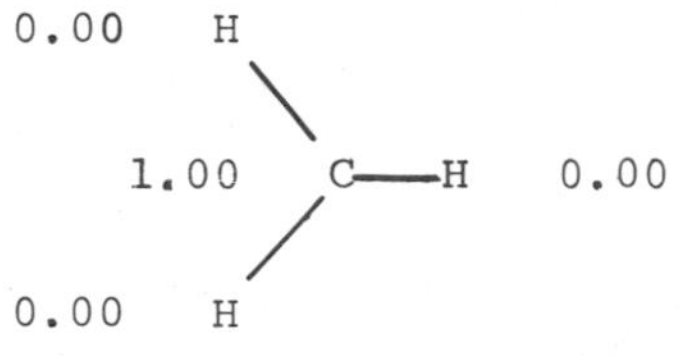

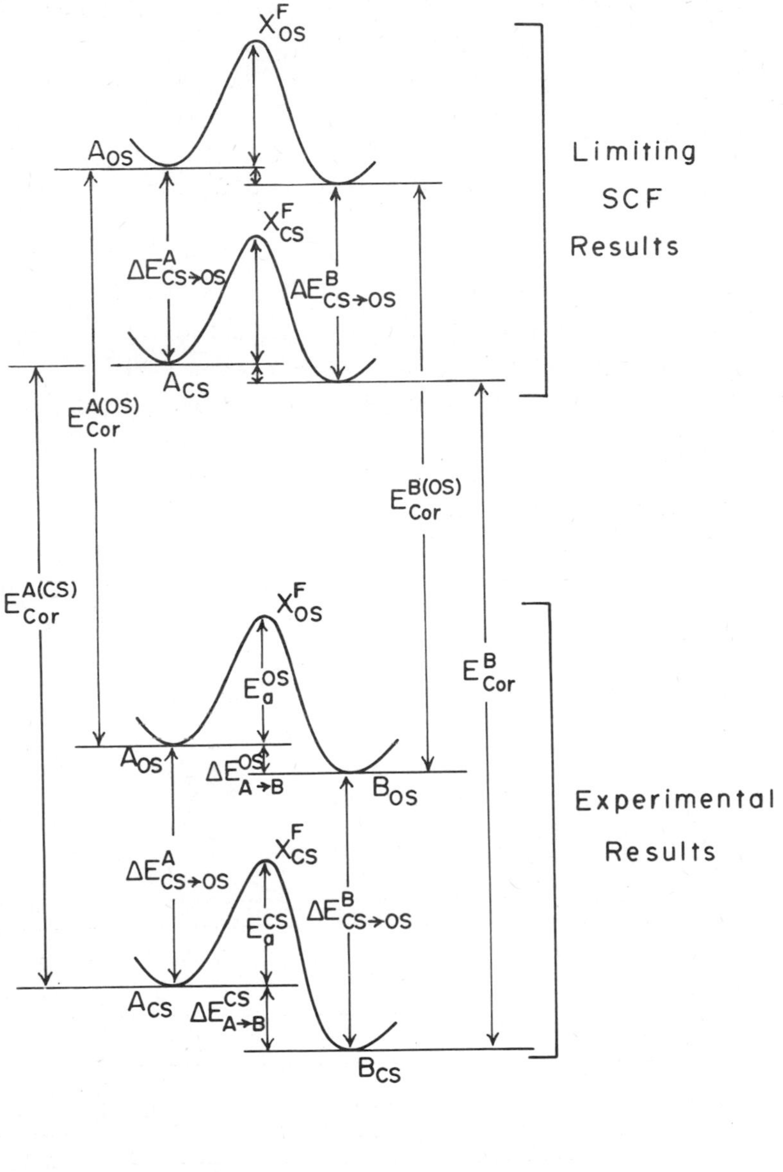

Figure XI-1. Experimental and limiting SCF results for energy differences. The effects of the correlation energy differences.

In contrast, the spin densities computed for CH_3 from a (spin polarized) UHF wavefunction show a noticeable delocalization of the spin density.

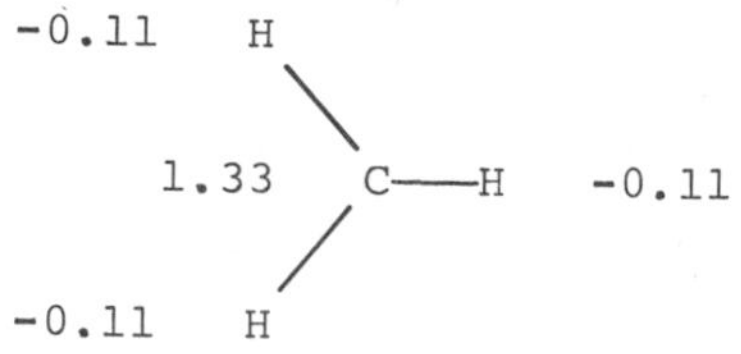

{XI-6}

In addition to more realistic spin densities, the UHF molecular orbitals have the traditional interpretation (i.e. Koopmans' theorem is applicable). However, a disadvantage is that the wavefunction is contaminated by other spin states. The UHF energies do not represent rigorous upper bounds and are strictly not physically realistic.

For all these reasons, it is understandable that both methods find extensive use and their applicability is determined by careful weighing of the advantages and disadvantages of one method against the other for a given problem. Some comparison between the RHF and UHF results for conformational changes in $H_2C=SiH_2$ is given in Table XI-1.

<u>TABLE XI-1</u>

A COMPARISON OF RHF AND UHF RESULTS FOR THE LOWEST TRIPLET STATE OF $H_2C=SiH_2$

	E_{PLANAR} (hartree)		$E_{PYRAMIDAL}$ (hartree)		$E_{PLANAR}-E_{PYR}$ (Kcal/mole)	
θ	RHF	UHF	RHF	UHF	RHF	UHF
0	-327.18387	-327.18985	-327.22187	-327.22613	23.85	22.77
90	-327.20174	-327.20698	-327.23179	-327.23538	18.86	17.82
E_0-E_{90} Kcal/mole	11.21	10.75	6.22	5.80		

2. Applications of Open Shell SCF Theories

    (a) Doublet States (Free Radicals and Molecular Ions)

    Systems that have doublet spin multiplicity may be
classified into three categories:

    (i)   Stable Species with an Odd Number of Electrons
    (ii)   Free Radicals and
    (iii)   Molecular Ions.

    In the first category are species with an odd number of
electrons that are known to be paramagnetic such as $\cdot NO$ and
$\cdot NO_2$. Others, such as transition metal complexes containing
a central metal with a single outer electron (e.g. Ag) would
form doublet complexes provided the ligands are   closed
shell molecules.

    In the second category are species that have a trans-
ient existence such as $\cdot CH_3$ or substituted carbon radicals as
exemplified by $\cdot CH_2-SH$. These radicals may be formed from
homolytic bond cleavage of any bond and the radical centres
(A and B) may involve atoms other than carbon.

$$A\!-\!B \;\rightarrow\; A\cdot \;+\; B\cdot \qquad\qquad \{XI-7\}$$

    In the third category are molecular ions generated from
closed shell molecules (M) either by the addition or removal
of a single electron. The energy differences

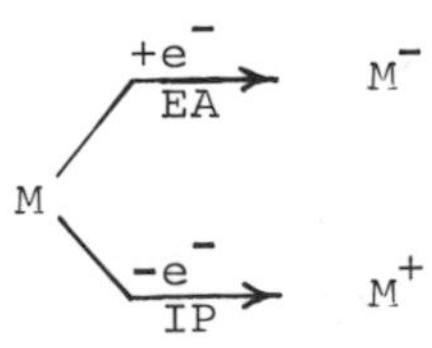

$$
\begin{array}{c}
\quad \xrightarrow[\;EA\;]{+e^{-}} \quad M^{-} \\[2mm]
M \\[2mm]
\quad \xrightarrow[\;IP\;]{-e^{-}} \quad M^{+}
\end{array}
\qquad\qquad \{XI-8\}
$$

associated with these two processes are called the electron
affinity (EA) and the ionization potential (IP) respectively.
Radical cations and radical anions may also be generated at
the electrodes in organic electrochemical reactions or by
various (usually inorganic) one electron oxidizing and

reducing agents. The process of molecular ionization is the basis of photoelectron spectroscopy or ESCA. The removal of an electron need not necessarily come from the outer (valence) shell. If it does a low energy (uv) ionization is involved and the method termed photoelectron spectroscopy. The high energy (x-ray) ionization spectroscopy is called ESCA. While the experimental techniques are different the basic theory that underlies these two methods is the same. The high energy excitation generates excited states of the molecular ion which may be generated by low energy ionization:

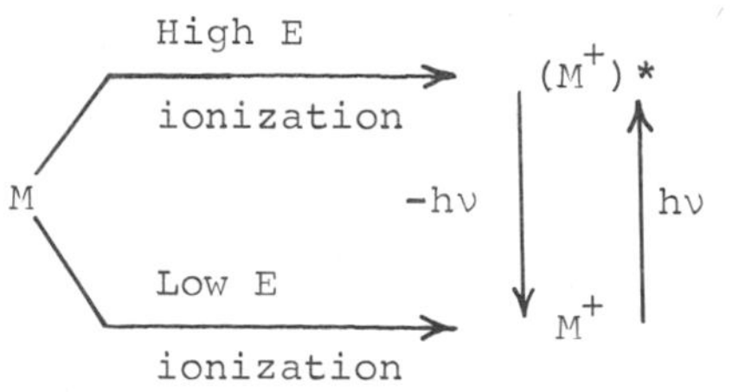

$$
\begin{array}{l}
M \quad \xrightarrow[\text{ionization}]{\text{High E}} \quad (M^+)^* \\[2em]
\qquad\qquad -h\nu \quad \Big\downarrow \quad \Big\uparrow\, h\nu \qquad \{XI\text{-}9\} \\[2em]
\quad \xrightarrow[\text{ionization}]{\text{Low E}} \quad M^+
\end{array}
$$

This interdependence of ground and excited states is illustrated for H_2S in Figure XI-2. Free radical formation and molecular ion generation are interrelated as shown by Figure XI-3 for the methane and ammonia families. The variations of the total energy with pyramidal angle are shown for four selected species from Figure XI-3 and reiterated in equation {XI-10}.

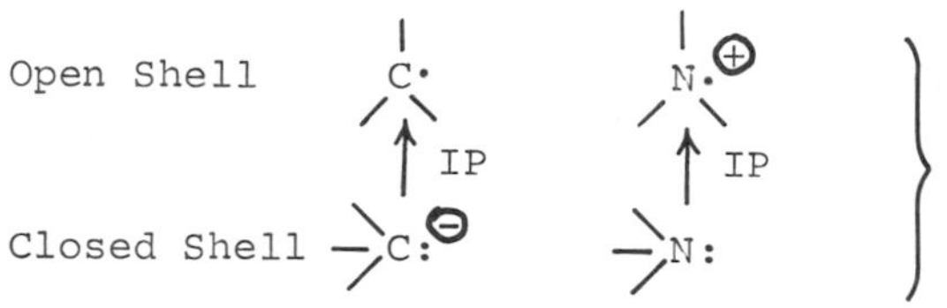

Open Shell

Closed Shell   {XI-10}

Note that here an electron pair is uncoupled both for the C and N series. Consequently the comparison of each pair yields adiabatic ionization potentials with a systematic error corresponding to the change in correlation energy as previously discussed. The situation is further aggravated in the case of the carbon series by the fact that basis sets optimized for the neutral atom are usually insufficient for the description of electron density on a negatively charged

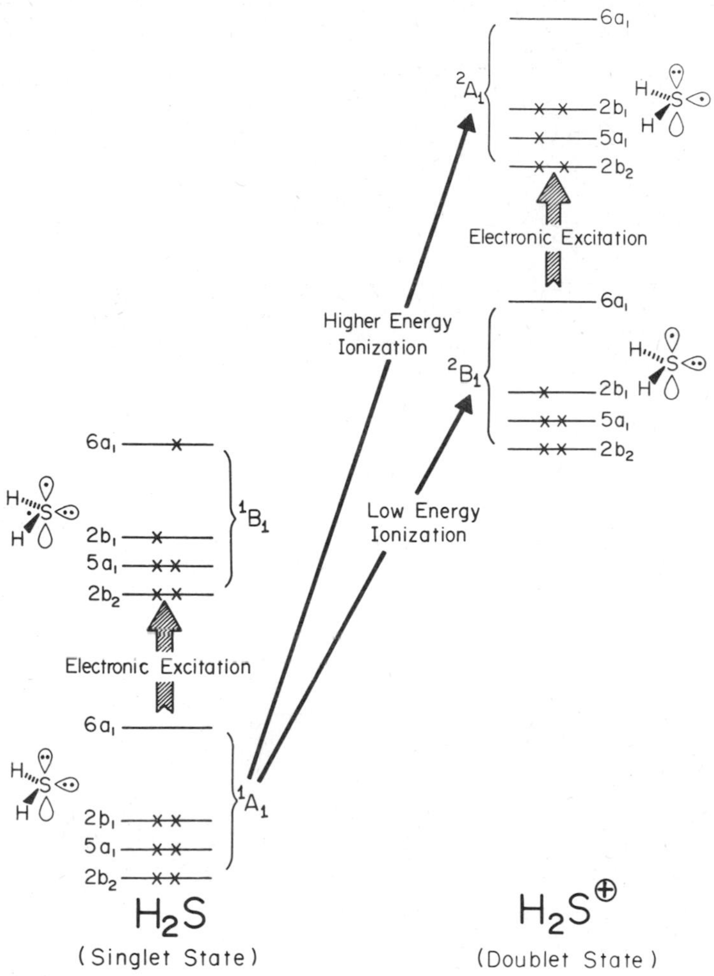

Figure XI-2. Ground and excited states of H_2S.

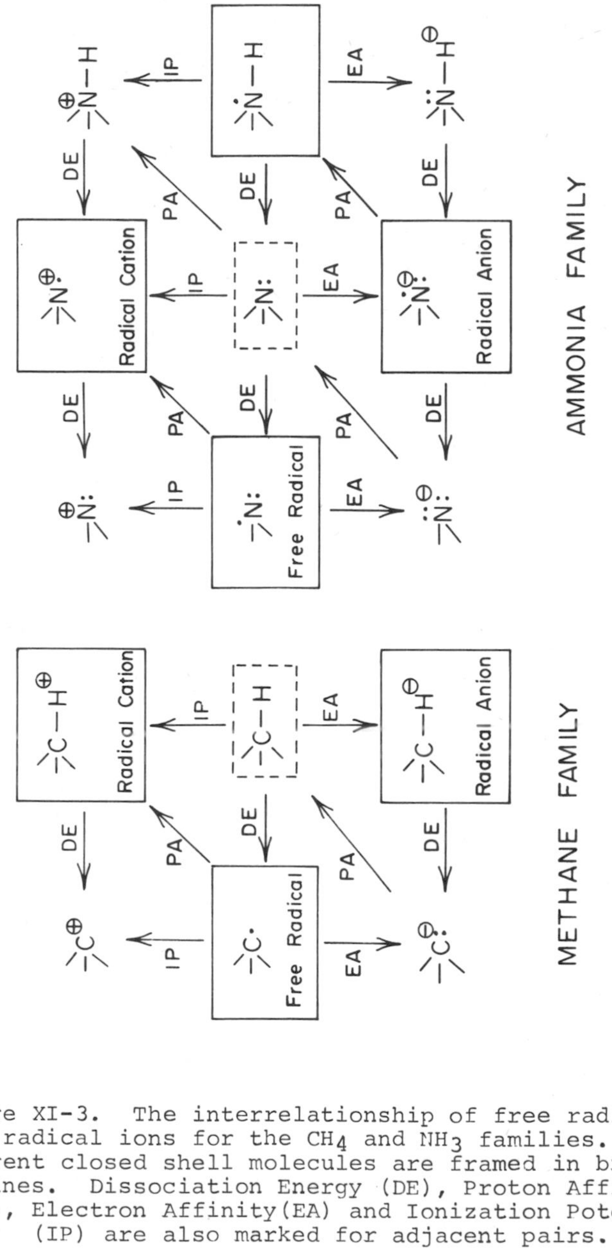

Figure XI-3. The interrelationship of free radicals and radical ions for the CH_4 and NH_3 families. The parent closed shell molecules are framed in broken lines. Dissociation Energy (DE), Proton Affinity (PA), Electron Affinity(EA) and Ionization Potential (IP) are also marked for adjacent pairs.

molecule. The combination of these two factors is thought to be responsible for the negative IP for the $:CH_3^- \rightarrow \cdot CH_3$ process which is apparent from Figure XI-4. However, the stereochemistries of the four species (pyramidal $:CH_3^-$ and $:NH_3$ planar CH_3 and $\cdot NH_3^+$) are expected to be correct due to the use of extensive, well balanced basis sets (R. E. Kari, Int. J. Quant. Chem.).

In general, the variation of energy with conformational change is expected to be correct for any free radical since the pairing scheme of the electrons usually remains unaltered. One such example is a carbon radical next to sulfur, $\cdot CH_2-SH$, which is traditionally thought to be stabilized by $d\pi-p\pi$ conjugation either

Electron-releasing conjugation:

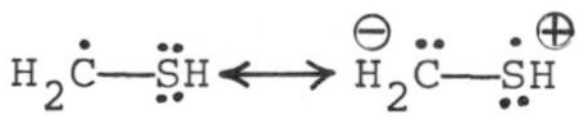

$$H_2\overset{\cdot}{C}-\overset{\cdot\cdot}{\underset{\cdot\cdot}{S}}H \longleftrightarrow H_2\overset{\ominus}{C}-\overset{\cdot\cdot}{\underset{\cdot\cdot}{S}}\overset{\cdot\, \oplus}{H} \qquad \{XI-11a\}$$

or Electron-sharing conjugation

$$H_2\overset{\cdot}{C}-\overset{\cdot\cdot}{\underset{\cdot\cdot}{S}}H \longleftrightarrow H_2C=\overset{\cdot}{\underset{\cdot\cdot}{S}}H \qquad \{XI-11b\}$$

Considering torsion about the C-S bond (θ) and inversion at the carbon atom (ϕ) as variable internal coordinates, one may generate a conformational energy surface of the type:

$$E = E(\theta,\phi) \qquad \{XI-12\}$$

(see Figure XI-5). Cross-sections in the above surface

$$E = E(\phi)$$
$$\{XI-13a,b\}$$
$$E = E(\theta)$$

are shown in Figure XI-6. The planar structure corresponds to the most favoured conformation and there is fairly low barrier to rotation about the C-S bond. The charge and spin distribution as computed within the UHF method are shown in $\{XI-14a\}$

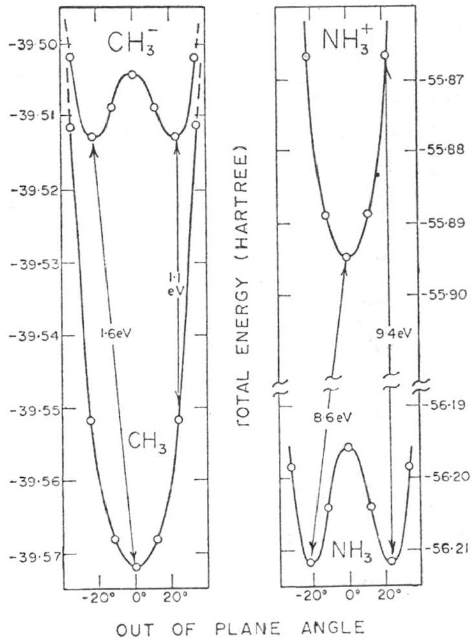

Figure XI-4. Conformational energy change with angle of pyramidal inversion for CH_3, $^-{:}CH_3$, $\cdot NH_3^+$, $:NH_3$.

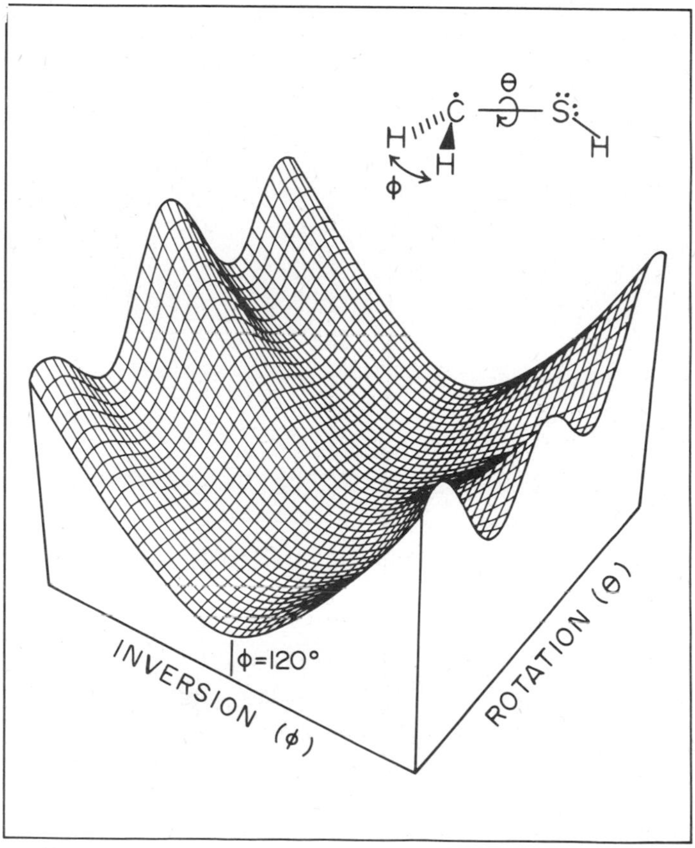

Figure XI-5. Rotation-inversion energy surface for H_2C-SH.

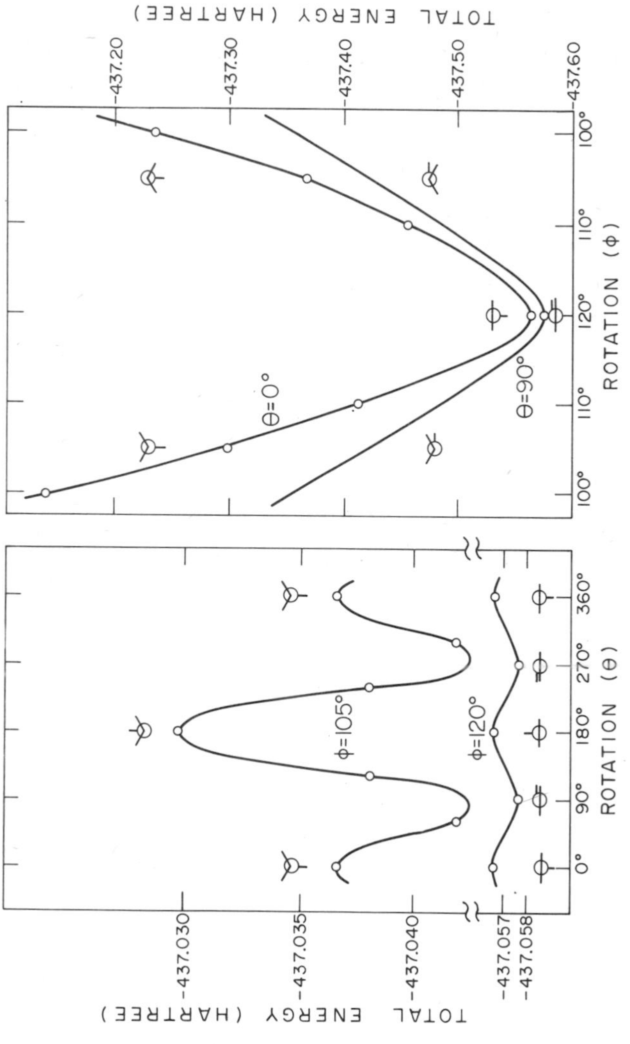

Figure XI-6. Rotation cross-section and inversion cross-section of the conformational energy surface of $H_2\dot{C}$-SH.

and {XI-14b} respectively

Charge Densities

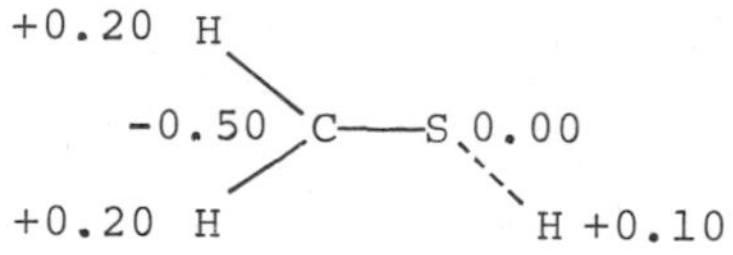

{XI-14a}

Spin Densities

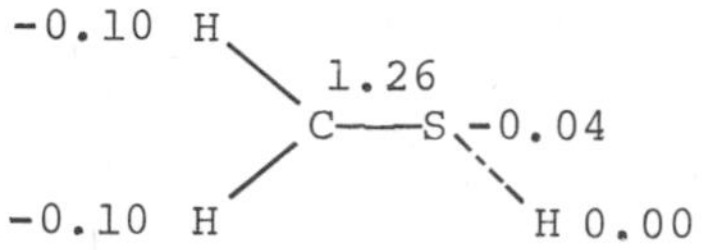

{XI-14b}

(b) Triplet States (Theoretical Stereochemistry and Theoretical Photochemistry)

The relative stabilities of the ground singlet (S_o) and the first excited triplet (T_1) state conformers of molecules with multiple bonds such as >C=X are of considerable interest. $H_2C=SiH_2$ and its derivatives are interesting examples. By analogy to ethylene the two lowest states might be expected to have the following most stable conformations:

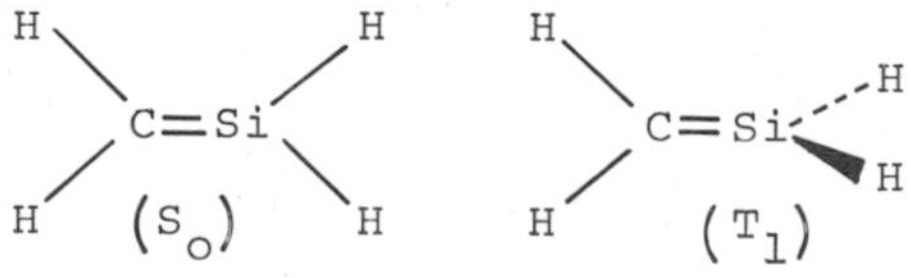

{XI-15}

Ab initio SCF calculations (RHF on S_o, UHF on T_1) could then be used to check this hypothesis.

As far as the relative energies of the S_o and T_1 states are concerned the difference in correlation energy between singlets and triplets presents a problem. As was pointed out at the beginning of this chapter, the S_o state, having one more electron pair, possesses a larger absolute correlation energy than the T_1 state. The difference between the two correlation energies may range between 0 and ∿30 Kcal mole^{-1}. Consequently a fairly large degree of uncertainty is involved in trying to establish the relative stabilities of the T_1 and S_o states.

To partially remove this difficulty the study could be calibrated against a molecule in which the T_1-S_o separation is known experimentally. Ethylene might lend itself to such a comparison since the relaxed T_1-S_o separation is known from spectroscopic observations and photochemical sensitization studies to be approximately 60 Kcal/mole. The results shown in Figure XI-7 indicates that $\Delta E(T_1$-$S_o)$ in ethylene is about 18 Kcal less than the experimental value although this figure is dependent on the basis set used. Since the valence electrons might be considered to be "looser" in the larger atom containing molecule, it might be expected that the valence pair correlation is smaller. This expectation is supported by the fact that the pair correlation energy in helium-like ions is inversely proportional to the "Size" of the electron pair [Theoret. Chim. Acta 37, 171 (1975)]. Thus one might estimate the correlation energy difference between T_1 and S_o of the H_2C=SiH_2 to be rather less than 18 Kcal mole^{-1} perhaps on the order of -10 Kcal mole^{-1}. With this estimate and the computed energy difference of +1.4 kcal mole^{-1}, the separation between T_1 and S_o in H_2C=SiH_2 and in particular its substituted derivatives could be so small that at least on the basis of thermodynamic considerations both states would be populated at room temperature.

The conformational study included the geometry optimization of four points of H_2C=SiH_2 using a minimal basis set:

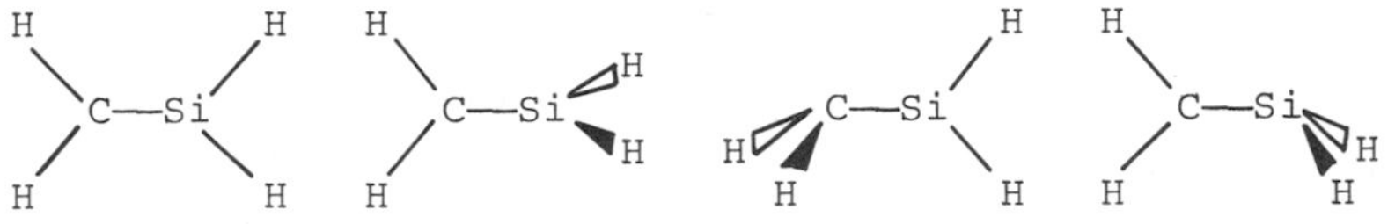

I (S_o) IV (T_1) III (S_o) II (T_1)

{XI-16a-d}

One dimensional cross-sections of the conformational energy hypersurface (see {XI-17})

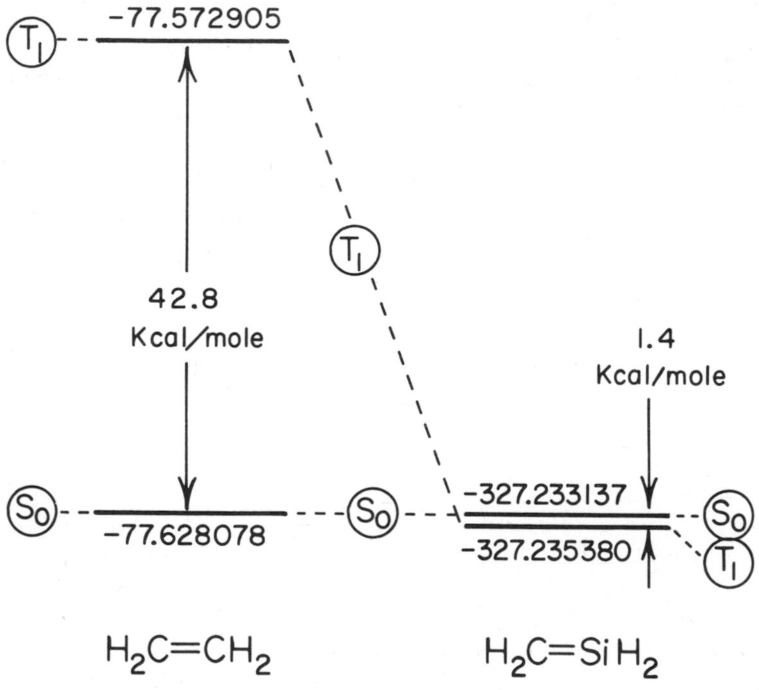

Figure XI-7. Thermodynamic stabilities of S_O and T_1 states of $H_2C{=}CH_2$ and $H_2C{=}SiH_2$ as computed with a minimal basis set.

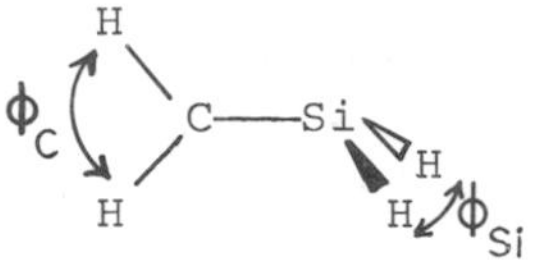

$$E = E(\theta, \phi_{Si}, \phi_C) \qquad \{XI-17\}$$

are shown in Figure XI-8. From this figure, it is apparent that structures I and II are minima and structures III and IV transition states.

Mulliken population analysis carried out on the most stable conformation of the singlet (S_o) and triplet (T_1) state of silaethylene resulted in the following charge distribution

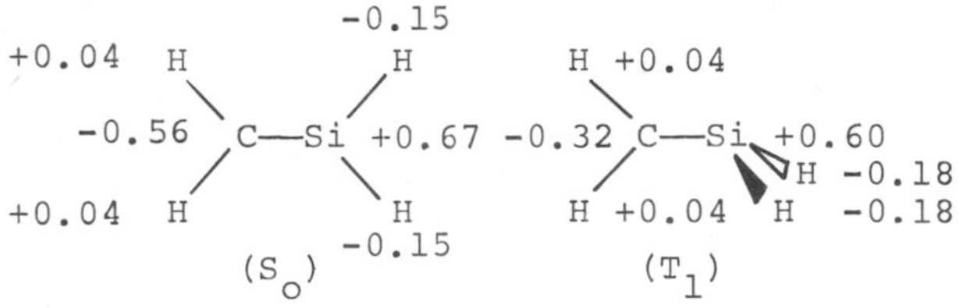

{XI-18}

Considering three "resonance structures" sometimes invoked in characterizing the reactivity of silaethylene

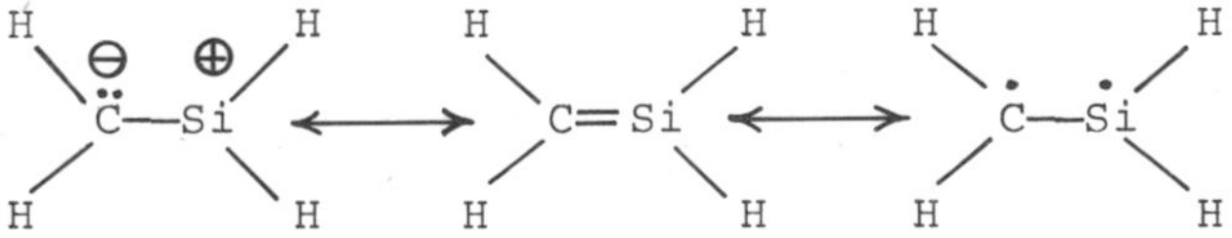

{XI-19a,b,c}

on the basis of the Mulliken population analysis the first two resonance contributors may be loosely associated with S_o while the second two resonance contributors may be more characteristic of T_1.

Another molecule in the $H_2C{=}X$ family in which the triplet-singlet separation as well as the variation with geometry change is of some interest is $H_2C{=}O$. Here the electron pairs are tighter than in $H_2Si{=}CH_2$, therefore the change in correlation energy between the singlet and triplet is expected to be greater, probably nearer the 30 Kcal/mole limit. More interesting than the T_1-S_o separation is the variation of total energy as a function of molecular

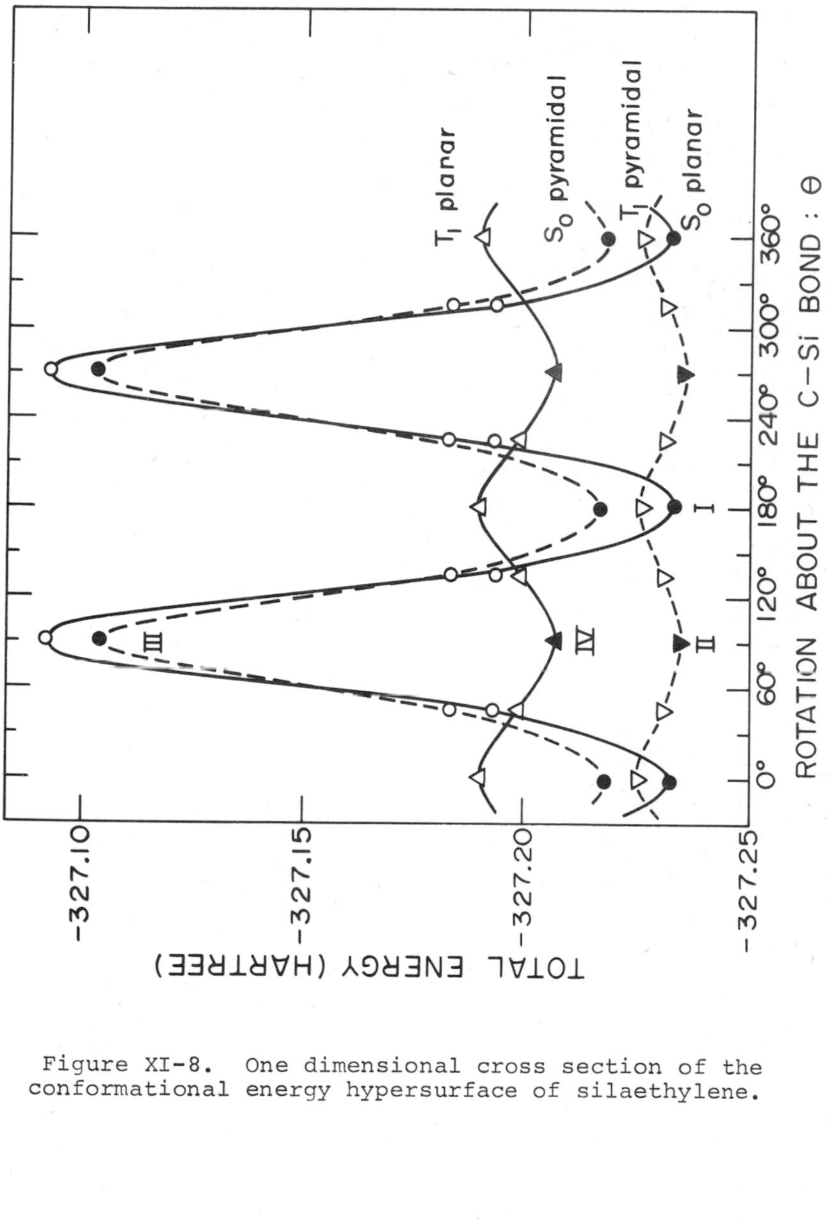

Figure XI-8. One dimensional cross section of the conformational energy hypersurface of silaethylene.

rearrangement. This leads to the domain of theoretical photochemistry. An interesting photochemical reaction of cyclic carbonyl compounds involves the following ring expansion

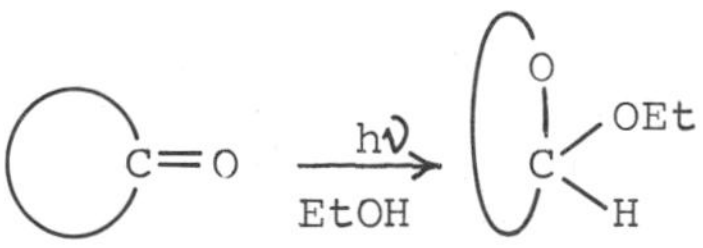

{XI-20}

The mechanism of the above reaction for strained rings is often rationalized by postulating an oxacarbene intermediate

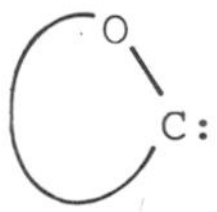

{XI-21}

The formation of this oxacarbene may be visualized to occur either via a concerted mechanism (direct alkyl migration from carbon to oxygen) or via a diradical intermediate which is generated from the excited carbonyl compound through a Norish type I homolytic carbon-carbon bond cleavage. The various possibilities are summarized in the following scheme

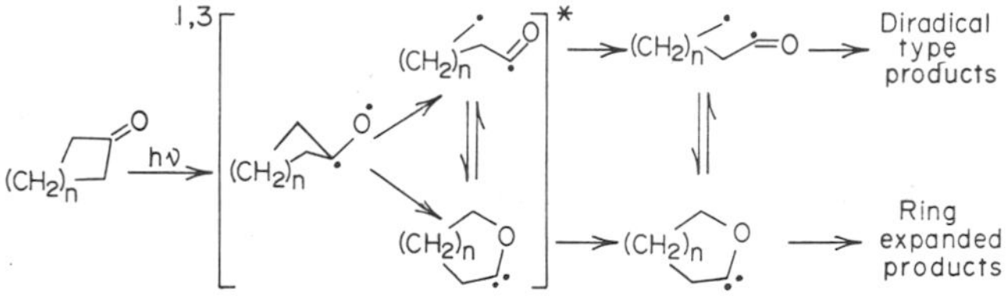

{XI-22}

The above problem is too large for ab initio computational study and so the mechanistic features of such a photochemical rearrangement may (hopefully) be simulated by the rearrangement of formaldehyde:

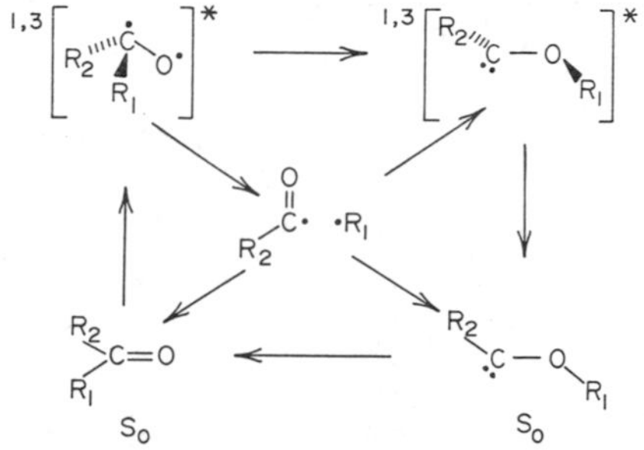

{XI-23}

SCF computations were carried out similar to those for the $H_2C{=}SiH_2$ system using RHF for the closed shell ground state (S_o) and UHF for the first excited triplet state (T_1). The thermodynamics of the two alternative mechanisms as shown in Figure XI-9 indicates a higher energy diradical intermediate.

The kinetic aspects of the reaction are illustrated in Figure XI-10 and would suggest that the diradical and concerted mechanism in the lowest triplet state could be competitive for formaldehyde.

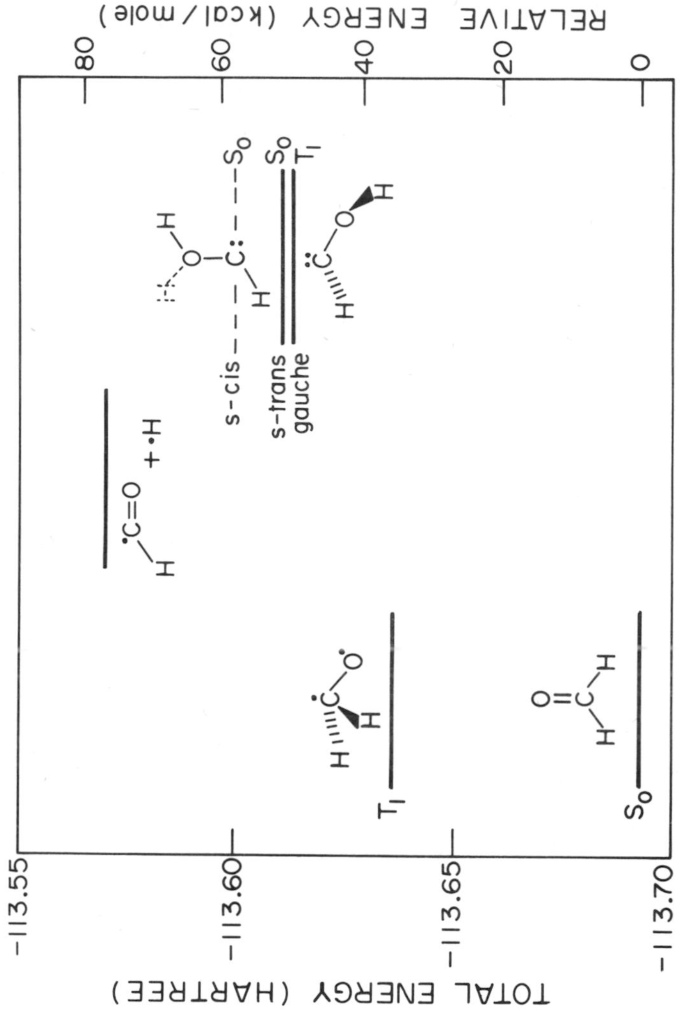

Figure XI-9. Thermodynamics of two alternative mechanisms for formaldehyde rearrangements.

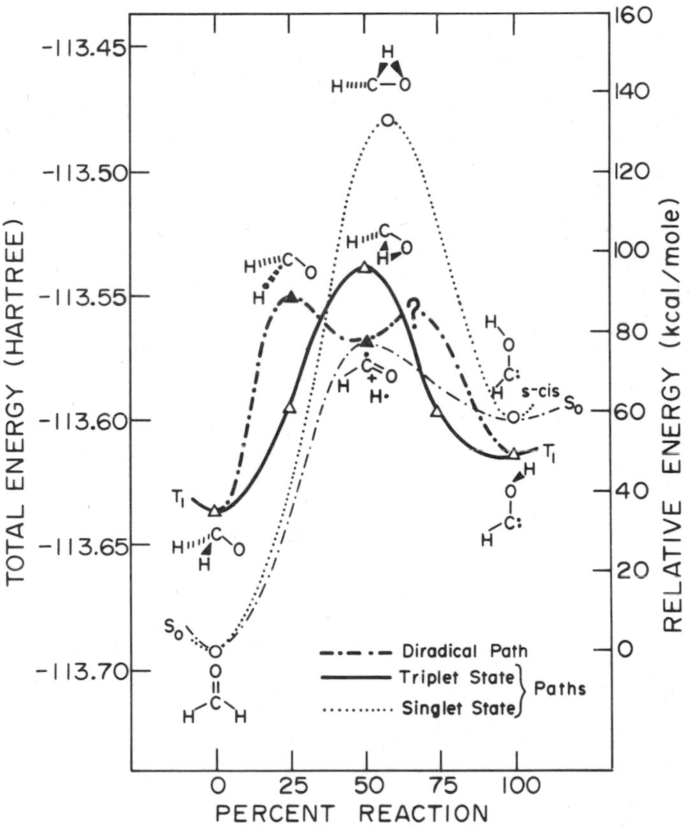

Figure XI-10. Kinetic aspects of the diradical and concerted mechanisms for the formaldehyde rearrangements.

SECTION D

PRACTICAL ASPECTS OF MO COMPUTATIONS

CHAPTER XII

BASIS SETS FOR MOLECULAR ORBITAL CALCULATIONS

1. Introduction

The use of a basis set of analytic functions in which to expand the unknown functions (the molecular orbitals) has been discussed previously at several points. Historically, the basis functions were chosen on the basis of their relationship to atomic orbitals hence the term LCAO-MO. It must be emphasized that the initial choice of a basis set will fix forever the reliability of the results of a calculation. There is no way to overcome a poor initial choice of basis functions. Moreover, the choice may be poor not only for the relatively obvious reason of a restriction to too few or too few types of basis functions but also for the more subtle reason of a lack of balance between the basis sets representing the various atomic centres in a molecule. The determination of carefully optimized and balanced basis sets is an extremely difficult and time consuming process but a very necessary one. With the increasing number of molecular calculations, experience is leading to fairly solid conclusions as to the sort of results which may be expected from various sizes and types of basis sets.

Historically, the basis functions were chosen by physical analogy to atomic orbitals. However, this physical analogy can lead to extreme but avoidable problems in mathematics and computer science. At present the emphasis in the choice of basis functions is on the computability of molecular integrals and particularly the three- and four-centre two electron integrals. The necessity of the rapid computation of integrals can be seen in Table XII-1 which presents the number of one and two electron integrals associated with a given number of basis functions. As the number of basis functions increases the number of two electron integrals (which are computationally difficult) increases roughly as N^4. Even for a relatively simple molecule such as ammonia a primitive Gaussian basis set size of say $N = 50$ would give only moderately good results and would involve on the order of a million integrals. Symmetry constraints can reduce the number of non-identical integral values but still

TABLE XII-1

THE NUMBER OF ONE (p) AND TWO (q) ELECTRON INTEGRALS ASSOCIATED WITH VARIOUS SIZE (N) BASIS SETS

N Size of Basis Set	p* One Electron Integrals	q* Two Electron Integrals
10	55	1,540
20	210	22,155
30	465	108,345
40	820	336,610
50	1275	814,725
100	5050	12,751,250
200	20100	202,015,050
300	45150	1,019,261,250

*$p = \frac{N(N+1)}{2}$ and $q = \frac{p(p+1)}{2}$

the sheer number of integrals involved in a molecular calculation requires that basis functions be chosen with the computational ease of the integrals foremost in mind. Thus, the more "physically" realistic exponential type functions or Slater type orbitals (STO) have been replaced by the more mathematically and computationally convenient Gaussian type functions (GTF) for many purposes. As mentioned the basic advantage of Gaussian type functions is due to the fact that the product of any two gaussians is also a gaussian with its centre on a line between the centres of the two original gaussian functions, as illustrated in Figure XII-1.

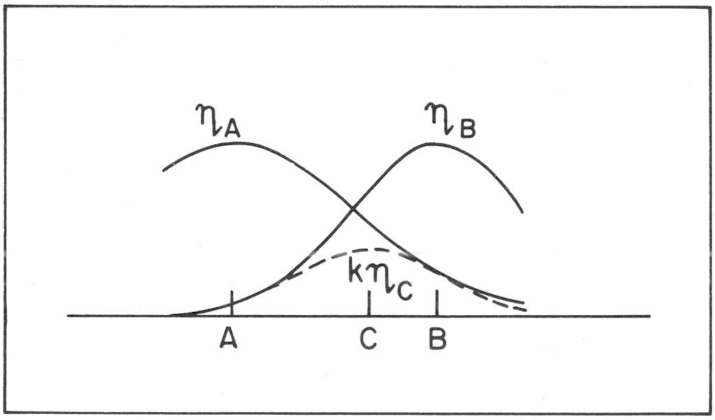

Figure XII-1. An illustration of the theorem that the product of two gaussians is also a gaussian.

Consequently, all integrals have explicit analytical expressions and may be evaluated rapidly. A similar theorem does not exist for STO. The principal disadvantages of GTF are their smooth behaviour (lack of a cusp) at the nucleus and their too rapid (e^{-r^2} rather than e^{-r}) decrease at large distances (cf. Figure XII-2). This improper asymptotic

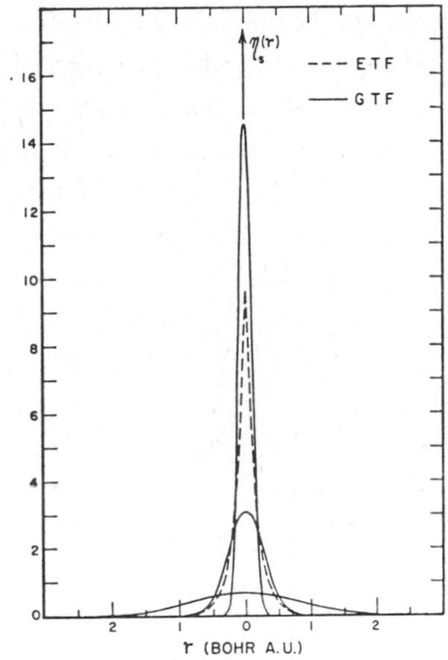

Figure XII-2. A comparison of a
1s-STO and 1s-GTF as a function of r.

behaviour requires the use of a larger number of GTF than STO for equivalent accuracy. However, the much greater speed per integral with GTF as opposed to STO allows for this greater total number of integrals.

2. Slater Type Orbitals

Prior to the early 1960's almost all molecular orbital calculations were performed (or were visualized as being performed) with STO. Today for large molecules of general geometry GTF would almost certainly be the choice but STO would still be preferred for diatomic molecules (in which, of course, no three- or four-centre integrals appear if the basic functions are centred only on the nuclei) and linear polyatomics (for which very sophisticated numerical methods dependent on the high symmetry of the systems exist for the multi-centre two electron integrals). The minimal basis set of STO is also of great conceptual importance. All semi-empirical calculations were originally related to such a basis set and many of the "concepts" so frequently used in organic chemistry (hybridization, resonance structures, etc.) can be easily visualized only in such a basis.

The functions are analogous to the actual solutions of the Schrödinger equation for the hydrogen atom but possess different nodal properties. The functions are given by

$$\eta_{n\ell m}(r,\theta,\phi) = A_n r^{n-1} e^{-\zeta r} Y_{\ell m}(\theta,\phi) \qquad \{XII-1\}$$

where n, ℓ and m are the principal, azimuthal and magnetic quantum numbers respectively; r, θ and ϕ have their usual meaning in a radial coordinate system; A_n is a normalizing factor given by

$$A = (2\zeta)^{n+1/2} \{(2n)!\}^{-1/2} \qquad \{XII-2\}$$

and $Y_{\ell m}$ are the unnormalized spherical harmonics which introduce the required angular dependence.

$$Y_{\ell m} = (-1)^{(m+|m|)/2} (4\pi)^{-1/2} \left\{\frac{(2\ell+1)(\ell-|m|)!}{(\ell+|m|)!}\right\}^{1/2} . P_\ell^{|m|}(\cos\theta) e^{im\phi}$$

$$\{XII-3\}$$

where $P_\ell^{|m|}$ are given by

$$P_\ell^{|m|}(\cos\theta) = \frac{\sin^{|m|}\theta(2)!}{2^\ell \ell!(\ell-m)!} \{\cos^{(\ell-|m|)}\theta - \frac{(\ell-|m|)(\ell-|m|-1)}{2(2\ell-1)} . \cos^{(\ell-|m|-2)}\theta$$

$$+ \frac{(\ell-|m|)(\ell-|m|-1)(\ell-|m|-2)(\ell-|m|-3)}{2.4.(2\ell-1)(2\ell-3)} . \cos^{(\ell-|m|-4)}\theta$$

$$- \ . \ . \ .$$

$$+ \frac{(-1)^P(\ell-|m|)(\ell-|m|-1) \ldots (\ell-|m|-2p+1)}{2.4 \ . \ . \ .(2p)(2\ell-1)(2\ell-3) \ldots (2\ell-2p+1)} . \cos^{(\ell-|m|-2p)}\theta$$

$$- \ . \ . \ .\} \qquad \{XII-4\}$$

TABLE XII-2

SLATER TYPE ATOMIC FUNCTIONS

n	ℓ	m	δ	Designation	Function
1	0	0	0	1s	$(\zeta^3/\pi)^{1/2}e^{-\zeta r}$
2	0	0	0	2s	$(\zeta^5/3\pi)^{1/2}re^{-\zeta r}$
2	1	0	0	$2p_z$	$(\zeta^5/\pi)^{1/2}ze^{-\zeta r}$
2	1	±1	1	$2p_x$	$(\zeta^5/\pi)^{1/2}xe^{-\zeta r}$
2	1	±1	-1	$2p_y$	$(\zeta^5/\pi)^{1/2}ye^{-\zeta r}$
3	0	0	0	3s	$(2\zeta^7/45\pi)^{1/2}r^2e^{-\zeta r}$
3	1	0	0	$3p_z$	$(2\zeta^7/15\pi)^{1/2}zre^{-\zeta r}$
3	1	±1	1	$3p_x$	$(2\zeta^7/15\pi)^{1/2}xre^{-\zeta r}$
3	1	±1	-1	$3p_y$	$(2\zeta^7/15\pi)^{1/2}yre^{-\zeta r}$
3	2	0	0	$3d_{z^2}$	$(\zeta^7/18\pi)^{1/2}(3z^2-r^2)e^{-\zeta r}$
3	2	±1	1	$3d_{xz}$	$(2\zeta^7/3\pi)^{1/2}xze^{-\zeta r}$
3	2	+1	-1	$3d_{yx}$	$(2\zeta^7/3\pi)^{1/2}yze^{-\zeta r}$
3	2	+2	1	$3d_{x^2-y^2}$	$(\zeta^7/6\pi)^{1/2}(x^2-y^2)e^{-\zeta r}$
3	2	+2	-1	$3d_{xy}$	$(2\zeta^7/3\pi)^{1/2}xye^{-\zeta r}$
4	0	0	0	4s	$(\zeta^9/315\pi)^{1/2}r^3e^{-\zeta r}$
4	1	0	0	$4p_z$	$(\zeta^9/105\pi)^{1/2}zr^2e^{-\zeta r}$
4	1	±1	1	$4p_x$	$(\zeta^9/105\pi)^{1/2}xr^2e^{-\zeta r}$
4	1	±1	-1	$4p_y$	$(\zeta^9/105\pi)^{1/2}yr^2e^{-\zeta r}$
4	2	0	0	$4d_{z^2}$	$(\zeta^9/252\pi)^{1/2}(3z^2-r^2)re^{-\zeta r}$
4	2	±1	1	$4d_{xz}$	$(\zeta^9/21\pi)^{1/2}xzre^{-\zeta r}$
4	2	±1	-1	$4d_{yz}$	$(\zeta^9/21\pi)^{1/2}yzre^{-\zeta r}$
4	2	±2	1	$4d_{x^2-y^2}$	$(\zeta^9/84\pi)^{1/2}(x^2-y^2)re^{-\zeta r}$
4	2	±2	-1	$4d_{xy}$	$(\zeta^9/21\pi)^{1/2}xyre^{-\zeta r}$
4	3	0	0	$4f_{z^3}$	$(\zeta^9/180\pi)^{1/2}z(5z^2-3r^2)e^{-\zeta r}$
4	3	±1	1	$4f_{xz^2}$	$(\zeta^9/120\pi)^{1/2}x(5z^2-r^2)e^{-\zeta r}$
4	3	±1	-1	$4f_{yz^2}$	$(\zeta^9/120\pi)^{1/2}y(5z^2-r^2)e^{-\zeta r}$
4	3	+2	1	$4f_{z(x^2-y^2)}$	$(\zeta^9/12\pi)^{1/2}z(x^2-y^2)e^{-\zeta r}$
4	3	±2	-1	$4f_{xyz}$	$(\zeta^9/3\pi)^{1/2}xyze^{-\zeta r}$
4	3	+3	1	$4f_{x^3}$	$(\zeta^9/72\pi)^{1/2}y(y^2-3x^2)e^{-\zeta r}$
4	3	±3	-1	$4f_{y^3}$	$(\zeta^9/72\pi)^{1/2}x(x^2-3y^2)e^{-\zeta r}$
5	0	0	0	5s	$(2\zeta^{11}/14175\pi)^{1/2}r^4e^{-\zeta r}$

The n, ℓ and m quantum numbers are subject to the restrictions

$$n = 1,2,3, \ldots$$

$$\ell = 0,1,2, \ldots (n-1)$$

$$m = -\ell, -(\ell-1) \ldots -1,0,1, \ldots (\ell-1),\ell$$

Substitution of these values will yield complex functions for $m \neq 0$, therefore linear combinations of those functions having the same n, ℓ and $|m|$ must be taken if real functions are desired. This is permissible since all functions with the same n and ℓ are degenerate in the atom. A fourth parameter, δ, is introduced where $\delta = +1$ denotes the positive linear combination and $\delta = -1$ the negative. The analytical expressions for the first 33 Slater type functions are given in Table XII-2 together with their n, ℓ, m and δ values and their usual shorthand designations.

A minimal basis set of STO is given by one such function for every atomic orbital occupied in the separated atoms (e.g. C 1s, 2s, 2p, $2p_x$, $2p_y$). The choice of the orbital exponents ζ (zeta) then must be made. ζ were once usually chosen on the basis of a set of empirical rules (Slater's Rules) but now variationally optimized ζ values for the free atoms are used. Table XII-3 collects these minimal basis or single zeta orbital exponents for H and C. In principle, the exponents should be reoptimized for the molecule, i.e. not only the linear parameters (the coefficients, $\underline{\underline{C}}$, in the LCAO-MO expansion) but also the non-linear parameters (orbital exponents, ζ) should be variationally optimized. In practice, the optimization of ζ values in the molecular environment is frequently prohibitively expensive.

TABLE XII-3

OPTIMUM FREE ATOM ζ VALUES FOR H AND C

	H	C
1s	1.0000	5.6727
2s	--	1.6083
2p	---	1.2107

A minimal basis set which retains the concept of a STO but replaces the actual STO by a linear expansion of GTF is termed an STO-NG basis set. Such a basis will be discussed further in the next section as it is fundamentally a gaussian basis set despite its name.

A considerable improvement on the minimal basis set is given by the double zeta basis set in which free atom occupied AO are represented by two STO with orbital exponents ζ and ζ'. The optimum free atom values of ζ and ζ' for C are collected in Table XII-4. Any basis set beyond the double zeta level may be considered an extended basis set. Of particular

TABLE XII-4

OPTIMUM FREE ATOM ζ AND ζ' VALUES FOR C

	C
1s	7.4831
1s'	5.1117
2s	1.8366
2s'	1.1635
2p	2.7238
2p'	1.2549

S. Huzinaga and C. Arnau, J. Chem. Phys. 53, 451 (1970)

importance is the addition of functions with higher ℓ values, i.e. (d, f, ...) on first row atoms and p on H which are termed polarization functions. The exponents of the polarization functions should in principle be optimized for the molecule in question but are usually not optimized in the course of a molecular calculation for economic reasons. Alternatively, optimization of polarization functions for atoms would necessitate the consideration of anionic species or excited states since by definition polarization functions are unoccupied in the atom and this approach usually is not used. It appears that calculated total energies are not too sensitive to small variations in the exponents of the polarization function and thus a reasonable choice might be made on the basis of small molecule calculations. For STO a ζ (2p) of 2.0 for H and ζ (3d) = 2.0 for C, are probably reasonable values.

3. Gaussian Type Functions

As has been discussed previously, in recent years most large scale MO calculations have used GTF. The radial dependence of a gaussian function may be written

$$\eta = r^n e^{-\alpha r^2}$$

Angular dependence may be introduced by a spherical harmonic factor $Y_{\ell m}(\theta,\phi)$ as was done with STO. However, for gaussians the angular dependence is frequently introduced as

$$Cx^p y^q z^s e^{-\alpha r^2}$$

where p, q and s are integers thus yielding <u>cartesian gaussians</u>. If p = s = 0 and q = 1 the function is of p_y symmetry. In addition to the cartesian gaussians the <u>gaussian lobe</u> method which uses only s type gaussians but floats the centres of these functions away from the atomic centres to simulate s, p, d, ... orbitals may also be used (cf. Figure XII-3).

As mentioned previously, widely used minimal basis sets which consist of gaussian functions are the STO-NG basis sets internal to the GAUSSIAN 70 program package. Calculations have indicated that N must be approximately 4 (i.e. STO-4G) to obtain values for most properties close to the actual <u>minimal basis set STO</u> results. N=3 (i.e. STO-3G) is also widely used for large molecules. The basis sets in GAUSSIAN 70 have a special constraint ($n\alpha_s = n\alpha_p$) on the orbital exponents which substantially reduces the two electron integral evaluation time without too serious a loss in variational

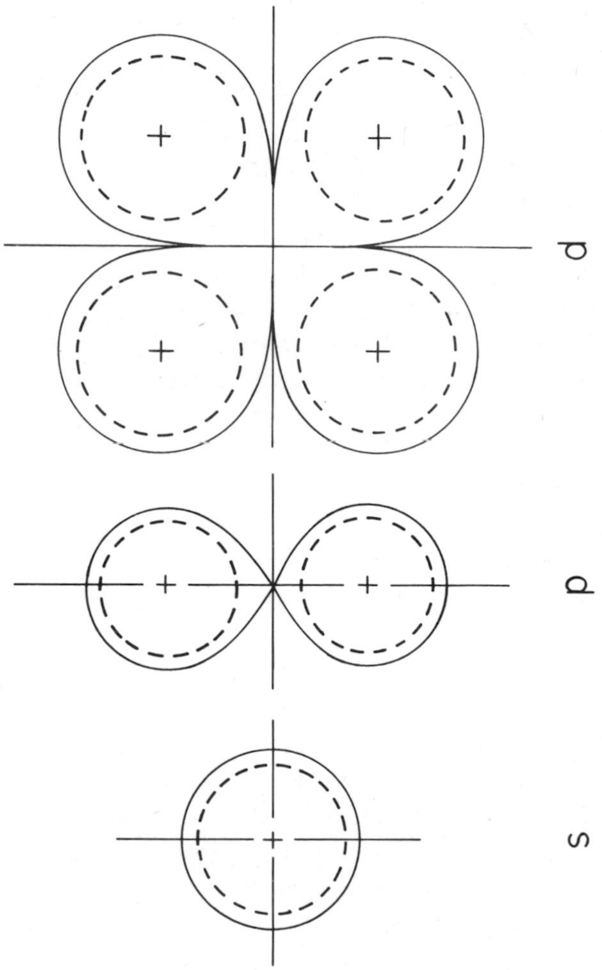

Figure XII-3. The off-centre (x) floating of gaussian lobe functions (broken lines) to simulate s, p, d, ... orbitals (solid lines).

freedom. The STO-3G and STO-4G basis sets for H and C are collected in Table XII-5. Standard scale factors chosen on the basis of a series of small molecule calculations are employed and incorporated in the exponents given.

The next basis set which might be considered is a "split valence" basis in which the core orbitals are represented as in a minimal basis set while the valence orbitals are made more flexible by allowing two functions for their representation, e.g. for H, (1s 1s') and for C, (1s 2s 2s' $2p_{x,y,z}$ $2p'_{x,y,z}$). The N-31G basis sets in GAUSSIAN 70 are of this type and are illustrated in Table XII-6 by the 4-31G basis for C and the 31G basis for H. The 4 indicates the number of gaussians in the STO-NG minimal basis representation of the core while 31G indicates that each valence orbital is represented by a total of 4 gaussians split into two functions one 3 term gaussian expansion and one a single gaussian. The basis sets again employ the $n\alpha_s = n\alpha_p$ exponent constraint and standard molecular scale factors.

In working with primitive GTF(PGTF) the large numbers of functions involved rapidly leads to a problem in the SCF routine. GTF allow fast integral evaluation due to the analytic nature of the techniques, however, 3 or 4 GTF are required to reach the same level of representation as a single STO and the SCF step can become prohibitively time consuming due to the larger size of the various matrices to be manipulated. Thus, contracted gaussian type functions (CGTF), linear combinations of gaussians with fixed exponents were introduced. Only the coefficients in each SCF orbital of the contracted functions are variationally determined in the SCF procedure. Of particular interest is a set of PGTF contracted to a basis set of approximately double zeta quality. A widely used contraction due to Dunning of a 9s 5p basis set of Huzinaga for carbon and 4s for hydrogen is given in Table XII-7.

As with STO beyond a double zeta basis set lies a _triple zeta basis_ set although in practice it is generally accepted that the most important functions to be added to the double zeta sp basis are polarization functions. Reasonable

TABLE XII-5

STO-3G AND STO-4G MINIMAL BASIS SETS FOR H AND C FROM GAUSSIAN 70

	STO	STO-3G	STO-4G
		Exponents	
H	1s	3.42525D+00	8.02142D+00
		6.23913D-01	1.46782D+00
		1.68855D-01	4.07777D-01
			1.35337D-01
C	1s	7.16168D+01	1.67716D+02
		1.30451D+01	3.06899D+01
		3.53051D+00	8.52600D+00
			2.82970D+00
	2sp*	2.94125D+00	6.87386D+00
		6.83483D-01	1.48804D+00
		2.22290D-01	4.83819D-01
			1.85818D-01
H,C	1s	1.54329D-01	5.67524D-02
		5.35328D-01	2.60141D-01
		4.44635D-01	5.32846D-01
			2.91625D-01
C	2s	-9.99672D-02	-6.22071D-02
		3.99513D-01	2.97680D-05
		7.00115D-01	5.58855D-01
			4.97767D-01
C	2p	1.55916D-01	4.36843D-02
		6.07684D-01	2.86379D-01
		3.91957D-01	5.83575D-01
			2.46313D-01

*$n\alpha_s = n\alpha_p$ orbital exponent constraint

TABLE XII-6

SPLIT VALENCE SHELL 4-31G
GAUSSIAN BASIS SETS
FOR H AND C FROM GAUSSIAN 70

	STO	Exponents 31G	Expansion Coefficients
H	1s	1.87311D+01	3.34946D-02
		2.82539D+00	2.34727D-01
		6.40121D-01	8.13757D-01
	1s'	1.61278D-01	1.00000D+00

	STO	Exponents 4-31G	Expansion Coefficients
C	1s	4.86967D+02	1.75983D-02
		7.33711D+01	1.22846D-01
		1.64135D+01	4.33782D-01
		4.34498D+00	5.61418D-01
			s
	2sp	8.67353D+00	-1.17489D-01
		2.09662D+00	-2.13994D-01
		6.04651D-01	1.17450D+00
			p
			6.40203D-02
			3.11203D-01
			7.52748D-01
			s or p
	2sp'	1.83558D-01	1.00000D+00

TABLE XII-7

9s5p GAUSSIAN BASIS SET FOR C
CONTRACTED TO 4s2p.
4s GAUSSIAN BASIS SET FOR H
CONTRACTED TO 2s.

Hydrogen s-Set		
Exponents	"2s"-Coefficients	
13.3615	0.032828	
2.0133	0.231208	s
0.4538	0.817238	
0.1233	1.000000	s'

Carbon s-Sets		
Exponents	"4s"-Coefficients	
4232.6100	0.002029	
634.8820	0.015535	
146.0970	0.075411	s
42.4974	0.257121	
14.1892	0.596555	
1.9666	0.242517	
5.1477	1.000000	s'
0.4962	1.000000	s"
0.1533	1.000000	s'"

Carbon p-Sets		
Exponents	"2p"-Coefficients	
18.1557	0.018534	
3.9864	0.115442	p
1.1429	0.386206	
0.3594	0.640089	
0.1146	1.000000	p'

values for the gaussian exponents α are given by $\alpha(2p) \approx 1.0$ for hydrogen and $\alpha(3d) \approx 0.8$ for carbon.

4. Some Remarks on Basis Set Optimization and Quality

Since the basis set fundamentally determines the reliability of the results which are obtained in any molecular orbital calculation, it is important to consider, at least briefly, two aspects of basis sets, their optimization and the measure of their quality. The term "basis set optimization" almost always refers to the variational determination with respect to the energy of a set of function exponents for a free atom. Note that even for the atom there is no guarantee that the exponents so obtained will also yield the best values for other properties (other than the energy) such as the moments of the charge distribution. Moreover, optimum atomic exponents are certainly not optimum molecular exponents although there is a general belief that the use of carefully optimized atomic exponents in molecular calculations is preferable.

The term "quality" used with respect to a basis set is of course a relative one. No absolute standard of basis set quality exists. Only when the predicted value of a particular molecular property is calculated can a statement like "Basis Set A is 'better', i.e. of higher quality than Basis B" be made. Only after the fact, i.e. after the molecular orbital calculation has been carried out can the initial basis set's quality truly be judged. However, certain conclusions based on previous experience may serve as guides in the selection of a basis set.

As to the size of a basis set and the types of functions to be included, it is clear that a minimal basis set even if carefully optimized is not capable of yielding reliable values for sensitive one-electron properties such as the dipole moment. However, such a basis set is likely to make relatively accurate predictions of at least the gross features of molecular geometry (planar vs pyramidal, linear vs bent, staggered vs eclipsed) and fairly good predictions of trends in bond lengths and angles. At the split valence or double zeta basis set level energy differences between various molecular structures are likely to be given more accurately. However, most one electron properties are unlikely to be given adequately by any basis set which neglects polarization functions. It should also be recalled (Chapter VIII) that the energetics of certain processes such as homolytic bond dissociation or electronic excitation will simply not be described properly at a single configuration (Hartree-Fock) level.

Having said something regarding the size and type of basis set, it is necessary to discuss a much more subtle problem, that of the balance of the basis set. When two or more atomic basis sets are brought together to form a molecular basis set how can one be certain that all the constituent parts of a molecule are described equally well (or equally poorly), i.e. that the basis set is balanced. If all atomic bases sets were carefully optimized to a given degree of accuracy then presumably the combination of such bases would be likely to be close to balanced. However, it is difficult to define precisely the degree of optimization with certain of the optimization methods due to the possibility of a relatively flat minimum in the exponent space or worse the possibility of the numerical procedure being trapped in a local minimum. Recently, the use of gradient techniques in which the energy is optimized with respect to all exponents by requiring that all $(\partial E/\partial \ln \alpha_i)$ approach zero $(\sim 10^{-5})$ (Figure XII-4) has provided a clearly defined quantitative measure of the quality of an atomic basis set, namely, the gradient length.

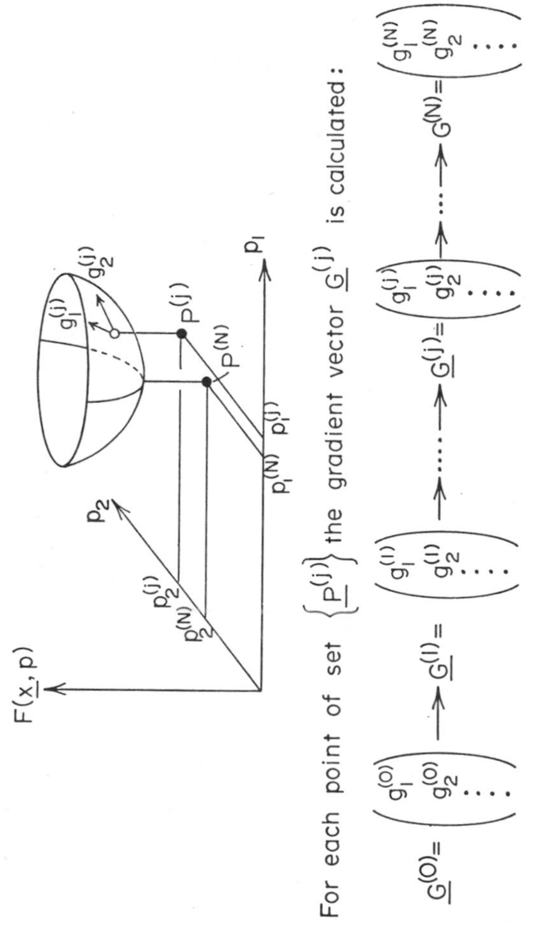

Figure XII-4. Illustration of the gradient method. ($P_i = \ln\alpha_i$). Note gradients decrease to zero at the minimum.

CHAPTER XIII

INFORMATION ON SELECTED COMPUTER PROGRAMS

1. The Quantum Chemistry Program Exchange (QCPE)

QCPE serves as a clearing house for information and programs for quantum and other chemists. All inquiries and general correspondence should be directed to:

Quantum Chemistry Program Exchange
Chemistry Department
Room 204
Indiana University
Bloomington, Indiana

QCPE is an excellent source of well tested, well documented computer programs to perform many of the sorts of calculations that have been discussed. Most of these programs are available in high level languages (usually FORTRAN IV), as punched cards (if necessary) or normally as card images on magnetic tape and often in versions for both CDC or IBM installations.

2. The Consolidated Index to QCPE Programs

The complete descriptions of programs available through QCPE are now contained in two separate publications. The publications are the 1974 QCPE Catalog (Vol. X) and the 1975 Supplement to this catalog. In order to provide a complete index to both of these publications we have included a consolidated index to both publications in the 1975 Supplement.

The index is organized into major headings which describe, in a broad sense, the types of programs which QCPE can offer. Under each of these headings the appropriate programs are identified by number and title. In the case of those programs which are not contained in the 1974 Catalog, but are only described in the 1975 Supplement, an asterisk (*) will precede the QCPE program number. For those programs which are not preceded by an asterisk, the reader is referred to the 1974 Catalog for the description of the program.

For those who find themselves with only the 1975 Supplement, the 1974 Catalog can be obtained at no charge from QCPE.

QCPE announces new programs on at least a quarterly basis in its newsletter. Practically speaking, new programs are announced even more often by means of special announcements. In order to have the most up-to-date information on QCPE's holdings, it is necessary to have the most recent QCPE Newsletter as well as the 1975 Supplement and the 1974 Catalog.

Matrix, Algebraic and Arithmetic Utility

*255. BIORTH: Subroutine for Matrix Biorthogonalization.
235. INVSQR: Inverse Square Root Matrix.
211. SERORT: Lowdin Symmetric Orthogonalization by Modified Series Method.
216. LOWDIN: Orthogonalization of Functions by Lowdin's Method.
85. LOWDIN: Orthogonalization of Normalized Functions by Lowdin's Method.
67. SIMPLEX: Minimum of a Function of Several (N) Constrained Variables.
66. STEPIT: Minimum of a Function of Several (N) Variables.
61. VD01A: Minimum of a Function of a Single Variable.
60. VA04A: Minimum of Function of Several Variables.
59. VA02A: Minimize Sum of Squares.
53. FACT: Factorial.
51. FACTOR: Factorial.
28. SCHMIT: Schmidt Orthogonalization.
2. DPSMIT: Schmidt Orthogonalization.
1. SCHMIT: Schmidt Orthogonalization.

Expansion and Special Functions

*254. GAMMAF: Subrouting to Compute the Incomplete Gamma Function.
207. ERFGEN; Calculation of Normalized Repeated Error Integrals.
58. EIGSER: Series Expansion of Matrix Eigenvalues and Eigenvectors.

Integral Calculations and Related Programs

*252. DERIC: One and Two Center Electron Repulsion Integrals Over Complex STO's.
218. PROJR: Overlap Calculation Between Two N-Electron Antisymmetric Wave Functions.
202. POESTO: Polyatomic One-Electron Slater-Type-Orbital Integral Program.
170. LIGAND: Field Integrals Using SCF Wave Functions.
182. DIATH2: A Computer Aid for Studying the Quantum Mechanics of the Chemical Bond.
153. A Program to Calculate Diatomic Overlap Integrals Between Slater Type Orbitals.

138. DIATOM: Diatomic Molecular Integral Program.
131. Integral Evaluation Routines Package.
119. OVERLAPII: Overlap Intergal Between Slater-Type Orbitals.
118. One ELECTRON IV: One Electron Integrals Between Linear Combinations of Slater-Type Orbitals.
117. One Electron II: One Electron Integrals Between Slater-Type Orbitals.
106. INT1: Calculation of Two-Center Exchange Integrals Occurring in Molecular Problems.
99. INTEG: One Center Integrals Over Slater Orbitals.
96. CSUBK: Integrals Over Spherical Harmonics.
87. ABAC: Three-Center Integrals of Type (ab/ac)
86. AABC: Three-Center Integrals of Type (aa/bc).
82. INTE: Moment Integral Calculations.
45. WPMAB: Calling Program for Three-Center Integral Program ABAC (QCPE 23).
44. WPMAA: Calling Program for Three-Center Integral Program AABC (QCPE 22).
43. STO2L: Store Two-Electron Integrals.
25. PS3: Four-Center Integrals Over Three S-Orbitals and One P-Orbital.
24. S4: Four-Center Integrals Over S-Orbitals.
23. ABAC: Three-Center Integrals of Type (ab/ac).
22. AABC: Three-Center Integrals of Type (aa/bc).
10. STWC: One-Electron Integrals Over Elliptical Orbitals.
9. STVYWC: One-Electron Integrals Over Elliptical Orbitals.

Eigenvalues and Eigenvectors

*264. CBORIS: Eigenvalues and Eigenvectors of a Complex Hermitean Matrix.
*253. CEIGEN: Eigenvalues and Eigenvectors of a Complex Hermitean Matrix.
180. NESBET: Eigenvalue and Eigenvector by Modified Nesbet Method.
114. EIGQR: Eigenvalues and Eigenvectors of a Real Symmetric Matrix
97. CEIG: Eigenvalues and Eigenvectors for (H-λS) C=O.
93. NESBET: Eigenvalues and Eigenvectors of Large Matrices.
63. MDG1: Eigenvalues and Eigenvectors of Large Matrices.
62. GIVENS: Eigenvalues and Eigenvectors by the Givens Method.

Symmetry Analysis and Related Numerical Quantities

*268. 3J/6J: Magnitudes of 3J and 6J Symbols.
*257. Symmetry Determination Program.
122. ROT: Rotation Group Representations.
90. RACAH: Racah Coefficients, 6j Symbols, etc.
55. ROT: Rotation Group Representations.
52. COUPLE: Vector Coupling Coefficients.
38. TERMS: Calculation of All Terms Arising from a

Symmetry Analysis and Related Numerical Quantities (Continued)

Given Electronic Configuration.
37. PROP: Calculation of Exact Eigenfunctions of Spin and Angular Orbital Momentum.
15. CKCOE: Condon-Shortley Coefficient.

Ab Initio Molecular Orbital Calculations and Related Programs

*263. SAMOS: Simulated Ab Initio Molecular Orbital System.
*251. POTLSURF: A Program to Compute the Potential Energy Surface Between a Closed-Shell Molecule and an Atom.
241. PHANTOM: Ab Initio Quantum Chemical Programs for CDC 6000 and 7000 Series Computers.
239. SMALLØBE: Ab Initio Gaussian Lobe SCF Closed Shell Program.
238. POLYATOM: Version II (IBM 360).
237. GAUSFIT: Gaussian Fit to Slater Orbitals.
236. GAUSSIAN 70: Ab Initio SCF-MO Calculations on Organic Molecules.
233. OEDM: One-Electron Diatomic Molecules.
199. POLYATOM: (Version 2) System of Programs for Quantitative Theoretical Chemistry.
181. PROJECT: Eigenfunctions of Spin and Orbital Angular Momentum by the Projection Operator Technique.
173. HEDIAG: Diatomic Molecule Symmetry Eigenfunctions by Direct Diagonalization.
145. Columbia University Molecular Properties Program.
125. LOCOSY: Floating Spherical Gaussian Orbital Model Calculations.
92. IBMOL: LCAO-MO-SCF Method for Molecules of General Geometry.
47. POLYATOM: Program Set for Non-Empirical Molecular Calculations.
46. SYMPRO: Symmetry Projection Program.

CNDO and INDO Molecular Orbital Calculations

*274. CNINDO-DYNAM: CNDO and INDO Molecular Orbital Program with Dynamic Data Storage.
*272. GSPCLIO: Modified PCILO.
*267. ORLOC: Localized Molecular Orbitals by the Edmiston, Ruedenberg, Trindle, Sinanoglu Method.
*249. VSS: Isoenergy Curve of Electronic Distributions.
242. CNINDO: With Bond Density Calculation.
240. CINDOM: CNDO and INDO Molecular Orbital Program (FORTRAN IV) for Medium Sized Computers.
224. FINITE: Finite Perturbation Calculation of NMR Coupling Constants.
223. CNINDO: CDC 6000 Series Version of QCPE 141.
221. PCILO: CDC 6600 Version.
220. PCILO: (Perturbation Configuration Interaction Using Localized Orbital) Method in the CNDO Hypothesis.
198. LMO: Localized Molecular Orbitals.
191. LOCAL: Molecular Orbital Localization Program.
185. Mulliken Method Calculations.

CNDO AND INDO Molecular Orbital Calculations (Continued)

174. CNDO/S-CI: Molecular Orbital Calculations with the Complete Neglect of Differential Overlap and Configuration Interaction.
144. Closed Shell Dual Purpose INDO/CNDO Semi-Empirical SCF.
142. CNINDO: CNDO and INDO Molecular Orbital Program (CDC FORTRAN 63).
141. CNINDO: CNDO and INDO Molecular Orbital Program (FORTRAN IV).
100. CNDOTWO: SCF-LCAO-MO with Complete Neglect of Differential Overlap.
91. CNDO/2: Molecular Calculations with Complete Neglect of Differential Overlap.

Huckel, Extended Huckel, Pi Electron and Related Calculations

*256. EHT-SPD: Extended Huckel Program for Atoms Through Fourth Row.
246. MIEHM: Modified Iterative Extended Huckel Method.
200. DEWARPI: SCF-LCAO-MO Calculations by Dewar's Pi Electron Method.
184. POLAHUC: Pi Polarizabilities of Non-Saturated Organic Molecules.
167. PEP-Pi Electron Program.
148. CMGMO: IBM 360 Compatible Version of OMEGAMO (QCPE 110).
143. LCAO-MO-SCF-CI Program for Aromatic System.
132. HUCKEL: Huckel Diagonalization-Revised.
124. UHFOCK: Unrestricted Hartree-Fock Program.
110. OMEGAMO: Huckel Calculations in which the Input Matrix is Modified by Charge Densities and Bond Lengths.
95. EXTHUC: Extended Huckel Theory.
81. HMO-II: Huckel Molecular Orbital Theory.
77. SCPOPEN: Open-Shell SCF-LCAO-MO.
76. POPLE PI: Pople Pi Electron Program.
72. SE18: Numerical Solution of Two-Dimensional Schroedinger Equations.
71. SCFOLO: Closed-Shell SCF-LCAO-MO.
70. HUCKEL: Diagonalization.
64. EXTHUC: Extended Huckel Theory Calculations.
48. EXTHUC: Extended Huckel Theory Calculations (CDC FORTRAN 63 Version of QCPE 30).

Other Molecular Orbital Calculations Based on Approximate Methods

*269. THERMO: Standard-State Thermodynamic Functions of Polyatomic Molecules.
228. MINDO/2': GEOMETRY Optimization Program- SIMPLEX.
227. Atomic Partial Charges Using Del RE's Theory.
217. OPTMO: A Molecular Orbital Program for Locating Equilibrium Geometries Employing the MINDO/2 or the Extended Huckel Methods.
166. TDHF: Time-Dependent Hartree-Fock Calculation in the Pariser-Parr-Pople Model.

Other Molecular Orbital Calculations Based on Approximate Methods (Continued)

137. MINDO: Molecular Orbital Calculations by the MINDO Method.
108. VIB: Numerical Solution of the One-Dimensional Radial Equation.
32. AMO-S: Alternant Molecular Orbital Program.

Scattering Programs

*273. A+BC: General Trajectory Program.
*248. CTAMYM: Modification of the Atomic-Diatomic Quasi-Classical Trajectory Program CLASTR.
229. CLASTR: Monte Carlo Quasi-Classical Trajectory Program.
187. Coupled Channel Scattering Matrices.

Crystallography Programs

*259. SYMPRJ/SYMPW: Symmetry Projection Programs.
245. PREDEN/CRYDEN: MO Calculations on Infinite Three-Dimensional Crystals.
222. MADE: Coulombic Lattice Energy of an Ionic Crystal.
216. Crystal-Field Splitting Calculations for Trivalent Lanthanide Ions (main program and 11 subroutines).
215. DIHEDR: Subroutine to Calculate the Value and the Sign of the Dihedral Angle of 4 Atoms Bonded L1-L2-L3-L4.
214. GPTHEORY: Molecular Point Group Symmetry Properties.
201. CRYSMO: CNDO-MO Calculations on Hydrogen Bonded Molecular Crystals.
163. IRREP: Description of a Fortran Program for the Calculation of Irreducible Representations of Finite Groups.
149. CARTCORD: Cartesian Coordinates for Lattice Atoms in a Crystal.
46. SYMPRO: Symmetry Projection Program.

Nuclear Magnetic Resonance (NMR) and Related Programs

232. NMR-LAOCN-4A: NMR Analysis by Least Squares Fit.
224. FINITE: Finite Perturbation Calculation of NMR Coupling Constants.
205. KOMBIP: A Complete Program for Generation of Lorentzian/Gaussian Line-Shape and "Stick"-Plot NMR Spectra.
188. UEAITR: NMR Analysis.
165. DNMR2: A Computer Program for the Calculation of Complex Exchange-Broadened NMR Spectra. Modified Version for Spin Systems Exhibiting Magnetic Equivalence or Symmetry.
154. POWPAT: Powder Patterns and Spectra for NMR Transitions.
152. ASSIGN: Computer Assignment Technique for NMR Spectra.
151. ICOSHED: Spheravr, NMRnarro.

Nuclear Magnetic Resonance (NMR) and Related Programs (Cont.)

150. MAGNSPEC3
140. DNMR: The Computation of Complex Exchange-Broadened NMR Spectra.
127. NMREN1: Nuclear Magnetic Resonance Energy Level Program.
126. NMRIT-IV: Nuclear Magnetic Resonance Iteration Program.
111. LAOCN3: Program for the Analysis of High-Resolution NMR spectra.
36. NMPRLT: Nuclear Magnetic Resonance Plotting Program.
35. NMREN2: Nuclear Magnetic Resonance Energy Level Program.
34. NMREN1: Nuclear Magnetic Resonance Energy Level Program.
33. NMRIT: Nuclear Magnetic Resonance Iteration Program.

Electron Spin Resonance (ESR) and Electron Paramagnetic Resonance (EPR) Programs

*265. SIMI4/SIMI4A: Simulation of Powder EPR Spectra.
243. General EPR Parameter Fitting Program.
210. ESRSPEC2: Second Order Electron Spin Resonance (ESR) Spectra.
197. ESRCON: Least Squares Fitting of Isotropic Multiline ESR Spectra.
192. Fields: EPR Transition Fields and Transition Probabilities.
160. ESSP2: Electron Spin Resonance Spectrum Simulation.
134. PARA: (Revised)-General EPR Fitting Program.
133. EPR: (Revised)-General EPR Spectrum Calculation.
128. ATLAS: Synthesis and Plotting of Hyperfine Patterns of Electron Spin Resonance Spectra.
83. ERSPEC: First Order Electron Spin Resonance Spectra.
69. PARA: General EPR Parameter Fitting.
68. EPR: General Spectrum Calculation.

Programs Developed to be Used as Education Tools

213. HYDRGN: Quantum Mechanical Bonding in the Hydrogen Molecule.
196. TITRATE: A Program for Titration Simulation.
182. DIATH2: A Computer Aid for Studying the Quantum Mechanics of the Chemical Bond.

Mass Spectroscopy Programs

*271. FIMS: Least squares Fitted Fractional Abundance of Isotopes.
*266. High Ionizing Voltage, Low Resolution Mass Spectrometric Analysis of Gas Oil Aromatic Fractions.
*262. MASSPEC: Mass Spectroscopy Isotope Program.

Spectroscopy, Instrumental Analysis and Related Programs

*270. ASROTOR: The Asymmetric Rotor Energy Levels Computer Program.
*260. NOMULT: Number of Multiplets in a Transition Array.
*247. QCFF/PI: A Program for the Consistent Force Field Evaluation of Equilibrium Geometries and Vibrational Frequencies of Molecules.
212. DECON: For Removing the Broadening of Spectra Due to Instrumental Effects.
204. PERTEN: One-Dimensional Vibrational Energy Levels by Perturbation.
164. MOSTR: Molecular Parameter for Diatomic Molecules Calculated from Infra-Red Spectral Data in the 2 to 15 Micron Region.
113. VIBROT: Vibration-Rotational Spectroscopic Constants Calculation.
107. MOSCOW: Kronig Kramers Transform Program.
74. POT: One-Dimensional Vibrational Energy Levels and Amplitudes.

Rate Constants, Kinetics and Related Programs

244. TGAP: Estimation of Gas-Phase Thermokinetic Parameters.
234. RRKM: General Program for Unimolecular Rate Constants.
195. Subroutine FIRST and SECOND for Rate Constant Evaluation
179. ACTEN: Arrhenius Activation Energies and Frequency Factors.
168. PARACT: Rate Constant Calculations from Differential Scanning Calorimetric Thermograms by the Bernoulli Equation.
162. DATAR: Rate Parameters from Differential Scanning Calorimetric Thermograms.
80. LSKINI: Least Squares Treatment of First Order Rate Data.
79. ACTENG: Activation Energy Calculation.

Calculation of Cartesian Coordinates of Atoms in Molecules

*250. ATCOOR: Calculation of Cartesian Coordinates.
226. COORD/1130: Calculations of Atomic Coordinates of Molecular Systems.
186. COORD: Time Sharing Version of QCPE 136.
169. CARCOR: Calculate Cartesian Coordinates for All Atoms in a Molecule.
136. COORD: Atomic Cartesian Coordinates for Molecules.
135. MBLD: Standard Geometric Models and Cartesian Coordinates of Molecules.
130. CORCAL: Molecular Atomic Coordinates from Bond Lengths and Bond Angles.

Normal Coordinate Analysis and Related Programs

*275. MODFOR: Normal Coordinate Analysis Solution by a Systematic Variation in the Wilson F-Matrix.
*258. Isotope Effects V-I.

Normal Coordinate Analysis and Related Programs (Continued)

*247. QCFF/PI: A Program for the Consistent Force Field Evaluation of Equilibrium Geometries and Vibrational Frequencies of Molecules.
208. COMPLY: Least Square Refinement of a Compliant Field.
178. XYZ: Geometry of Molecules.
177. NORCRD: Long XYZ Version.
176. NORCRD: Short XYZ Version.
78. GF MATRIX: Secular Determinant Solution in Normal Coordinate Analysis.
57. GNTCP: General Normal Coordinate Treatment.

Programs of General Utility

231. NEPROP: Subroutine for Numerical Propagation of Uncertainties.
230. FITIT: Non-Linear Least Squares Fit.
225. MOLAREA: Calculation of the Surface Area of a Non-Spherical Molecule or Molecular Cavity in a Fluid from the Van der Waals' Radii of Component Atoms.
219. PRINT: Plotting and Representation in Three Dimensions by Line Printer.
206. SUPERCIS: Urey-Bradley Five Atom Configuration.
203. FDBVM: Numerical Eigenvalues and Matrix Elements for the Quantum Mechanical Radial Equation.
194. CONVERT: IBM 360 FORTRAN Programs to CDC 6600 FORTRAN Conversion.
193. Self-Consistent Charge and Configuration Calculations on Inorganic Compounds.
190. APER: Seventh Order Nondegenerate Double Perturbation Calculations.
175. PLOT: Printer Plotting Subroutine.
157. AFIT: Molecular Orbital Calculations by the Jacobi Method.
147. FRS3: FORTRAN Real Sort Version 3.
139. Perturbation Calculations.
115. Linear Least Squares Fitting of Weighted Data.

3. Abstracts for Selected Programs

A. Coordinate Calculation

136. COORD - Atomic Cartesian Coordinates for Molecules

by M. J. S. Dewar, Department of Chemistry, University of Texas, Austin, Texas and N. C. Baird, Department of Chemistry, University of Western Ontario, London, Ontario, Canada.

This program calculates the cartesian coordinates in molecules, given the bond lengths, bond angles and some dihedral angles.

Explicit input instructions are contained in the COMMENT cards in the program source deck.

FORTRAN IV
Symbolic Cards: 250

169. CARCOR-Calculate Cartesian Coordinates for All Atoms in a Molecule

by A. Crackett, Physical Chemistry Laboratory, Oxford University, England.

This program calculates cartesian coordinates for all atoms in a molecule, given bond lengths, bond angles and a twist angle to specify the conformation.

FORTRAN IV (CDC 3600/modified from KDF9 FORTRAN)
Symbolic Cards: 1000

B. Semi Empirical Molecular Orbital Programs

167. PEP-Pi Electron Program

by U. Mueller Westerhoff, IBM Research Laboratory, San Jose, California.

The closed-shell π-SCF-LCAO-MO program PEP (Pi-Electron Program) performs calculations according to the general approximations of the Pariser-Parr-Pople method, including configuration interaction.

This is a development originally based on QCPE 71.

FORTRAN IV (IBM/360)
Symbolic Cards: 1250

200. DEWARPI: SCF-LCAO-MO Calculations by Dewar's PI Electron Method

N. Colin Baird, Department of Chemistry, University of Western Ontario, London, Ontario, Canada.

This program calculates bond distances, pi and sigma bonding energies, heats of formation, electron densities, etc. for both ground and excited states of organic molecules and ions by the pi electron method of Dewar et al.

FORTRAN IV
Symbolic Cards: 600

246. MIEHM: Modified Iterative Extended Huckel Method

by Jens Spanget-Larsen, Division of Theoretical Chemistry, Aarhus University, Aarhus, Denmark.

This complete program performs Iterative Extended Huckel level LCAO-MO calculations with one-center and two-center contributions included in the iterative procedure.

FORTRAN IV (CDC 6000 Series)
Symbolic Cards: 2000

256. EHT-SPD: Extended Huckel Program for Atoms Through Fourth Row

by P. Dibout, Laboratoire de Chimie Theorique, University of Rennes, Rennes, France.

Performs calculations according to the extended Huckel method (iterative or not) for molecules including elements from first to fourth row. Basis is constituted by valence orbitals (S, P and D Slater type) up to 5S, 5P and 4D.

Nature of the problem solved:

This program calculates according to the extended Huckel scheme, or iterative extended Huckel scheme:

- Energy Levels
- Electronic Energy
- Wave Functions
- Bond Orders
- Orbital and Atomic Charges
- Dipole moment including polarization term
- Coulomb interaction between atoms of the molecule
- Solvation Energy

Auxiliary Program: AMUNU

AMUNU creates a data set used to calculate overlap integrals. This data set is not altered by execution of EHT-SPD program.

FORTRAN IV (IBM 360/370)
Symbolic Cards: 2400

NOTE: The COMMENT cards and therefore much of the documentation for this program are in French.

141. CNINDO: CNDO and INDO Molecular Orbital Program (FORTRAN IV)

by Paul A. Dobosh, Carnegie-Mellon University, Pittsburgh, Pennsylvania.

This program written for the IBM System 360/67 digital computer for the calculation of CNDO and INDO molecular orbitals. The program is capable of computing CNDO wave functions for open and closed shell molecules containing H to F. The matrices in the program are large enough to allow molecules containing up to 35 atoms or 80 basis functions (whichever is smaller). One atomic orbital basis function is allowed for hydrogen (1s), four each to the elements Na through Cl (3s, $2p_x$, $3p_y$, $3d_{yz}$, $3d_{x^2-y^2}$, $3d_{xy}$).

FORTRAN IV (IBM 360/67)
Symbolic Cards: 2000

240. CINDOM: CNDO and INDO Molecular Orbital Program (FORTRAN IV) For Medium Sized Computers

by Roy E. Bruns, Instituto de Quimica, Universidade Estadual de Campinas, Campinas, SP, Brasil.

This program is a modification of QCPE 141 for the calculation of CNDO and INDO molecular orbitals using medium sized computers. The program is capable of computing CNDO wave functions for open and closed shell molecules containing elements H to Cl and INDO open and closed shell calculations for molecules containing H to F. The matrices are large enough to allow calculations for molecules containing up to 20 atoms and 50 basis functions. By modifying the dimension statements, calculations for molecules with up to 70 basis functions can be performed on computers with a storage capacity less than that required by QCPE 141 for identical-sized problems. For problems with basis sets larger than 70 this modification has no practical advantages over QCPE 141.

This program has been organized into a ROOT and four PHASES for use with computers with chain (overlay) facilities. This procedure is well described in the documentation.

IBM 360 FORTRAN
IBM 360/44 (128K)
Symbolic Cards: 2000

274. CNINDO-DYNAM: CNDO and INDO Molecular Orbital Program with Dynamic Data Storage

by Alice Chung-Phillips, Department of Chemistry, Miami University, Oxford, Ohio 45056

The program is a modification of QCPE 141. Modifications include: (1) conversion of the old main routine and part of subroutine INTGRL to the new main routine and subroutine DYNAMI: (2) replacement of the old subroutine COEFFT with the new functions Y and Z and subroutine FIND: and (3) correction of certain ill-defined limits for DO lopes in the old subroutines HUCKOP, SCFOPN and OPRINT.

The special feature of this program is dynamic data storage. In the subprograms, the arrays originally stored in COMMON/ARRAYS/, INFO/, /GAB/, and /INFO1/ of CNINDO are converted to dynamic arrays, or arrays with variable dimensions. The integer variables used to specify these dimensions are NATOMS, the number of atoms, and N, the number of atomic orbitals. In the main program, the dynamic arrays are placed end to end into a single storage block. The individual dimensions of the arrays are no longer declared explicitly. Instead, the size of the storage block, NTOTAL, is specified. NTOTAL is an integer constant and may be changed before each execution. NTOTAL also represents the maximum storage allocated for all the dynamic arrays. The storage actually occupied by the dynamic arrays in the storage block, ITOTAL, is equal to M, M+1, or M+2 words, where M=NATOMS(10+2NATOMS)+N(15+6N). With this formula a user may estimate ITOTAL for the molecules involved and assign a proper number of words to NTOTAL provided NTOTAL≥ITOTAL.

In the QCPE version of CNINDO-DYNAM, NTOTAL is chosen as 25000 words. The size of the storage block is defined by the statements: "INTEGER*4 A", "REAL*8 B", "DIMENSION A (25000)", DIMENSION B (12500)", and "EQUILVALENCE (A(1),B(1))". Note that the number of elements in B is one-half of that in A. To change the amount of storage allocated for the dynamic arrays, simply change the numbers in the two DIMENSION statements described here and in the statement "NTOTAL= 25000". These changes require very little effort since all three statements concerned appear in the main routine only.

The program no longer has specific limits for NATOMS and N. (In CNINDO the respective limits are 35 and 80). The present limit is for ITOTAL. This limit is just NTOTAL, a number the user sets up to suit his own needs. For NTOTAL=25000 words, the limiting molecule is somewhere between C_9H_{20} and $C_{10}H_{22}$. Also, the program requires 192 K bytes of core storage without overlay and 154 K bytes with overlay. An overlay deck is provided.

Documentation of the program appears as comment statements in the main routine and subroutine DUMMY. The latter is to be removed for execution. Two test decks are provided: ethane (CNDO CLSD) and allyl radical (INDO OPEN).

The method of dynamic data storage is described in the article entitled "Dynamic Data Storage in FORTRAN IV", by A. Chung-Phillips and R. W. Rosen. The method is recommended for all general interest programs. The cindo-DYNAM program is discussed in a note entitled, "Dynamic Data Storage for Molecular Orbital Calculations", by A. Chung-Phillips. Both papers are being submitted for publication (ACP, January, 1974).

FORTRAN IV (IBM 360 and 370)
Symbolic Cards: 2000

261. CNDO/2-3R CNDO for Third Row Elements

by H. L. Hase and A. Schweig, Fachbereich Physikalische Chemie, Universitaet Marburg, Marburg/Lahn, West Germany

This program is an extension of the usual CNINDO program (QCPE 141) to elements of the third row. Germanium, arsenic, selenium and bromine are included. The method of extension and the choice of parameters have been described elsewhere [H. L. Hase, A. Schweig, Theoret. Chim. Acta 31, 215 (1973)].

The following subroutines of the CNINDO program have been modified: BLOCK DATA, INTGRL, SS, HUCKCL, SCFCLO, CPRINT, HUCKOP, SCFOPN, OPRINT. The subroutine COEFFT has been substituted by the subroutines COEFF, COEEFS, BINOM.

FORTRAN IV (IBM 360/370)
Symbolic Cards: 1800

174. CNDO/S-CI Molecular Orbital Calculations with the Complete Neglect of Differential Overlap and Configuration Interaction

by J. Del Bene, H. H. Jaffe', R. L. Ellis and G. Kuehnlenz, University of Cincinnati, Cincinnati, Ohio

Program performs molecular orbital calculation for molecules in close shell electronic configurations, using the CNDO/S method described by J. De. Bene and H. H. Jaffe', J. Chem. Phys. 48, 1807 (1968) with the following revisions:

1. The two-center two-electron repulsion integrals are calculated by the Mataga approximation.

2. The CI matrix is symmetry blocked.

Up to 31 centers and 100 basis members can be employed and the program requires a computing system with an overlay facility.

FORTRAN IV
Symbolic Cards: 2000

137. MINDO - Molecular Orbital Calculations by the MINDO Method

by N. Colin Baird, Department of Chemistry, University of Western Ontario, London, Ontario, Canada

This program executes SCF-LCAO-MO calculations for the valence electrons of organic molecules in their ground states at equilibrium bond distances, primarily to estimate total bonding energies. Detailed input instructions for the program are contained in the comment cards.

FORTRAN IV
Symbolic Cards: 700

279. MINDO/3: Modified Intermediate Neglect of Differential Overlap

by M. J. S. Dewar, H. Metiu, P. J. Student, A. Brown, R. C. Bingham, D. H. Lo, C. A. Ramsden, H. Kollmar, P. Weiner, P. K. Bischof, Department of Chemistry, University of Texas at Austin, Texas

MINDO/3 represents the endpoint in a series of developments spanning many years and having MINDO/1 and MINDO/2 as its direct predecessors. The purpose of these efforts has been to develop a calculational tool of sufficient accuracy and reliability to permit chemists calculational access to important information which is experimentally inaccessible.

This system automatically calculates the geometry and energy of a molecule by minimizing the energy with respect to all geometrical parameters. Heat of atomization is found by subtracting the energies of the component atoms. Heat of atomization is then converted to heat of formation using experimental values for heats of formation of gaseous atoms.

This system has options for open-shell calculations (radicals and triplets) by the "half-electron" method and for inclusion of CI with the lowest doubly excited configuration (for biradical-like species) together with the DFP geometry program.

In the DFP geometry program minimization of a real valued function of an N-component real vector is accomplished by the Davidon-Fletcher-Powell algorithm. (Computer Journal, Vol. 6, p. 163.)

This system is parameterized to handle the following atoms: H, B, C, N, O, F, Si, P, S and Cl.

Pertinent recent references: R. C. Bingham, M. J. S. Dewar, D. H. Lo, JACS, Vol. 97, No. 6, p. 1285.

See also: JACS, Vol. 97, No. 6, p. 1294.
JACS, Vol. 97, No. 6, p. 1302.
JACS, Vol. 97, No. 6, p. 1307.

FORTRAN IV (CDC 6600)
Symbolic Cards: Approximately 7000

220. PCILO (Perturbation Configuration Interaction using Localized Orbital) Method in the CNDO Hypotheses

by P. Claverie, J. P. Daudey, S. Diner, Cl. Giessner-Prettre, M. Gilbert, J. Langlet, J. P. Malrieu, U. Pincelli and B. Pullman, Laboratoire de Chimie Quantique, Institut de Biologie Physicochimique, 13, rue P. et M. Curie, Paris 53, France.

PCILO is an automatic program intended for the calculation of electronic ground state properties of a molecular system in the framework of semi-empirical methods for all valence electrons. In its present form it calculates the electronic ground state energy and the one particle density-matrix in an approximation which lies beyond the SCF one using the CNDO II hypotheses for the integrals. Due to its speed and accuracy, compared to the SCF calculations by the CNDO II method of Pople and Segal, PCILO is specially fitted for conformation analysis. The number of valence electrons of the system (limited here to 120) can easily reach 200 or 250, the computing time remaining reasonable. The only limitation comes from the precision of the computer.

PCILO was born during the study of the perturbation treatment of configuration interaction using SCF localized orbitals. This gives a definition of its technical framework. From a conceptual point of view PCILO is equivalent to a method looking at the molecule as an assembly of "two-centers, two-electron molecules" (chemical bonds) in the interaction; the interaction being treated by perturbation theory in an antisymmetrized basis. Classical theoretical chemistry and the studies on the localization of SCF orbitals provide a large justification for this conception: the chemical formula is a very good order approximation for the study of a molecule.

The method relies on four fundamental steps:

1. Building of bonding and antibonding orbitals orthogonalized in some way.

2. Bonding orbitals are used to construct a Slater determinant which is the zeroth order wavefunction for the molecule.

3. Antibonding orbitals allow the construction of Slater determinants corresponding to excited configurations. These configurations are characterized by the bonds involved in the excitations, and can be classified according to the two different possibilities: excitation with or without electron transfer from one bond to another.

4. In the basis of all these determinates the molecular hamiltonian is represented by a "configuration interaction matrix". The lowest eigenvalue and the corresponding eigenvector are calculated by a Rayleigh-Schrodinger perturbation series. Ground state properties are obtained along this way. Excited states could be studied in an analogous way but technical difficulties due to degeneracy appear and such a calculation is not included in the standard version of PCILO. The total molecular hamiltonian is split according to the Epstein-Nesbet procedure, which is equivalent to taking the diagonal part of the interaction configuration matrix as representation of the non-perturbed hamiltonian and the non-diagonal part as perturbation. The absence of diagonal terms in the perturbation insures a faster convergence of the series. At the same time the first order correction to the energy is zero.

FORTRAN IV (IBM 360/75)
Symbolic Cards: 2700

263. <u>SAMOS, Simulated Ab Initio Molecular Orbital System</u>

by Brian O'Leary, Department of Chemistry, University of Alabama, Birmingham, Alabama

B. J. Duke, Department of Chemistry, University of Lancaster, Bailrigg, Lancaster, England

J. E. Eilers, Department of Chemistry, SUNY, College of Rockport, New York.

This offering consists of four major programs all of which are part of this system for accomplishing Simulated Ab Initio Molecular Orbital (SAMO) calculations. The four components of this system are the following:

1. SAMO Method for Closed-Shell Molecules.

2. SAMO Method for Open-Shell Radicals, Using the Spin Unrestricted Formalism.

3. SAMO Method for Polymers.

4. General Library Service Program for the SAMO system.

The detailed description of each segment follows:

1. SAMOM

This program can be used to evaluate the molecular orbitals for the molecular ground state of closed-shell molecules, using the SAMO technique. The resulting wave function is a simulation of the one that would be obtained using the usual Roothaan *ab initio* LCAOMO method $\underline{F}\ \underline{C} = \underline{S}\ \underline{C}\ \underline{\lambda}$.

The elements of the Fock matrix F are transferred from *ab initio* results on smaller "pattern" molecules. A hybrid orbital basis set, constructed from Gaussian orbitals, is used throughout. The overlap matrix $\underline{S}$ is evaluated exactly.

Reference: J. Eilers and D. Whitman, JACS 95, 2067 (1973).

2. SAMOU

This program can be used to evaluate the molecular orbitals for a particular class of open-shell radicals using the SAMO technique and the spin-unrestricted formalism. The resulting wavefunction is a simulation of the one that would be evaluated using the *ab initio* unrestricted Hartree-Fock (UHF) method.

$$\lambda = A\ \{U_1\alpha U_2\alpha\ \ldots\ U_m\alpha V_1\beta V_2\beta\ \ldots\ V_n\beta\}\quad m > n$$

With the different orbitals for different spins U_i and V_j satisfying the equations

$$\underline{F}^{\alpha}\ \underline{C}^{\alpha} = \underline{S}^{\alpha}\underline{C}^{\alpha}\ \underline{\lambda}^{\alpha}$$

The elements of the Fock matrices $\underline{F}^{\alpha}$ and $\underline{F}^{\beta}$ are transform *ab initio* UHF results on some "pattern" radicals and closed shell restricted Hartree-Fock results on other "pattern" molecules. A hybrid orbital basis set, constructed from Gaussian orbitals, is used throughout. The overlap matrix S is evaluated exactly. The radical must be such that the odd electron is essentially localized in a distinct part of the molecule.

Reference: B. J. Duke, B. O'Leary and J. Eilers, J. Chem. Soc. (Farad. Trans. II) in press.

3. SAMOP

This program evaluates the band structure of polymers with translational symmetry in one dimension using the SAMO method. This method is an economical way of simulating the results that would be obtained by an *ab initio* restricted Hartree-Fock closed shell LCAOMO procedure.

(Reference - method for molecules - J. Eilers and D. Whitman, JACS **95**, 2067 (1973)).

Reference - polymers - B. J. Duke and B. O'Leary, Chem. Phys. Lett. **20**, 459 (1973).

4. SAMOL

The SAMO method depends on the transferability of Fock matrix elements over hybrid basis orbitals in LCAOMO *ab initio* wavefunctions. The method uses such matrix elements for small "pattern" molecules to simulate (by transferability) the Fock matrix for larger molecules. Since the total number of matrix elements to be considered can be very large, this process of "transferring" them from the pattern molecule should be made as automatic as possible. This program aims to do this by producing, from a series of libraries of Fock elements for small molecules, the Fock element data in a form suitable for input by the program SAMOM and SAMOP.

The program uses two techniques. In the first, each matrix element for the pattern molecule is tagged automatically with a number of identifiers. A search for the large molecule then attempts to find matrix elements with the tags required for that molecule from the pattern molecule libraries. This approach is particularly suitable for large molecules of high symmetry which are to be simulated from a small number of small pattern molecules. The second technique, suitable for large molecules of low symmetry simulated from a larger number of pattern molecules is simpler but requires more thought from the user. This second technique is programmed only to give data suitable for SAMOM.

FORTRAN IV (IBM 360/370)
Symbolic Cards: Approximately 10000

3. C. Ab Initio Molecular Orbital Programs

239. SMALLØBE: Ab Initio Gaussian Lobe SCF Closed-Shell Program

by D. D. Shillady, Chemistry Department, Virginia Commonwealth University, Richmond, Virginia

This program was designed with speed and low core requirements in mind. It offers a subminimal gaussian lobe basis for geometry optimization of closed shell species and two minimal basis sets roughly comparable to 4G nuclear-centered gaussians. The program can be restarted in several ways to piecemeal a long run and it will run in less than 135K bytes of 360/40 storage. A simple property package offers population analysis and dipole moment. Data input is extremely simple and up to 12 atoms, 35 contracted orbitals or 108 primitive gaussian spheres can be treated. The generation of s, p and d STØ mimics is completely automatic including scaling to single zeta values.

IBM 360 FORTRAN
Symbolic Cards: 1300

92. IBMOL - LCAO-MO-SCF Method for Molecules of General Geometry

by D. R. Davis and E. Clementi, IBM Research Laboratory, San Jose, California

Program computes molecular wave functions for closed shell systems using Gaussian type basis functions. Up to 800 Gaussian functions may be used, with all gaussians through f-type permitted. The program operates under the IBSYS Overlay structure and it is written almost entirely in FORTRAN IV.

IBM 7090 FORTRAN IV and MAP
Symbolic Cards: About 10000

199. POLYATOM (Version 2) System of Programs for Quantitative Theoretical Chemistry

by D. B. Neumann, National Bureau of Standards, Washington D. C., H. Basch, Scientific Laboratory, Ford Motor Company, Dearborn, Michigan, R. l. Kornegay, Bell Telephone Laboratories, Murray Hill, New Jersey, L. C. Snyder, Bell Telephone Laboratories, Murray Hill, New Jersey, J. W. Moskowitz, New York University, Washington Square College, New York, C. Hornback, New York University, Washington Square College, New York, S. P. Liebmann, Department of Chemistry, University of Pennsylvania.

The POLYATOM system of computer programs has been written to make quantitative wave mechanical descriptions of molecules. These programs employ a gaussian basis set to compute single determinant SCF wavefunctions and corresponding properties in an a-prior style which includes all electrons and computes all integrals. The documation describes the basic philosophy and structure of POLYATOM (Version 2). It includes an account of most of the subroutines and several examples or applications giving input and output for the programs.

Programming Language: FORTRAN IV with auxiliary routines in assembly language for the GE 635 and the CDC 6600

Symbolic Cards: 16000

238. POLYATOM (Version II) (IBM 360)

by T. D. Metzgar and J. E. Bloor, Chemistry Department, The University of Tennessee, Knoxville 37916

This program is POLYATOM (Version 2) modified to enable it to be run on an IBM 360/65 in double precision. In its present form the SCF part requires a maximum of 268K core. However we have used it with slow core so that a maximum of 256K core and 32K LCS is required. The molecular properties package requires 468K or 224K and 256K LCS.

The data preparation and description of the program are as in the original POLYATOM (Version 2) write-up, or as listed in the program.

Test data for HF is in the original write-up. In addition a calculation on Cl_2 is supplied.

The double Hamiltonian open shell (PA 42) part has not been tested out.

In addition to the normal POLYATOM documentation which is pertinent to this program a short check out report is included as well as the Cl_2 test output.

IBM 360 FORTRAN
Symbolic Cards: Approximately 16000

241. PHANTOM: Ab Initio Quantum Chemical Programs for CDC 6000 and 7000 Series Computers

by D. Goutier and R. Macaulay, Department de Chemie, Université de Montréal, P. Q. Canada and A. J. Duke, Center for Computer Studies, The University of Leeds, Leeds, Great Britain.

This is an extensively modified and somewhat expanded version of POLYATOM Version II.

This system is very CDC oriented and for that reason it will be distributed on magnetic tape only in a special format which is used on CDC computers for maintaining large program libraries. This format is the UPDATE format.

The POLYATOM documentation is still appropriate to this system. In addition much new documentation which relates specifically to using this system as contained in UPDATE format is also included.

In testing, this system has proven very flexible and convenient to use.

FORTRAN IV (CDC 6000 17000 Series Computers)
Symbolic Cards: Approximately 16000

236. GAUSSIAN 70: Ab Initio SCF-MO Calculations on Organic Molecules

by W. J. Hehre, Department of Chemistry, University of California, Irvine, California 92664.

W. A. Lathan, Department of Chemistry, University of Rochester, Rochester, New York 14627.

R. Ditchfield, Department of Chemistry, Dartmouth College, Hanover, New Hampshire 03755.

M. D. Newton, Department of Chemistry, Brookhaven National Laboratory, Upton, L. I., New York 11973.

J. A. Pople, Department of Chemistry, Carnegie Mellon University, Pittsburgh, Pennsylvania 15213.

Gaussian 70 is a system of FORTRAN programs written for the IBM 360/370 and overlayed so as to require a maximum of 226K bytes of storage.

Two series of bases are built into the program: the minimal STO-NG[1] and extended 4-31G[2] sets. In addition provision has been made for the input of arbitrary sets of (s and p) functions subject to the overall size restrictions:

Maximum number of atoms 35.

Maximum number of atomic orbitals 75.

Maximum number of Gaussian functions per atomic orbital 6.

Maximum total number of Gaussians 240.

s and p functions only.

The system has been designed with flexibility and ease of use in mind, as well as overall program efficiency and it enables the user to perform such tasks as geometry and basis set optimization and potential surface scanning with little more input than is normally required for a single calculation.

Input and operating details are extensively documented in COMMENT cards within the program. In addition three sample inputs and associated outputs are provided for initial testing.

Primarily because of the novel design of the integral evaluation package, *ab initio* calculations on moderate size molecules (up to 10 or 12 heavy atoms) with the minimal STO-3G basis are only about 20 times more costly than with currently available semi-empirical schemes (CDNO, INDO, MINDO). Thus for those who now use QCPE 141 and related programs and who are not working on exceptionally large molecules, this system may well be worth the effort to examine thoroughly.

1. W. J. Hehre, R. F. Stewart and J. A. Pople, J. Chem. Phys. 51, 2657 (1969); W. J. Hehre, R. Ditchfield, R. F. Stewart and J. A. Pople, ibid. 52, 2191 (1970).

2. R. Ditchfield, W. J. Hehre and J. A. Pople, ibid. 54, 724 (1971); W. J. Hehre and W. A. Lathan, ibid. 56, 5255 (1972).

In addition applications of the STO-3G and 4-31G basis to theoretical problems in organic chemistry are well illustrated in the series of papers from the group of Professor Pople, "Molecular Orbital Theory of the Electronic Structure of Organic Compounds" appearing in The Journal of the American Chemical Society from 1970 to the present time.

FORTRAN IV (IBM 360/370)
Symbolic Cards: 13270

3. D. Localization of Molecular Orbitals

191. LOCAL: Molecular Orbital Localization Program

by Bernard Tinland, University of Lyon, Villeurbanne, France.

This subroutine performs localization of molecular orbitals in the ZDO approximation using the technique of Edmiston and Ruedenberg (Rev. Mod. Phys. 35, 457 (1963)). It is in convenient form to be added without modification to the CNINDO program (QCPE 141). The

CNINDO program must be modified slightly to include this subroutine. The necessary modification is given in the operating instructions.

IBM 360/75 FORTRAN IV (G Level)
Symbolic Cards: 200

198. LMO: Localized Molecular Orbitals

by T. G. M. Dolph, M. J. Shulta and Dr. Keith F. Purcell, Department of Chemistry, Kansas State University, Manhattan, Kansas 66502.

This program obtains localized molecular orbitals from CNDO or INDO canonical wavefunctions by maximizing the intra-orbital repulsion energy. The program outputs localized eigenvectors, the transformation matrix (canonical MO's), localized molecular orbital repulsion energy, 1/2* bond index, active charges, atom charges and total active charges.

FORTRAN IV (IBM 360/50)
Symbolic Cards: 800

267. ORLOC: Localized Molecular Orbitals by the Edmiston, Ruedenberg, Trindle, Sinanoglu Method

by Paul M. Kuznesof, Instituto de Quimica, Universidade Estadud de Campinas, Brasil

ORLOC is a subroutine which performs the two-dimensional unitary transformations on pairs of molecular orbitals (i,j) with systematic variations of the indices i,j. The iteration is terminated when the programmed degree of localization is achieved. Output is the set of localized orbitals, the final transformation matrix, and a new calculation of the charge density-bond order matrix as a check.

The program has the following advantages over QCPE 191:

1. It is faster due to the different iteration scheme.

2. It does not require assignment of a new matrix. Therefore, core memory requirements are practically unchanged from the original CNINDØ.

3. Test input is provided (cyclopropane) to be run with the adapted CNINDØ. The output localized orbitals should agree with those in J. Mol. Structure 8, 333 (1971).

FORTRAN IV (IBM)
Symbolic Cards: 150

3. E. Geometry Optimization Methods

OPTMO: A Molecular Orbital Program for Locating Equilibrium Geometries Employing the MINDO/2 or the Extended Huckel Methods

by Andrew Komornicki and James McIver, Department of Chemistry, SUNY at Buffalo, New York.

OPTMO is a multi-purpose program written to provide equilibrium geometries and energies on a variety of molecular systems. The program has provisions to work within either the MINDO/2 or the Extended Huckel framework. Depending on the options chosen the user may request either one function evaluation, or a series of function evaluations which will lead to the equilibrium geometry predicted by the method employed. A function evaluation involves the calculation of the energy and the gradient for a particular molecular geometry. The program has been written in such a way as to provide maximum efficiency with the minimum amount of user input. The MINDO/2 portion is based on the work of Dewar and coworkers and the parametrization employed is from the original papers. The Extended Huckel method of Hoffman has also been taken from the original references as have been the empirical parameters. Since within the past few years Hoffman has advocated the use of a hydrogen exponent of 1.3 rather than 1.0, this has also been implemented. In its present form the MINDO/2 method includes parameters for the atoms of carbon, hydrogen, oxygen and nitrogen. The Extended Huckel method is parametrized for hydrogen, boron, carbon and nitrogen.

FORTRAN IV (CDC 6400)
Symbolic Cards: 2000

228. MINDO/2: Geometry Optimization Program: SIMPLEX

by Michael J. S. Dewar and Patrick J. Student, Department of Chemistry, University of Texas at Austin, Texas.

The program performs geometry optimization of the MINDO/2' energy using the SIMPLEX algorithm. (See references below for more information concerning the algorithm.)

This program is in FORTRAN for a CDC 6600 and uses approximately 32K of core. The program now handles up to 30 atoms, 60 orbitals and 50 variables. The program requires N^2 SCF evaluations where N is the number of variables to be optimized.

The sample problems and documentation with this program provide some insight into its versatility. Included are sample input for Ethylene and Bicyclobutane.

The Bicyclobutane example is especially interesting because its calculation contains parameters which are dependent on previously defined parameters and this situation can be handled, in general, by the inclusion of a user supplied subroutine which handles this problem. For bicyclobutane the subroutine DEPVAR is included with the program.

The documentation also indicates how the systematic change in a parameter, which is defined as the reaction coordinate, can be handled in studying a reaction.

References: SIMPLEX:

J. A. Nelder and R. Mead, Comp. J., 308 (1964).

A. Brown, M. J. S. Dewar, H. Metiu, P. Student and J. Wasson, Proc. Roy. Soc., to be published.

FORTRAN IV (CDC 6600)
Symbolic Cards: Approximately 2800

CHAPTER XIV

CLOSING REMARKS

This chapter consists of a number of policy statements which border on philosophy and therefore are highly subjective. Anyone who has performed a series of computations will have already at least partially developed his own set of prejudices. Consequently, the present set of policies are presented largely to serve as illustrations for those not yet committed to any particular "theoretical religion".

1. The Choice of the Problem

Since a practicing experimental organic chemist will have already chosen a research problem, the title seems to suggest that only pure theoreticians must search for a chemical problem. Such an implication is false. More often than not the experimental problem is in some way unsuitable for computational study, usually because it involves too many variables, i.e. too many or too large molecules. Another problem involving some simplification is necessary. The choice of the 'reduced' problem is usually done with the 'full' problem in mind, retaining what are (on intuitive grounds) believed to be the essential features, for example, the reactive functional groups, while eliminating as many other variables as possible.

For example, in Chapter XI the oxacarbene ring expansion reaction of some cyclic ketones was discussed:

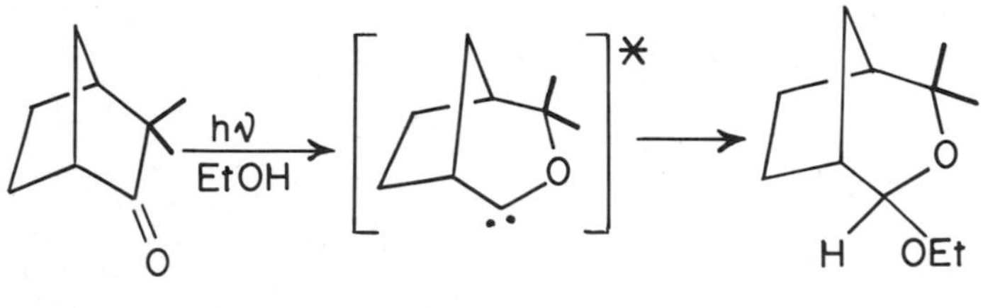

{XIV-1}

This large problem was simulated by the rearrangement to oxacarbene of formaldehyde

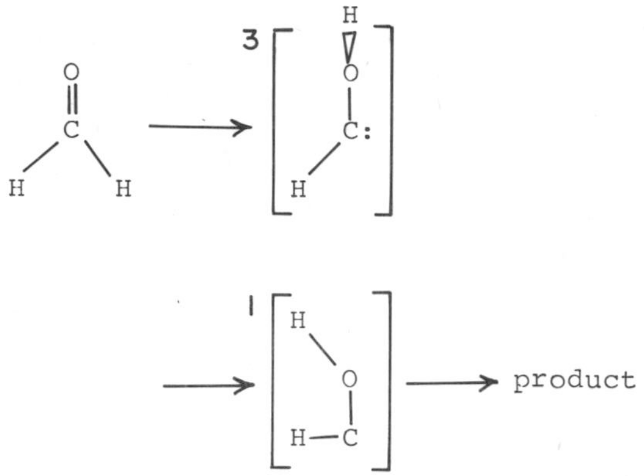

{XIV-2}

Such a 'model' system may be treated rigorously by ab initio methods. A great deal has been understood about the nature of the reaction by just such a study. However, even if the reaction surface was explored reasonably well, it does not mean that hydrogen migration from carbon to oxygen in an excited state of any carbonyl compound can faithfully mimic all the details of alkyl migration from carbon to oxygen.

In order to look at the details of alkyl migration from C to O in excited carbonyl compounds one needs another "model" compound that is structurally slightly closer to the original problem. In the present case acetaldehyde might be used instead of formaldehyde to simulate the photochemical reaction

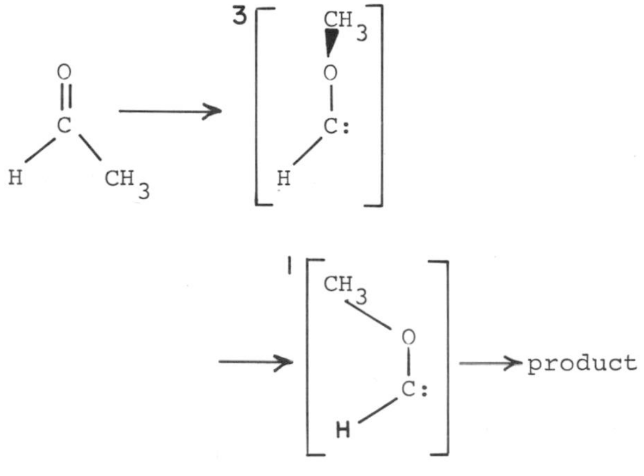

{XIV-3}

of interest. However the "ring strain" factor which is apparently involved in a reaction like {XIV-1} has so far been completely disregarded. A much more expensive but again more realistic study of cyclic ketones of different ring size might be performed:

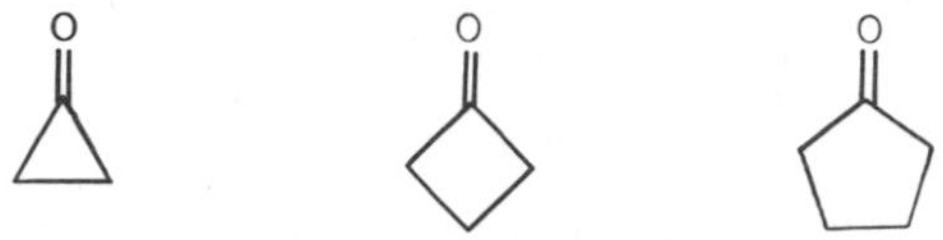

{XIV-4}

In the unattainable limit of the infinite computer budget one would finally return to the actual system {XIV-1} and handle it by *ab initio* calculations.

Although most of the above arguments have been made in terms of the limited resources available, it should be noted that even if one could directly perform calculations on a system like that given in {XIV-1} the results might be very difficult to interpret conceptually without the other stages of the calculation as outlined above.

2. The Choice of the Theoretical Approach

The choice of the problem must be correlated with the choice of the theoretical approach. To do calculations on the molecules which an organic chemist typically uses in his experiments requires a different theoretical approach than that which would be used for the study of a small molecule. Considering the all electron or all valence electron approaches only, there exists at least three categories of theoretical research tools:

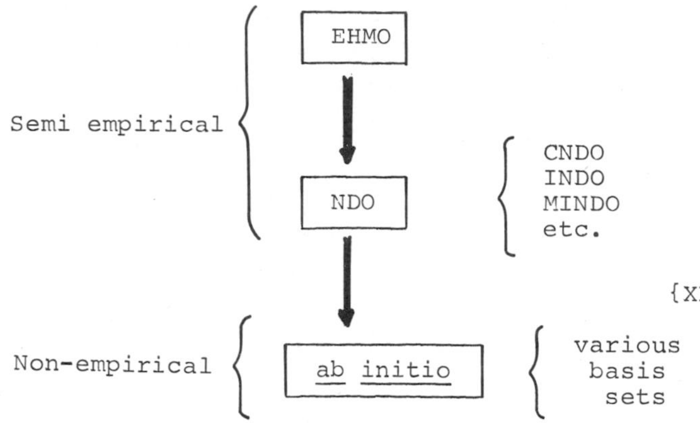

{XIV-5}

Although research schools exist which are essentially dedicated entirely to one approach to the exclusion of all others, such a philosophy seems somewhat fanatic or dogmatic. It would appear more reasonable to suggest that each method has its own domain (not necessarily with very well defined boundaries) of optimum applicability and thus the choice of the theoretical approach should be done in conjunction with the choice of the chemical problem of interest.

3. The Choice of the Program

There was a time, not long ago (it was after all only in the early 1960's that the first correct ab initio calculations on polyatomic molecules with more than about four electrons appeared in the literature), when the computer program had to be written and tested before any problem oriented calculations could begin. Today, thanks to the excellent work of the people at the Quantum Chemistry Program Exchange (QCPE) and to the numerous unselfish program authors, accurate, well tested and well documented programs are readily available. Consequently, in conjunction with the previous two choices, we can usually select a program from the QCPE Catalog. Note, however, that any general program will not, by definition, be the most efficient for any given particular problem. Unless the problem is at the very limits of computational possibility, this fact need not be of too great a concern. Nevertheless, the possibility of (perhaps even relatively minor) changes to existing programs to suit a particular set of problems at hand should not be overlooked. Of course the computational and man effort required to correctly modify and test a program must be balanced against the number of times it will be used.

There are several programs in all three categories given above ({XIV-5}). However, in a reflection of the bias mentioned at the very beginning of this chapter attention will be paid to the *ab initio* programs. At present, perhaps the most popular (among organic chemists) *ab initio* program is GAUSSIAN 70. Its built-in features such as geometry optimization, geometry scanning, standard internal basis sets and its relatively high efficiency and speed, are all attractive features if one is interested chiefly in calculations on small or medium-size organic molecules. Many of the calculations used as examples in these notes were performed with GAUSSIAN 70. Theoreticians would probably tend to prefer IBMOL4 or POLYATOM 2, or various modifications of these programs, since they allow for greater flexibility in basis sets, in stepwise use of the program, in the size of the problem treated, etc. For example, the study of the role of d and f polarization functions in sulfur compounds [I. G. Csizmadia, General and Theoretical Aspects of the Thiol Group, in "The Chemistry of the Thiol Group", edited by S. Patai, John Wiley and Sons, 1974, pp. 1-109] involved the combined use of the IBMOL and POLYATOM programs. However, it may be useful to carry out preliminary studies with the aid of GAUSSIAN 70 before attempting more sophisticated computations. A combination of various computer programs could be regarded as almost a practical necessity for the well-rounded theoretical organic chemist.

4. The Choice of the Basis Set

For semi-empirical MO computations invariably Slater type orbitals (STO) are used since generally only overlap integrals are specifically computed. For non-empirical MO computations, with the exception of linear molecules, Gaussian type functions (GTF) are used almost exclusively.

Chapter XII dealt with the basis set problem in some detail. However, a few points bear repetition. At present, no universally accepted Gaussian basis sets are available. Computations done at different laboratories are in general not directly comparable unless done in the same basis sets. The following references listed in chronological order may be consulted for details regarding a number of Gaussian basis sets:

1. I. G. Csizmadia, M. C. Harrison, J. W. Moskowitz and B. T. Sutcliffe, Progr. Rept. MIT-SSMTG 49, 78 (1963) [see also: Theoret. Chim. Acta 6, 191 (1966)].
2. S. Huzinaga, J. Chem. Phys. 42, 1293 (1965).
3. J. L. Whitten, J. Chem. Phys. 44, 359 (1966).
4. H. Basch, N. B. Robin, N. A. Kuebler, J. Chem. Phys. 47, 1201 (1967).
5. A. Veillard, Theor. Chim. Acta 12, 405 (1968).
6. S. Huzinaga, Y. Sakai, J. Chem. Phys. 50, 1371 (1968).
7. C. Salez, A. Veillard, Theor. Chim. Acta 11, 441 (1968).
8. H. Basch, C. J. Hornback, J. W. Moskowitz, J. Chem. Phys. 51, 1311 (1969).
9. D. R. Whitman, C. J. Hornback, J. Chem. Phys. 51, 398 (1969).
10. S. Huzinaga, C. Arnau, J. Chem. Phys. 52, 2224 (1970).
11. A. J. H. Wachters, J. Chem. Phys. 52, 1033 (1970).
12. B. Roos, P. Siegbahn, Theor. Chim. Acta 17, 209 (1970).
13. R. Ditchfield, W. J. Hehre, J. A. Pople, J. Chem. Phys. 52, 5001 (1970); 54, 724 (1971) [GAUSSIAN 70].
14. R. Roos, A. Veillard, G. Vinot, Theor. Chim. Acta 20, 1 (1971).
15. T. A. Claxton, N. A. Smith, Theor. Chim. Acta 22, 378 (1971).

16. W. J. Hehre, R. Ditchfield, J. A. Pople, J. Chem. Phys. 56, 2257 (1972) [GAUSSIAN 70].

17. K. Ruedenberg, R. C. Raffenetti, R. D. Bardo, in Proceedings of the 1972 Boulder Seminar Research Conference on Theoretical Chemistry, edited by D. W. Smith (Wiley, New York, 1973), p. 164.

18. R. C. Raffenetti, J. Chem. Phys. 59, 5963 (1973).

19. R. C. Raffenetti, K. Ruedenberg, J. Chem. Phys. 59, 5978 (1973).

20. R. D. Bardo, K. Ruedenberg, J. Chem. Phys. 59, 5956 (1973); 59, 5966 (1973); 60, 918 (1974); 60, 932 (1974).

21. B. M. Rode, Chem. Phys. Letters 27, 264 (1974).

22. R. E. Kari, P. G. Mezey, I. G. Csizmadia, J. Chem. Phys. 63, 581 (1975).

23. P. G. Mezey, I. G. Csizmadia, O. P. Strausz, Can. J. Phys. 53, 2512 (1975).

24. P. G. Mezey, M. H. Lien, K. Yates, I. G. Csizmadia, Theor. Chim. Acta 40, 75 (1975).

25. R. E. Kari, P. G. Mezey, I. G. Csizmadia, J. Chem. Phys. 64, 632 (1976).

26. P. G. Mezey, R. E. Kari, I. G. Csizmadia, J. Chem. Phys. in press.

Recall that a given set of primitive gaussians may be contracted in different ways

1	single	zeta	(minimal)
1 1/2	sesqui	zeta	(split valence shell)
2	double	zeta	
2 1/2	ses	zeta	
3	triple	zeta	
	⋮	⋮	

{XIV-6}

Clearly the contracted basis set, over which the SCF calculation is performed is of major importance. However, even a generous (say double zeta) contraction scheme cannot mask any short comings in the primitive gaussian set.

This fact recalls to mind the question of the balance of molecular basis sets. Suppose one wishes to investigate a heteronuclear bond such as

$$\overset{\delta+}{S}\text{—}\overset{\delta-}{O} \qquad \{XIV-7\}$$

with a uniquely defined dipole moment. If one over represents the S atom with respect to the oxygen the computed charge distribution {XIV-8} as well as the dipole moment may be opposite to that of the experimental {XIV-7}

$$\overset{\delta-}{S}\text{—}\overset{\delta+}{O} \qquad \{XIV-8\}$$

In contrast, if one over represents the oxygen atom with respect to the sulfur atom the direction of polarization will not be opposite to the experimentally observed one {XIV-7} but it could be exaggerated to such a degree that the calculations show an exothermic bond cleavage in a heterolytic fashion:

$$S^{2+} \quad O^{2-} \qquad \{XIV-9\}$$

Consequently it is of great importance that the orbital representation of the constituent atoms be in balance with each other within the molecule.

5. The Choice of the Geometry

The best approach is to carry out a full geometry variation for all molecules in any study. This requires the generation of an energy hypersurface (cf. Chapter IX):

$$E = E(q_1, q_2, \ldots, q_{3N-6}) \qquad \{XIV-10\}$$

Such a hypersurface includes all possible structures.

However, such a general approach is very time consuming (both in terms of human time and computer time) and may not be practical. Often, an educated guess for the molecular geometry based on experimental results or previous calculations on similar species is a necessity. Such an educated guess is of course also useful as a reasonable starting point for full geometry optimization. The following publications may be of some use in making such an educated guess:

1a H. J. M. Bowen et al., Tables of interatomic distances and configuration in molecules and ions. Chem. Soc. (London) Spec. Publ. 11 (1958).

1b Tables of interatomic distances and configuration in molecules and ions. Supplement 1956-1959. Spec. Publ. 18 (1965).

2 G. Herzberg. Molecular Spectra and Molecular Structure III. Electronic spectra and electronic structure of polyatomic molecules. Van Nostrand, Toronto (1966).

3a W. G. Richards, T. E. H. Walker and R. K. Hinkley. A Bibliography of ab initio molecular wave functions. Clarendon Press, Oxford (1971).

3b W. G. Richards, T. E. H. Walker, L. Farnell and P. R. Scott. Bibliography of ab initio molecular wave functions. Supplement for 1970-1973. Clarendon Press, Oxford (1974).

6. The Choice of the Computer

Often there is no choice. Even if there exists a choice one would probably prefer to work at his own institution and if there is only one computer that predetermines this choice.

Many programs are available from QCPE both for various IBM (International Business Machine) and for various CDC (Control Data Corporation) computers. If the available machine is different from the one the program was developed to run on, the program must be adapted to the local environment. If the program is written in a high level language such as

FORTRAN, the adaptation procedure may not be all that difficult. However, if some of the routines are written in assembler or worse, machine-level language adaptation will normally be quite difficult. Any novice is well advised to secure all available help from the local computing centre.

7. Ready to Begin

Having nearly finished these notes you should be nearly ready to begin. After all the most successful and perhaps the only way to learn something is by doing it. The policies outlined in this chapter will perhaps be of interest until (probably fairly quickly) you develop your own.

It is strongly recommended that the novice tackle a few simple problems first, even if they have been treated by others previously.

For example, if the experimental problem involves a secondary alkyl iodide such as

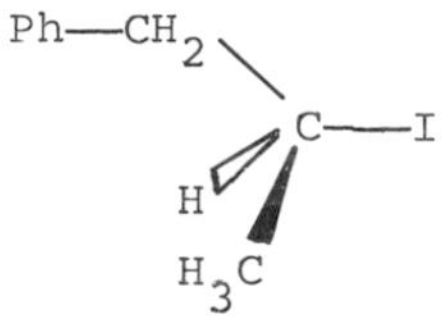

{XIV-11}

one could aim to simulate it by

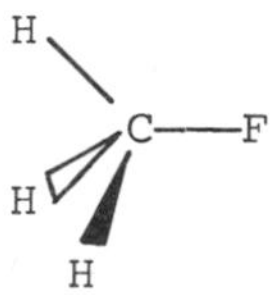

{XIV-12}

and a good place to begin would be

H——F {XIV-13}

However, more problems will arise later. Once the problem is fairly well defined it may become desirable to compute a good portion of the energy hypersurface {XIV-10} that describes the process under investigation. Such a project could easily require the computation of over 200 points and sometimes as many as 1000 or more.* Once this "production" stage is reached the computational cost per point is of considerable importance. The cost of computer time for a molecule of chemical interest may be high for even a single point if a sufficiently good quality basis set is used. For this reason, it is not unheard of for a computing chemist to travel vast distances to use a lower cost computer. Teaming up for such gigantic projects is not uncommon and thus the otherwise oppressive computational load may be divided among several laboratories. Of course prime computer time (Monday to Friday, 9 to 5) is hardly ever used for such production work and clearly most "number crunching" jobs are done on the "graveyard shift", on weekends, and on holidays.

Once started, you will not be alone! In the mid-1960's it was a rare thing for an organic chemist to be involved in theoretical computation, but in the mid-1970's it is becoming increasingly more common. The time will come when it will be as routine as NMR spectroscopy and every organic chemist will carry out both theoretical and experimental investigations of the same phenomenon.

---

*See Figure XIV-1 for an example of a simple chemical reaction, $H_2 + H^+ \rightleftharpoons H_3^+$.

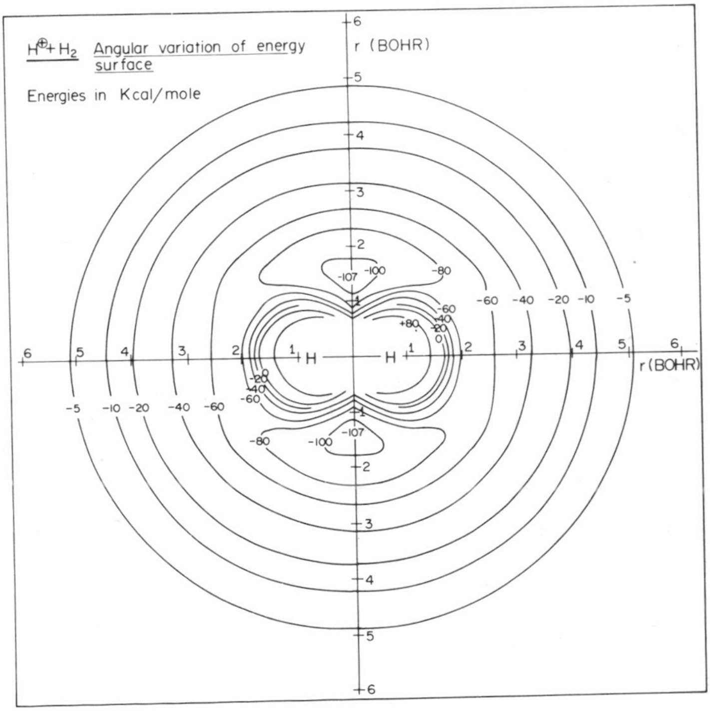

Figure XIV-1. The relaxed cross section of the hypersurface for the reaction $H_2 + H^+ \rightleftarrows H_3^+$. This hypersurface calculation involved 210 individual points each computed by a full CI (cf. Ch. VIII) in an sp Gaussian basis set. Each point required ∿20 min on an IBM 7094 [Can. J. Chem., 47, 4097 (1969), J. Chem. Phys. 52, 6205 (1970)].

SECTION E

APPENDIX

CHAPTER XV

DETAILED FORMALISM OF ROOTHAAN'S SCF THEORIES

1. Roothaan Closed Shell Restricted Hartree-Fock Formalism

Recalling equation {IV-53}

$$E = 2 \sum_{p=1}^{M} h_{pp} + \sum_{p=1}^{M} \sum_{q=1}^{M} (2J_{pq} - K_{pq}) \qquad \{XVA-1\}$$

where

$$J_{pq} = \langle \phi_p | \hat{J}_q | \phi_p \rangle$$

$$= \int \phi_p^*(1) \left[\int \phi_q^*(2) \frac{1}{r_{12}} \phi_q(2)\, d\tau_2 \right] \phi_p(1)\, d\tau_1 \qquad \{XVA-2a\}$$

$$K_{pq} = \langle \phi_p | \hat{K}_q | \phi_p \rangle$$

$$= \int \phi_p^*(1) \left[\int \phi_p^*(2) \frac{1}{r_{12}} \phi_q(2)\, d\tau_2 \right] \phi_q(1)\, d\tau_1 \qquad \{XVA-2b\}$$

as in Chapter IV.

Rewriting {XVA-1}

$$E = 2 \sum_{p=1}^{M} \int \phi_p^* \left[\hat{h} + \sum_{q}^{M} (2\hat{J}_q - \hat{K}_q) \right] \phi_p\, d\tau \qquad \{XVA-3\}$$

and we require E to be stationary, i.e.

$$\delta E = 0 \qquad \{XVA-4\}$$

subject to the constraint that the orbitals ϕ_p remain orthonormal, i.e.

$$\int \phi_p^* \phi_p d\tau = 1$$

$$\int \phi_p^* \phi_q d\tau = 0 \quad p \neq q \qquad \{XVA-5a,b\}$$

Consider an arbitrary variation in E caused by an infinitesimal variation in the MO, $\delta\phi_p$. Then from {XVA-3}

$$\delta E = 2 \sum_{p=1}^{M} \int \delta\phi_p^* \{\hat{h} + \sum_{q}^{M} (2\hat{J}_q - \hat{K}_q)\} \phi_p d\tau \qquad \text{\{XVA-6\}}$$

$$+ 2 \sum_{p=1}^{M} \int (\delta\phi_p) \{\hat{h}^* + \sum_{q}^{M} (2\hat{J}_q^* - \hat{K}_q^*\} \phi_p^* d\tau$$

Defining

$$\{\hat{h} + \sum_{q}^{M} (2\hat{J}_q - \hat{K}_q)\} = F \qquad \text{\{XVA-7\}}$$

then

$$\delta E = 2 \sum_{p=1}^{M} \int \delta\phi_p^* \hat{F} \phi_p d\tau + 2 \sum_{p=1}^{M} \int \delta\phi_p \hat{F}^* \phi_p^* d\tau \qquad \text{\{XVA-8\}}$$

The orthonormality restriction on the ϕ_p leads to the restriction

$$\int \delta\phi_p^* \phi_q d\tau + \int \delta\phi_p \phi_q^* d\tau = 0 \qquad \text{\{XVA-9\}}$$

on the variation in the ϕ_p.

A standard mathematical technique for such a constrained extreme value problem is the use of undetermined Lagrangian multipliers. Introducing the Lagrangian multipliers $-2\varepsilon_{qp}$ into {XVA-9} and performing the summation gives

$$-2 \sum_{p}^{M} \sum_{q}^{M} \varepsilon_{qp} \int \delta\phi_p^* \phi_q d\tau - 2 \sum_{p}^{M} \sum_{q}^{M} \varepsilon_{pq} \int \delta\phi_p \phi_q^* d\tau = 0 \qquad \text{\{XVA-10\}}$$

Comparing {XVA-10} with {XVA-6} the two equations are of the same form and may be added to give:

$$\delta E = 2 \sum_{p}^{M} \int \delta\phi_p^* \{\hat{F}\phi_p - \sum_{q}^{M} \phi_q \varepsilon_{qp}\} d\tau$$

$$+ 2 \sum_{p}^{M} \int \delta\phi_p \{\hat{F}^*\phi_p^* - \sum_{q}^{M} \phi_q^* \varepsilon_{pq}\} d\tau = 0 \qquad \{XVA\text{-}11\}$$

The two terms are the complex conjugate of one another. Therefore, both must be zero.

$$2 \sum_{p}^{M} \int (\delta\phi_p^*) \{\hat{F}\phi_p - \sum_{q}^{M} \phi_q \varepsilon_{qp}\} d\tau = 0$$

$$2 \sum_{p}^{M} \int (\delta\phi_p) \{\hat{F}^*\phi_p^* - \sum_{q}^{M} \phi_q^* \varepsilon_{pq}\} d\tau = 0 \qquad \{XVA\text{-}12a,b\}$$

Now $\delta\phi_p^* \neq 0$ and $\delta\phi_p \neq 0$ thus

$$\hat{F}\phi_p - \sum_{q}^{M} \phi_q \varepsilon_{qp} = 0$$

$$\{XVA\text{-}13a,b\}$$

$$\hat{F}^*\phi_p^* - \sum_{q}^{M} \phi_q^* \varepsilon_{pq} = 0$$

or

$$\hat{F}\phi_p = \sum_q^M \phi_q \varepsilon_{qp}$$

$$\hat{F}^*\phi_p^* = \sum_q^M \phi_q^* \varepsilon_{pq} \qquad \{XVA\text{-}14a,b\}$$

Taking the complex conjugate of {XVA-14b} and subtracting it from {XVA-14a} gives

$$0 = \sum_q^M \phi_q \, (\varepsilon_{qp} - \varepsilon_{pq}^*) \qquad \{XVA\text{-}15\}$$

which implies

$$\varepsilon_{qp} - \varepsilon_{pq}^* = 0 \qquad \{XVA\text{-}16\}$$

or

$$\varepsilon_{qp} - \varepsilon_{qp}^{\dagger} = 0 \qquad \{XVA\text{-}17\}$$

Thus a matrix with the elements ε_{pq} is a Hermitian matrix, the two equations {XVA-14a,b} are equivalent and only one needs to be solved.

$$\hat{F}\phi_p = \sum_q^M \phi_q \varepsilon_{qp} \qquad \{XVA\text{-}18\}$$

Since the $\underline{\underline{\varepsilon}}$ matrix is Hermitian it may be diagonalized leading to M equations each of the form

$$\hat{F}\phi_p = \phi_p \varepsilon_p \qquad \{XVA\text{-}19\}$$

F is the Fock operator and the ε_p originally introduced as undetermined Lagrangian multipliers may be interpreted as "orbital energies" in the Koopmans' theorem sense (Chapter VI-3).

2. Roothaan Open Shell Restricted Hartree-Fock Formalism

Roothaan's open shell restricted Hartree-Fock approach reduces the open shell equations for a large number of cases to a pseudoeigenvalue form by absorbing off diagonal multipliers (which cannot in general be eliminated simply by a unitary transformation due to the coupling of the closed and open shell) into the effective Hamiltonian, i.e. by redefining the Fock operator*.

Roothaan's method treats the case where

$$\phi = (\phi_c, \phi_o) \qquad \{XVA-20\}$$

ϕ, set of MO

ϕ_c, set of closed shell MO

ϕ_o, set of open shell MO

and the orbitals are assumed orthonormal. The energy in a case suitable for the Roothaan method may be written as

$$E = 2 \sum_{k=1}^{n_c} H_k + \sum_{k=1}^{n_c} \sum_{\ell=1}^{n_c} (2J_{k\ell} - K_{k\ell}) + 2f \sum_{k=1}^{n_c} \sum_{m=1}^{n_o} (2J_{km} - K_{km})$$

$$+ f \,[2 \sum_{m=1}^{n_o} h_m + f \sum_{m=1}^{n_o} \sum_{n=1}^{n_o} (2aJ_{mn} - bK_{mn}) \qquad \{XVA-21\}$$

where

---

*The notation used in this section differs from that of section 1 k,ℓ; refer to closed shell orbitals

m,n; refer to open shell orbitals

n_c; number of closed shell orbitals

n_o; number of open shell orbitals

a,b; constants depending on spin state

f; the fractional occupancy of the open shell (0 < f < 1), i.e. the number of occupied open shell spin orbitals divided by the number of available open shell spin orbitals.

Coupling operators involving the off-diagonal Lagrangian multipliers, which as before preserve orthonormality, are now introduced.

<u>Coulomb and Exchange Coupling Operators</u>

(Hermitian) $$\hat{L}_i|\phi_i\rangle = \langle\phi_i|\hat{J}_o|\phi\rangle|\phi_i\rangle + \langle\phi_i|\phi\rangle\hat{J}_o|\phi_i\rangle \quad \{XVA\text{-}22a\}$$

$$\hat{M}_i|\phi_i\rangle = \langle\phi_i|\hat{K}_o|\phi\rangle|\phi_i\rangle + \langle\phi_i|\phi_i\rangle\hat{K}_o|\phi_k\rangle \quad \{XVA\text{-}22b\}$$

closed shell: $$\hat{L}_c = \sum_{k=1}^{n_c} \hat{L}_k \qquad \hat{M}_c = \sum_{k=1}^{n_c} \hat{M}_k \quad \{XVA\text{-}23a\}$$

open shell: $$\hat{L}_o = f\sum_{m=n_c+1}^{n_o} \hat{L}_m \qquad \hat{M}_o = f\sum_{m=n_c+1}^{n_o} \hat{M}_m \quad \{XVA\text{-}23b\}$$

total: $$\hat{L}_r = \hat{L}_c + \hat{L}_o \qquad \hat{M}_r = \hat{M}_c + \hat{M}_o \quad \{XVA\text{-}23c\}$$

The effects of these coupling operators on the orbitals are given by

$$\hat{L}_c-\hat{J}_o|\phi_k\rangle = \sum_{\ell}|\phi_\ell\rangle\langle\phi_\ell|\hat{J}_o|\phi_k\rangle \qquad \{XVA\text{-}24a\text{-}d\}$$

$$\hat{M}_c-\hat{K}_o|\phi_k\rangle = \sum_{\ell}|\phi_\ell\rangle\langle\phi_\ell|\hat{K}_o|\phi_k\rangle$$

$$\hat{L}_o-f\hat{J}_o|\phi_m\rangle = f\sum_{n}|\phi_n\rangle\langle\phi_n|\hat{J}_o|\phi_m\rangle$$

$$\hat{M}_o-f\hat{K}_o|\phi_m\rangle = f\sum_{n}|\phi_n\rangle\langle\phi_n|\hat{K}_o|\phi_m\rangle$$

and

$$\hat{L}_c|\phi_m\rangle = \sum_{k=1}^{n_c}|\phi_k\rangle\langle\phi_k|\hat{J}_o|\phi_m\rangle \qquad \{XVA\text{-}25a\text{-}d\}$$

$$\hat{M}_c|\phi_m\rangle = \sum_{k=1}^{n_c}|\phi_k\rangle\langle\phi_k|\hat{K}_o|\phi_m\rangle$$

$$\hat{L}_o|\phi_k\rangle = f\sum_{m=n_c+1}^{n_o}|\phi_m\rangle\langle\phi_m|\hat{J}_o|\phi_k\rangle$$

$$\hat{M}_o|\phi_k\rangle = f\sum_{n=n_c+1}^{n_o}|\phi_m\rangle\langle\phi_m|\hat{K}_o|\phi_k\rangle$$

i.e. the coupling operators transform open shell orbitals into closed shell orbitals and vice versa.

The variational problem is treated in an exactly analogous fashion to the closed shell case.

$$\delta E = 2 \sum_{k} [\langle\delta\phi_k|\hat{H}|\phi_k\rangle + \langle\phi_k|\hat{H}|\delta\phi_k\rangle]$$

$$+ \sum_{k,\ell} \{\langle\delta\phi_k|2\hat{J}_\ell-\hat{K}_\ell|\phi_k\rangle + \langle\delta\phi_\ell|2\hat{J}_k-\hat{K}_k|\phi_\ell\rangle\}$$

$$+ 2f \sum_{m} [\langle\delta\phi_m|\hat{H}|\phi_m\rangle + \langle\phi_m|\hat{H}|\delta\phi_m\rangle]$$

$$+ f^2 \sum_{m,n} \{\langle\delta\phi_m|2a\hat{J}_n-b\hat{K}_n|\phi_m\rangle + \langle\delta\phi_n|2a\hat{J}_m-b\hat{K}_m|\phi_n\rangle\}$$

$$+ f \sum_{k,m} \{\langle\delta\phi_k|2\hat{J}_m-\hat{K}_m|\phi_k\rangle + \langle\delta\phi_m|2\hat{J}_k-\hat{K}_k|\phi_m\rangle\}$$

{XVA-26}

which becomes

$$0 = \delta E = 2\sum_{k}\langle\delta\phi_k|\hat{H} + \sum_{\ell}(2\hat{J}_\ell-\hat{K}_\ell) + f\sum_{m}(2\hat{J}_m-\hat{K}_m)|\phi_k\rangle + \text{c.c.}$$

$$+ 2f\sum_{m} \langle\delta\phi_m|f(\hat{H} + \sum_{n} 2a\hat{J}_n-b\hat{K}_n) + f\sum_{k}(\hat{J}_k-\hat{K}_k)|\phi_m\rangle + \text{c.c.}$$

{XVA-27}

where c.c. abbreviates the complex conjugate of the expression immediately preceding it. {XVA-27} may be rearranged to give:

$$0 = \delta E = 2\sum_{k} \langle\delta\phi_k|\hat{H} + 2\hat{J}_c-\hat{K}_c + 2\hat{J}_o-\hat{K}_o|\phi_k\rangle + \text{c.c.}$$

$$+ 2\sum_{m} \langle\delta\phi_m|f(\hat{H} + 2\hat{J}_c-\hat{K}_c + 2a\hat{J}_o-b\hat{K}_o)|\phi_m\rangle + \text{c.c.}$$

{XVA-28}

and introducing the Lagrangian multiplier $-2\theta_{ij}$ for a given i,j and summing over all i,j

$$-2 \sum_{ij} \theta_{ji} \langle\delta\phi_i|\phi_i\rangle - 2 \sum_{ij} \theta_{ij} \langle\phi_i|\delta\phi_i\rangle = 0 \qquad \{XVA-29\}$$

This expression is added to that above for δE

$$0 = 2 \sum_k \langle\delta\phi_k| (\hat{H} + 2\hat{J}_c - \hat{K}_c + 2\hat{J}_o - \hat{K}_o)|\phi_k\rangle - \sum_j \phi_{jk}|\phi_j\rangle + c.c.$$

$$+ 2 \sum_m \langle\delta\phi_m| f(\hat{H} + 2\hat{J}_c - \hat{K}_c + 2a\hat{J}_o - b\hat{K}_o|\phi_m\rangle - \sum_m \theta_{jm}|\phi_j\rangle + c.c.$$

$$\{XVA-30\}$$

The expression should hold for any $\delta\phi_i$ hence:

$$(\hat{H} + 2\hat{J}_c - \hat{K}_c + 2\hat{J}_o - \hat{K}_o)|\phi_k\rangle = \sum_j \theta_{jk}|\phi_j\rangle$$

$$f(\hat{H} + 2\hat{J}_c - \hat{K}_c + 2a\hat{J}_o - b\hat{K}_o)|\phi_m\rangle = \sum_j \theta_{jm}|\phi_j\rangle \qquad \{XVA-31a,b\}$$

$$(\hat{H} + 2\hat{J}_c - \hat{K}_c + 2\hat{J}_o - \hat{K}_o)|\phi_k\rangle = \sum_\ell |\phi_\ell\rangle\theta_{\ell k} + \underline{\sum_n |\phi_n\rangle\theta_{nk}}$$

$$f(\hat{H} + 2\hat{J}_c - \hat{K}_c + 2a\hat{J}_o - b\hat{K}_o)|\phi_m = \underline{\sum_\ell |\phi_\ell\rangle \theta_{\ell m}} + \sum_n |\phi_n\rangle \theta_{nm}$$

$$\{XVA-32a,b\}$$

In the closed shell case, the orbitals can <u>always</u> be subjected to a unitary transformation which brings the matrix of Lagrangian multipliers into diagonal form. In the open shell case the available transformations may separately diagonalize the closed and open shell parts but <u>cannot in general eliminate the off-diagonal multipliers coupling the closed and open shells</u>. That is, it is <u>possible</u> to reduce to

$$\hat{Q}|\phi_k\rangle = \theta_k|\phi_k\rangle + \sum_n |\phi_n\rangle\theta_{nk}$$

$$\hat{R}|\phi_m\rangle = \theta_m|\phi_m\rangle + \sum_\ell |\phi_\ell\rangle\theta_{\ell m} \qquad \text{\{XVA-33a,b\}}$$

but off-diagonal multipliers remain. However, the <u>problem specified thus far</u> may be reduced to a convenient form by re-expressing the closed shell-open shell coupling terms with the aid of the coupling operators previously defined so as to absorb terms involving off diagonal multipliers into the left hand sides of the equations

$$(\hat{H} + 2\hat{J}_c - \hat{K}_c + 2\hat{J}_o - \hat{K}_o)|\phi_k\rangle = \sum_\ell |\phi_\ell\rangle\theta_{\ell k} + \sum_n |\phi_n\rangle\theta_{nk}$$

$$f(\hat{H} + 2\hat{J}_c - \hat{K}_c + 2a\hat{J}_o - b\hat{K}_o)|\phi_m\rangle = \sum_\ell |\phi_\ell\rangle\theta_{\ell m} + \sum_n |\phi_n\rangle\theta_{nm}$$

{XVA-32a,b}

Multiplying {XVA-32a} from the left by $\langle\phi_m|$ and {XVA-32b} from the left by $\langle\phi_k|$ gives

$$\langle\phi_m|(\hat{H} + 2\hat{J}_c - \hat{K}_c + 2\hat{J}_o - \hat{K}_o|\phi_k\rangle = \sum_\ell \langle\phi_m|\phi_\ell\rangle\theta_{\ell k} + \sum_n \langle\phi_m|\phi_n\rangle\theta_{nk}$$

$$f\langle\phi_k|(\hat{H} + 2\hat{J}_c - \hat{K}_c + 2a\hat{J}_o - b\hat{K}_o)|\phi_m\rangle = \sum_\ell \langle\phi_k|\phi_\ell\rangle\theta_{\ell m} + \sum_n \langle\phi_\ell|\phi_n\rangle\theta_{mn}$$

{XVA-34a,b}

Recalling that the orbitals are orthonormal and remembering that $\sum_\ell$ does not include m and $\sum_n$ does not include ℓ gives

$$\langle\phi_m|(\hat{H} + 2\hat{J}_c - \hat{K}_c + 2\hat{J}_o - \hat{K}_o|\phi_k\rangle = \theta_{mk}$$

$$f\langle\phi_k|(\hat{H} + 2\hat{J}_c - \hat{K}_c + 2a\hat{J}_o - b\hat{K}_o|\phi_m\rangle = \theta_{km} \qquad \text{\{XVA-35a,b\}}$$

Taking {XVA-35a} times (-f/(1-f)) and {XVA-35b} by (1/(1-f)) and adding gives

$$-f\langle\phi_m|2\,\frac{(1-a)}{(1-f)}\,\hat{J}_o - \frac{(1-b)}{(1-f)}\,\hat{K}_o|\phi_k\rangle = \theta_{mk} \qquad \text{\{XVA-36\}}$$

and defining

$$\alpha = \frac{1-a}{1-f}$$

and {XVA-37a,b}

$$\beta = \frac{1-b}{1-f}$$

$$\boxed{-f\langle\phi_m|2\alpha\hat{J}_o-\beta\hat{K}_o|\phi_k\rangle = \theta_{mk}} \qquad \text{\{XVA-38\}}$$

Employing {XVA-25a-d} results in

$$-\sum_n|\phi_n\rangle\theta_{nk} = \sum_n|\phi_n\rangle\langle\phi_n|2\alpha f\hat{J}_o-\beta f\hat{K}_o|\phi_k\rangle$$

{XVA-39a,b}

$$\underline{\underline{-\sum_n|\phi_n\rangle\theta_{nk}}} = (2\alpha\hat{L}_o-\beta\hat{M}_o)|\phi_k\rangle$$

$$\underline{\underline{-\sum_\ell|\phi_\ell\rangle\theta_{\ell m}}} = f\sum_\ell|\phi_\ell\rangle\langle\phi_\ell|(2\alpha\hat{J}_o-\hat{\beta}K_o)|\phi_m\rangle$$

{XVA-40a,b}

$$-\sum_\ell|\phi_\ell\rangle\theta_{\ell m} = f(2\alpha L_c-\beta M_c)|\phi_m\rangle$$

where $\sum_n|\phi_n\rangle\langle\phi_n|$ and $\sum_\ell|\phi_\ell\rangle\langle\phi_\ell|$ are "identity elements" and the underlined expressions are the off-diagonal terms which remained in {XVA-32a,b}. These equations allow the re-expression of the coupling terms in the closed and open shells in the desired manner (i.e. the off-diagonal lagrange mutlipliers have been absorbed into the operator)

$$\hat{\underline{F}}_c \underline{\phi}_c = \underline{\phi}_c \underline{\eta}_c \qquad \text{and} \qquad \hat{\underline{F}}_o \underline{\phi}_o = \underline{\phi}_o \underline{\eta}_o$$

{XVA41a,b}

where

$$\left.\begin{array}{ll} \underline{\eta}_c & \text{elements } \theta_{\ell k} \\ \underline{\eta}_o & \text{elements } \theta_{nm/f} \end{array}\right\} \text{Hermitian matrices}$$

and

$$\hat{F}_c = \hat{H} + 2\hat{J}_c - \hat{K}_c + 2\hat{J}_o - \hat{K}_o + 2\alpha\hat{L}_o - \beta\hat{M}_o$$

$$\hat{F}_o = \hat{H} + 2\hat{J}_c - \hat{K}_c + 2\alpha\hat{J}_o - b\hat{K}_o + 2\alpha\hat{L}_c - \beta\hat{M}_c$$

{XVA-42a,b}

The available arbitrary unitary transformations for the closed and open shells separately may be chosen so the matrices $\underline{\underline{\eta}}_c$ and $\underline{\underline{\eta}}_o$ become diagonal. Hence, there exists at least one set of orbitals satisfying:

$$\hat{F}_c|\phi_k\rangle = \eta_k|\phi_k\rangle \quad \text{closed shell}$$

$$\hat{F}_o|\phi_m\rangle = \eta_m|\phi_m\rangle \quad \text{open shell}$$

{XVA-43a,b}

In this case, the total energy can be expressed in terms of the η_i ("orbital energies") and the one electron integrals H_i (exactly analogous to the closed shell expression

$$E = \sum_i (H_i + \varepsilon_i)$$

$$E = \sum_k (H_k + \eta_k) + f \sum_m (H_m + \eta_m) \qquad \text{\{XVA-44\}}$$

The analogy to the closed shell case would be exact if $-\eta_k$

and η_m were approximately equal to the ionization potentials for the removal of an electron from

$$|\phi_k\rangle \quad \text{or} \quad |\phi_m\rangle \quad \text{(Koopmans' Theorem)}$$

Such a relationship is not in general true.